NONLINEAR FUNCTIONAL EVOLUTIONS IN BANACH SPACES

Nonlinear Functional Evolutions in Banach Spaces

by

Ki Sik Ha

Department of Mathematics,
Pusan National University,
Pusan, Republic of Korea

KLUWER ACADEMIC PUBLISHERS
DORDRECHT / BOSTON / LONDON

A C.I.P. Catalogue record for this book is available from the Library of Congress.

ISBN 1-4020-1091-5

Published by Kluwer Academic Publishers,
P.O. Box 17, 3300 AA Dordrecht, The Netherlands.

Sold and distributed in North, Central and South America
by Kluwer Academic Publishers,
101 Philip Drive, Norwell, MA 02061, U.S.A.

In all other countries, sold and distributed
by Kluwer Academic Publishers,
P.O. Box 322, 3300 AH Dordrecht, The Netherlands.

Printed on acid-free paper

Printed in the Netherlands.

To my wife
in commemoration of
her sixtieth birthday
and to my children

Contents

Preface

There are many problems in nonlinear partial differential equations with delay which arise from, for example, physical models, biochemical models, and social models. Some of them can be formulated as nonlinear functional evolutions in infinite-dimensional abstract spaces.

Since Webb (1976) considered autonomous nonlinear functional evolutions in infinite-dimensional real Hilbert spaces, many nonlinear analysts have studied for the last nearly three decades autonomous nonlinear functional evolutions, non-autonomous nonlinear functional evolutions and quasi-nonlinear functional evolutions in infinite-dimensional real Banach spaces.

The techniques developed for nonlinear evolutions in infinite-dimensional real Banach spaces are applied. This book gives a detailed account of the recent state of theory of nonlinear functional evolutions associated with accretive operators in infinite-dimensional real Banach spaces.

Existence, uniqueness, and stability for 'solutions' of nonlinear functional evolutions are considered. Solutions are presented by nonlinear semigroups, or evolution operators, or methods of lines, or inequalities by Bénilan.

This book is divided into four chapters.

Chapter 1 contains some basic concepts and results in the theory of nonlinear operators and nonlinear evolutions in real Banach spaces, that play very important roles in the following three chapters.

Chapter 2 deals with autonomous nonlinear functional evolutions in infinite-dimensional real Banach spaces.

Chapter 3 is devoted to non-autonomous nonlinear functional evolutions in infinite-dimensional real Banach spaces.

Finally, in Chapter 4 quasi-nonlinear functional evolutions are considered in infinite-dimensional real Banach spaces.

I thank very much the Seminar Team on Nonlinear Analysis led by Professor Ki-yeon Shin in Pusan National University, which gave me the first incentive to write this book.

I thank Dr. Liesbeth Mol, Senior Publishing Editor for Mathematics, Kluwer Academic Publishers, whom I met and discussed with about writing this book at the book corner of The Third World Congress of Nonlinear Analysts, Catania, Sicily, Italy, 2000.

I also thank the staff at Kluwer Academic Publshers for undertaking the publication of this book and for very pleasant cooperation.

Finally, I express my special thanks to my wife and my children for their ongoing encouragement of the writing of this book.

Ki Sik Ha

April 30, 2002

Chapter 1

NONLINEAR EVOLUTIONS

The purpose of this chapter is to state some basic concepts and to introduce well known results in the nonlinear operator theory and the theory of nonlinear evolutions in real Banach spaces that will play very important roles in the subsequent development of this book. In Section 1.1 we consider properties of accretive operators with one-sided directional derivatives and duality mappings. Section 1.2 contains the generation of nonlinear semigroups and solutions of autonomous homogeneous nonlinear evolutions. Section 1.3 deals with the existence for strong solutions, integral solutions, and limit solutions of autonomous non-homogeneous nonlinear evolutions. Section 1.4 is devoted to definitions and properties of generalized domains. Section 1.5 treats properties of evolution operators and their generation. In Section 1.6, Section 1.7, and Section 1.8 we discuss how to solve non-autonomous nonlinear evolutions by Crandall–Liggett, Crandall–Pazy, and Evans, respectively, whilst Section 1.9 deals with them by Pavel's method. Comments and notes for references are found in Section 1.10.

1.1. Accretive Operators

Throughout this book let X be a real Banach space with norm $\|\cdot\|$ and let X^* be the dual space of X.

A mapping $A : X \supset D(A) \to 2^X$ is called an operator of X with the domain

$$D(A) = \{x \in X \mid Ax \neq \phi\}$$

of A, the range

$$R(A) = \cup\{Ax \mid x \in D(A)\}$$

of A and the graph

$$G(A) = \{[x, y] \in X \times X \mid y \in Ax\}$$

of A.

$G(A)$ is identified with A.

For two operators A, $B : X \to 2^X$ and a constant $\lambda > 0$ we define operators $A + B$, A^{-1}, λA, $AB : X \to 2^X$ by

$$\begin{aligned}
A + B &= \{[x, y + z] \in X \times X \mid [x, y] \in A,\ [x, z] \in B\}, \\
D(A + B) &= D(A) \cap D(B); \\
A^{-1} &= \{[x, y] \in X \times X \mid [y, x] \in A\}, \\
D(A^{-1}) &= R(A); \\
\lambda A &= \{[x, \lambda y] \mid [x, y] \in A\}, \\
D(\lambda A) &= D(A); \\
AB &= \{[x, z] \in X \times X \mid \text{ there exists } y \in X \text{ such that } [x, y] \in A \\
&\qquad \text{and } [y, z] \in B\}, \\
D(AB) &= \{x \in X \mid x \in D(B),\ Bx \subset D(A)\}.
\end{aligned}$$

From these definitions we have

$$(A + B)x \equiv Ax + Bx = \{y + z \in X \mid y \in Ax,\ z \in Bx\}$$

for every $x \in D(A + B)$,

$$A^{-1}x = \{y \in X \mid x \in Ay\}$$

for every $x \in D(A^{-1})$,

$$(\lambda A)x \equiv \lambda(Ax) = \{\lambda y \in X \mid y \in Ax\}$$

for every $x \in D(\lambda A)$ and

$$\begin{aligned}
(AB)x \equiv A(Bx) &= \{z \in X \mid \text{there exists } y \in D(A) \\
&\qquad \text{such that } y \in Bx \text{ and } z \in Ay\}
\end{aligned}$$

for every $x \in D(AB)$.

Let X^* be the dual space of X. We denote by $x^*(x)$, or (x, x^*), or (x^*, x) the value of $x^* \in X^*$ at $x \in X$.

We define an operator $F : X \to 2^{X^*}$ by

$$F(x) = \{x^* \in X^* \mid x^*(x) = \|x\|^2 = \|x^*\|^2\} \tag{1.1.1}$$

for every $x \in X$. From the Hahn–Banach theorem $F(x) \neq \emptyset$.
The operator F is called the duality mapping of X.

A real Banach space X is said to be strictly convex if for every linearly independent x, $y \in X$

$$\|x + y\| \neq \|x\| + \|y\|.$$

A real Banach space X is said to be uniformly convex if

$$\lim_{n\to\infty} \|x_n - y_n\| = 0$$

for every x_n, $y_n \in X$ with $\|x_n\| = \|y_n\| = 1$, $n = 1, 2, \cdots$, and

$$\lim_{n\to\infty} \|x_n + y_n\| = 2.$$

It is known that a real Banach space is strictly convex if and only if

$$\|\lambda x + (1 - \lambda)y\| < 1$$

for every x, $y \in X$ with $x \neq y$, $\|x\| = \|y\| = 1$ and $0 < \lambda < 1$; a uniformly convex Banach space is strictly convex; a uniformly convex Banach space is reflexive; a real Hilbert space is uniformly convex.
By the defintion of the duality mapping we obtain the following:

PROPOSITION 1.1.1 *Let F be the duality mapping of X defined as in* (1.1.1). *Then:*

(1) $F(x)$ is a closed convex subset of X for every $x \in X$;

(2) $F(\lambda x) = \lambda F(x)$ for every $x \in X$ and $\lambda \in \mathbf{R}$;

(3) $(x^* - y^*, x - y) \geq 0$ for every x, $y \in X$, $x^* \in F(x)$ and $y^* \in F(y)$;

(4) $F(X) = X^*$ if X is reflexive;

(5) F is single-valued if X^* is strictly convex;

(6) F is uniformly continuous on every bounded subset of X if X^* is uniformly convex;

(7) F is the identity of X, that is, $F = I$ if X is a real Hilbert space.

To define accretive operators we need one-sided directional derivatives. We define them and list some properties of them as follows:
We define a functional $[\cdot, \cdot]_h : X \times X \to \mathbf{R}$ by

$$[y, x]_h = \frac{\|x + hy\| - \|x\|}{h}$$

for every $x, y \in X$ and $h \neq 0$. Then the limits

$$\lim_{h\to 0+} [y,x]_h \text{ and } \lim_{h\to 0-} [y,x]_h$$

exist for every $x, y \in X$. So we define one-sided directional derivatives $[\cdot,\cdot]_+, [\cdot,\cdot]_- : X \times X \to \mathbf{R}$ by

$$[y,x]_+ = \lim_{h\to 0+} [y,x]_h \tag{1.1.2}$$

and

$$[y,x]_- = \lim_{h\to 0-} [y,x]_h \tag{1.1.3}$$

for every $x, y \in X$.

Some properties of $[\cdot,\cdot]_+$ and $[\cdot,\cdot]_-$ are listed in the following:

LEMMA 1.1.1 *Let $[\cdot,\cdot]_+$ and $[\cdot,\cdot]_-$ be defined as in* (1.1.2) *and* (1.1.3), *respectively. Then we have*

(1) $[y,x]_- \leq [y,x]_+$ for every $x, y \in X$;

(2) $|\,[y,x]_\pm\,| \leq \|y\|$ for every $x, y \in X$;

(3) $|\,[y,x]_\pm - [z,x]_\pm\,| \leq \|y-z\|$ for every $x, y, z \in X$;

(4) $[y,x]_+ = -[-y,x]_- = -[y,-x]_-$ for every $x, y \in X$;

(5) $[\beta y, \alpha x]_\pm = \beta[y,x]_\pm$ for every $x, y \in X$ and $\alpha, \beta \geq 0$;

(6) $[\alpha x, x]_\pm = \alpha\|x\|$ for every $x, y \in X$ and $\alpha \in \boldsymbol{R}$;

(7) $[y+z,x]_+ \leq [y,x]_+ + [z,x]_+$ and $[y+z,x]_- \geq [y,x]_- + [z,x]_-$ for every $x, y, z \in X$;

(8) $[y+z,x]_+ \geq [y,x]_+ + [z,x]_-$ and $[y+z,x]_- \leq [y,x]_- + [z,x]_+$ for every $x, y, z \in X$;

(9) $[\alpha x + y, x]_\pm = \alpha\|x\| + [y,x]_\pm$ for every $x, y \in X$ and $\alpha \in \boldsymbol{R}$;

(10) $[\cdot,\cdot]_+$ is upper semicontinuous on $X \times X$ and $[\cdot,\cdot]_-$ is lower semicontinuous on $X \times X$;

(11) If $x(t) : [a,b] \to X$ has the right and left derivatives at $t = t_0 \in (a,b)$ then $\|x(t)\| : [a,b] \to \boldsymbol{R}$ has also the right and left derivatives at $t = t_0$ and

$$\frac{d^\pm\|x(t)\|}{dt} = \left[\frac{d^\pm x(t)}{dt}, x(t)\right]_\pm$$

at $t = t_0$.

Making use of the duality mapping F we define the semi-inner products $(\cdot,\cdot)_+, (\cdot,\cdot)_- : X \times X \to \mathbf{R}$ by

$$(y,x)_+ = \sup\{x^*(y) \mid x^* \in F(x)\} \tag{1.1.4}$$

and

$$(y,x)_- = \inf\{x^*(y) \mid x^* \in F(x)\} \tag{1.1.5}$$

for every x, $y \in X$, respectively, which are closely related to $[\cdot,\cdot]_+$ and $[\cdot,\cdot]_-$.

From Lemma 1.1.1 we now list the properties of $(\cdot,\cdot)_+$ and $(\cdot,\cdot)_-$ in the following lemma:

LEMMA 1.1.2 *Let* $(\cdot,\cdot)_+$ *and* $(\cdot,\cdot)_-$ *be defined as in* (1.1.4) *and* (1.1.5), *respectively. Then:*

(1) $(y,x)_+ = \|x\|[y,x]_+$ *and* $(y,x)_- = \|x\|[y,x]_-$ *for every* x, $y \in X$;

(2) $(y,x)_- \le (y,x)_+$ *for every* x, $y \in X$;

(3) $|(y,x)_\pm| \le \|x\|\|y\|$ *for every* x, $y \in X$;

(4) $|(y,x)_\pm - (z,x)_\pm| \le \|x\|\|y-z\|$ *for every* x, y, $z \in X$;

(5) $(y,x)_+ = -(-y,x)_- = -(y,-x)_-$ *for every* $x, y \in X$;

(6) $(\beta y, \alpha x)_\pm = \alpha\beta(y,x)_\pm$ *for every* x, $y \in X$ *and* α, $\beta \ge 0$;

(7) $(\alpha x, x)_\pm = \alpha\|x\|^2$ *for every* $x \in X$ *and* $\alpha \in \boldsymbol{R}$;

(8) $(y+z,x)_+ \le (y,x)_+ + (z,x)_+$ *and* $(y+z,x)_- \ge (y,x)_- + (z,x)_-$ *for every* x, y, $z \in X$;

(9) $(y+z,x)_+ \ge (y,x)_+ + (z,x)_-$ *and* $(y+z,x)_- \le (y,x)_- + (z,x)_+$ *for every* x, y, $z \in X$;

(10) $(\alpha x + y, x)_\pm = \alpha\|x\|^2 + (y,x)_\pm$ *for every* x, $y \in X$ *and* $\alpha \in \boldsymbol{R}$;

(11) $(\cdot,\cdot)_+$ *is upper semi-continuous on* $X \times X$ *and* $(\cdot,\cdot)_-$ *is lower semi-continuous on* $X \times X$;

(12) *If* $x(t) : [a,b] \to X$ *has the right and left derivatives at* $t = t_0 \in (a,b)$ *then* $\|x(t)\|^2 : [a,b] \to \boldsymbol{R}$ *has the right and left derivatives at* $t = t_0$ *and*

$$\frac{d^\pm \|x(t)\|^2}{dt} = 2\Big(\frac{d^\pm x(t)}{dt}, x(t)\Big)_\pm$$

at $t = t_0$.

We are now in a position to define accretive operators which play a role very important in the theory of nonlinear evolutions and nonlinear functional evolutions.

DEFINITION 1.1.1 *An operator $A : X \supset D(A) \to 2^X$ is said to be accretive in X if*

$$\|(x_1 - x_2) + \lambda(y_1 - y_2)\| \geq \|x_1 - x_2\|$$

for every $[x_i, y_i] \in A$, $i = 1, 2$ and $\lambda > 0$.

An operator A of X is said to be strongly accretive if

$$\|(x_1 - x_2) + \lambda(y_1 - y_2)\| \geq \alpha\|x_1 - x_2\|$$

for every $[x_i, y_i] \in A$, $i = 1, 2$, $\lambda > 0$ and some $\alpha \geq 1$.

An accretive operator $A : X \supset D(A) \to 2^X$ is said to be m-accretive in X if $R(I + \lambda A) = X$ for every $\lambda > 0$.

An accretive operator $A : X \supset D(A) \to 2^X$ is called maximal accretive in X if $[\xi, \eta] \in A$ for every $[\xi, \eta] \in X \times X$ satisfying

$$\|(x - \xi) + \lambda(y - \eta)\| \geq \|x - \xi\|$$

for every $[x, y] \in A$.

THEOREM 1.1.1 *Let $A : X \supset D(A) \to 2^X$ be an operator of X. Then the following (1)-(4) are equivalent:*

(1) A is an acretive operator of X;

(2) $[y_1 - y_2, x_1 - x_2]_+ \geq 0$ for every $[x_i, y_i] \in A$, $i = 1, 2$;

(3) $(y_1 - y_2, x_1 - x_2)_+ \geq 0$ for every $[x_i, y_i] \in A$, $i = 1, 2$;

(4) For every $[x_i, y_i] \in A$, $i = 1, 2$, there exists $x^* \in F(x_1 - x_2)$ such that $(y_1 - y_2, x^*) \geq 0$.

DEFINITION 1.1.2 *Let X be a real Hilbert space with inner product $(\cdot, \cdot)$ and norm $\|\cdot\| = \sqrt{(\cdot, \cdot)}$.*

An operator $A : X \supset D(A) \to 2^X$ is said to be monotone if

$$(y_1 - y_2, x_1 - x_2) \geq 0$$

for every $[x_i, y_i] \in A$, $i = 1, 2$.

A monotone operator $A : X \supset D(A) \to 2^X$ is called maximal monotone in X if $[\xi, \eta] \in A$ for every $[\xi, \eta] \in X \times X$ satisfying

$$(y - \eta, x - \xi) \geq 0$$

for every $[x, y] \in A$.

In view of Definition 1.1.1 and Definition 1.1.2 we can state the following properties:

PROPOSITION 1.1.2 *Let X be a real Banach space. Then:*

(1) An m-accretive operator of X is maximal accretive and closed in X;

(2) When A is accretive in X and $R(I + \lambda A) = X$ for some $\lambda > 0$ A is m-accretive in X;

(3) When X^* is uniformly convex, an m-accretive operator of X is demi-closed.

PROPOSITION 1.1.3 *Let X be a real Hilbert space. Then:*

(1) An operator A of X is monotone in X if and only if A is accretive in X;

(2) An operator A of X is maximal monotone in X if and only if A is m-accretive in X.

Let S be a non-empty subset of a real Banach space X. We define

$$\begin{aligned} S^0 &= \{x \in S \mid \|x\| = \inf_{y \in S} \|y\|\}, \\ \|S^0\| &= \inf_{x \in S} \|x\|. \end{aligned}$$

LEMMA 1.1.3 *Let S be a non-empty closed convex subset of X. Then:*

(1) If X is reflexive then S^0 is non-empty;

(2) If X is strictly convex then S^0 is singleton.

From Theorem 1.1.1 we list the properties of accretive operators which are useful in the next chapters as follows:

PROPOSITION 1.1.4 *Let $A : X \supset D(A) \to 2^X$ be an operator in X. We define operators J_λ, $A_\lambda : D(J_\lambda) \subset X \to 2^X$ by*

$$J_\lambda = (I + \lambda A)^{-1}, \quad A_\lambda = \frac{1}{\lambda}(I - J_\lambda)$$

for every $\lambda > 0$.

J_λ and A_λ are called the resolvent and the Yosida approximant of A, respectively.

Let $A + \alpha I$ be accretive in X for some $\alpha \in \boldsymbol{R}$. Then:

(1) For every $\lambda > 0$ with $\lambda\alpha < 1$ $J_\lambda : D(J_\lambda) \subset X \to X$ is single-valued with $D(J_\lambda) = R(I + \lambda A)$ and $R(J_\lambda) = D(A)$;

(2) *For every* $x_i \in D(J_\lambda)$, $i = 1, 2$,

$$\|J_\lambda x_1 - J_\lambda x_2\| \leq (1 - \lambda\alpha)^{-1}\|x_1 - x_2\|;$$

(3) *For every* $\lambda > 0$ *with* $\lambda\alpha < 1$, $x \in D(J_\lambda) \cap D(A)$ *and* $[x, y] \in A$

$$\|J_\lambda x - x\| \leq (1 - \lambda\alpha)^{-1}\|y\|,$$

and thus $\lim_{\lambda \to 0+} J_\lambda x = x$ *for every* $x \in D(J_\lambda) \cap \overline{D(A)}$;

(4) *For every* λ, $\mu > 0$ *with* $\lambda|\alpha| < 1$, *and* $x \in D(J_\lambda^n)$, $n = 1, 2, \cdots$

$$\|J_\lambda^n x - x\| \leq n(1 - \lambda|\alpha|)^{-n+1}\|J_\lambda x - x\|;$$

(5) *For every* λ, $\mu > 0$ *with* $\lambda\alpha < 1$, $\mu\alpha < 1$, *and* $x \in D(J_\lambda)$

$$\frac{\mu}{\lambda}x + \frac{\lambda - \mu}{\lambda}J_\lambda x \in D(J_\mu)$$

and

$$J_\mu\Big(\frac{\mu}{\lambda}x + \frac{\lambda - \mu}{\lambda}J_\lambda x\Big) = J_\lambda x;$$

(6) *For every* $\lambda > 0$ *with* $\lambda\alpha < 1$ $A_\lambda : D(A_\lambda) \subset X \to X$ *is single-valued with* $D(A_\lambda) = D(J_\lambda)$ *and* $R(A_\lambda) = R(J_\lambda)$;

(7) *For every* $\lambda > 0$ *with* $\lambda\alpha < 1$ $A_\lambda + \alpha(1 - \lambda\alpha)^{-1}I$ *is accretive in* X;

(8) *For every* $\lambda > 0$ *with* $\lambda\alpha < 1$ *and* $x_i \in D(A_\lambda)$, $i = 1, 2$

$$\|A_\lambda x_1 - A_\lambda x_2\| \leq \frac{1}{\lambda}(1 + (1 - \lambda\alpha)^{-1})\|x_1 - x_2\|;$$

(9) *For every* $\lambda > 0$ *with* $\lambda\alpha < 1$ *and* $x \in D(A_\lambda)$ $A_\lambda x \in AJ_\lambda x$;

(10) *For every* $\lambda > 0$ *with* $\lambda\alpha < 1$ *and* $x \in D(A_\lambda) \cap D(A)$

$$\|A_\lambda x\| \leq (1 - \lambda\alpha)^{-1}\|A^0 x\|,$$

where $A^0 : X \supset D(A^0) \to 2^X$ *is defined by* $A^0 x = (Ax)^0$ *for every* $x \in D(A^0)$ *and* $\|A^0 x\| = \inf_{y \in Ax}\|y\|$;

(11) *For every* λ, $\mu > 0$ *with* $\mu \leq \lambda$, $\lambda\alpha < 1$, $\mu\alpha < 1$, *and* $x \in D(A_\lambda) \cap D(A_\mu)$

$$(1 - \lambda\alpha)\|A_\lambda x\| \leq (1 - \mu\alpha)\|A_\mu x\|;$$

(12) *Let* $\lambda > 0$, $\mu > 0$ *with* $\lambda\alpha < 1$, *and* $\mu\alpha(1 - \lambda\alpha)^{-1} < 1$. *If* $R(I + \lambda A) = X$ *then* $R(I + \mu A_\lambda) = X$;

(13) *Let $\lambda > 0$ with $\lambda\alpha < 1$. If $R(I + \lambda A) = X$ for every $x \in X$*

$$\lim_{n\to\infty} \left(I + \lambda A_{\frac{1}{n}}\right)^{-1} x = (I + \lambda A)^{-1} x.$$

PROPOSITION 1.1.5 *Let the dual space X^* of X be uniformly convex and let $A : D(A) \subset X \to X$ be m-accretive in X. Then:*

(1) *For every $x \in X$ and $\{x_n\} \subset D(A)$ with $x = \lim_{n\to\infty} x_n$ and that $\|Ax_n\|$ is bounded in X we have $x \in D(A)$ and $w-\lim_{n\to\infty} Ax_n = Ax$;*

(2) *$w-\lim_{\lambda\to 0+} A_\lambda x = Ax$ for every $x \in D(A)$, where $w-\lim$ denotes the weak limit.*

1.2. Nonlinear Semigroups

In this section we are concerned with nonlinear semigroups which have a connection closed to solutions of nonlinear evolutions and nonlinear functional evolutions associated with accretive operators in X.

We now introduce the definition of nonlinear semigroups and their generators.

DEFINITION 1.2.1 *Let X be a real Banach space with norm $\|\cdot\|$. Let X_0 be a closed subset of X. Let $S(t) : X_0 \to X_0$ be a one-parameter nonlinear operator for every $t \geq 0$.*

The family $\{S(t) \mid t \geq 0\}$ is called a nonlinear semigroup on X_0 of type α if:

(1) *$S(0)x = x$ for every $x \in X_0$;*

(2) *$S(t + s)x = S(t)S(s)x$ for every $x \in X_0$ and $t, s \geq 0$;*

(3) *$\lim_{t\to 0+} S(t)x = x$ for every $x \in X_0$;*

(4) *For every $x, y \in X_0$, $t \geq 0$ and for some $\alpha \in \boldsymbol{R}$*

$$\|S(t)x - S(t)y\| \leq e^{\alpha t} \|x - y\|.$$

When $\alpha = 0$ $\{S(t) \mid t \geq 0\}$ is called a nonlinear semigroup on X_0 of contractions , or simply a nonlinear semigroup on X_0.

Let $\{S(t) \mid t \geq 0\}$ be a nonlinear semigroup on X_0 of type α. We define an operator $A_0 : D(A_0) \subset X_0 \to X$ by

$$\begin{aligned} D(A_0) &= \left\{x \in X_0 \;\middle|\; \lim_{t\to 0+} \frac{1}{t}(x - S(t)x) \text{ exists}\right\}, \\ A_0 x &= \lim_{t\to 0+} \frac{1}{t}(x - S(t)x), \quad x \in D(A_0). \end{aligned}$$

This operator A_0 is called an infinitesimal generator, or simply a generator, of $\{S(t) \mid t \geq 0\}$.

Nonlinear semigroups are generated by accretive operators. We state the Crandall–Liggett theorem:

THEOREM 1.2.1 (Crandall–Liggett [26]) *Let* $A : X \supset D(A) \to 2^X$ *be an operator and* $\alpha \in \boldsymbol{R}$ *such that* $A + \alpha I$ *is accretive in* X *and*

$$R(I + \lambda A) \supset \overline{D(A)}$$

for every $\lambda > 0$ *with* $\lambda\alpha < 1$. *Then the limit*

$$\lim_{n\to\infty} \left(I + \frac{t}{n}A\right)^{-n} x$$

exists for every $x \in \overline{D(A)}$ *uniformly in* t *on every bounded interval of* $[0, \infty)$.

Let $t \geq 0$. *We define a one-parameter operator* $S(t) : \overline{D(A)} \to \overline{D(A)}$ *by*

$$S(t)x = \lim_{n\to\infty} \left(I + \frac{t}{n}A\right)^{-n} x$$

for every $x \in \overline{D(A)}$. *Then* $\{S(t) \mid t \geq 0\}$ *is a nonlinear semigroup on* $\overline{D(A)}$ *of type* α *and*

$$\|S(t)x - S(s)x\| \leq (e^{2|\alpha|t+s} + e^{4|\alpha|t})|t - s|\|A^0 x\|$$

for every $x \in D(A^0)$ *and* $t,\ s \geq 0$.

Moreover,

$$S(t)x = \lim_{\lambda\to 0+} (I + \lambda A)^{-[\frac{t}{\lambda}]} x$$

for every $t \geq 0$ *and* $x \in \overline{D(A)}$, *where this convergence is uniform on every bounded interval of* $[0, \infty)$ *and* $[t/\lambda]$ *denotes the largest integer* $\leq t/\lambda$ *for* $0 < \lambda \leq t$.

In this case we say that $\{S(t) \mid t \geq 0\}$ is generated by $-A$.

Some results on nonlinear semigroups and nonlinear compact semigroups is presented as follows, which will be used in the next chapters.

THEOREM 1.2.2 *Let* $A : X \supset D(A) \to 2^X$ *be an m-accretive operator of* X *and let* $\{S(t) \mid t \geq 0\}$ *be the nonlinear semigroup generated by* $-A$. *Then:*

(1) For every $x \in \overline{D(A)}$, $t \geq 0$ *and* $[\xi, \eta] \in A$

$$\|S(t)x - \xi\| \leq \|S(s)x - \xi\| + \int_s^t [-\eta, S(\tau)x - \xi]_+ d\tau;$$

(2) For every $x \in \overline{D(A)}$ *and* $\lambda > 0$

$$\|J_\lambda x - x\| \leq \frac{4}{\lambda} \int_0^\lambda \|S(\tau)x - x\| d\tau.$$

Let $A : X \supset D(A) \to 2^X$ be an m-accretive operator of X and let $\{S(t) \mid t \geq 0\}$ be the nonlinear semigroup on $\overline{D(A)}$ of contractions generated by $-A$.

$\{S(t) \mid t \geq 0\}$ is called a compact semigroup if $S(t)$ is a compact operator for every $t > 0$. $\{S(t) \mid t \geq 0\}$ is called an equicontinuous semigroup if $\{S(t)x \mid x \in B\}$ is equicontinuous at every $t > 0$ for every bounded subset B of X.

We have Brézis' theorem on compact semigroups.

THEOREM 1.2.3 (Brézis [16]) *Let* $A : X \supset D(A) \to 2^X$ *be an* m*-accretive operator of* X *and let* $\{S(t) \mid t \geq 0\}$ *be the nonlinear semigroup on* $\overline{D(A)}$ *of contractions generated by* $-A$*. Then the following (1) and (2) are equivalent:*

(1) $\{S(t) \mid t \geq 0\}$ *is a compact semigroup;*

(2) $J_\lambda = (I + \lambda A)^{-1}$ *is a compact operator for every* $\lambda > 0$ *and* $\{S(t) \mid t \geq 0\}$ *is an equicontinuous semigroup.*

Now we introduce definitions and existence theorems for solutions of autonomous homogeneous nonlinear evolutions of the type

$$\begin{cases} \dfrac{dx}{dt}(t) + Ax(t) \ni 0, & t \geq 0 \\ x(0) = x_0, \end{cases} \tag{1.2.1}$$

where $A : X \supset D(A) \to 2^X$ and $x_0 \in X$.

DEFINITION 1.2.2 *1. A continuous function* $x(t) : [0, \infty) \to X$ *is called a (classical) solution of* (1.2.1) *on* $[0, \infty)$ *if:*

(1) $x(t)$ *is continuously differentiable on* $[0, \infty)$*;*

(2) $x(t)$ *satisfies* (1.2.1).

2. A continuous function $x(t) : [0, \infty) \to X$ *is called a strong solution of* (1.2.1) *on* $[0, \infty)$ *if:*

(1) $x(0) = x_0$*;*

(2) $x(t)$ *is absolutely continuous on every compact interval of* $[0, \infty)$*;*

(3) $x(t)$ *is differentiable a.e.* $t \in (0, \infty)$*;*

(4) $x(t)$ *satisfies*

$$\frac{dx}{dt}(t) + Ax(t) \ni 0$$

a.e. $t \in (0, \infty)$.

We consider autonomous homogeneous nonlinear evolutions of the type

$$\begin{cases} \dfrac{dx}{dt}(t) + A_\lambda x(t) = 0, & t \geq 0 \\ x(0) = x_0 \end{cases} \tag{1.2.2}$$

associated with the Yosida approximant A_λ of an operator A for every $\lambda > 0$ and for some $x_0 \in X$.

THEOREM 1.2.4 *Let* $A : X \supset D(A) \to 2^X$ *be an operator and* $\alpha \in \boldsymbol{R}$ *such that* $A + \alpha I$ *is accretive in* X *and*

$$R(I + \lambda A) \supset \overline{\mathrm{co}} D(A)$$

for every $\lambda > 0$ *with* $\lambda\alpha < 1$. *Let* $\{S(t) \mid t \geq 0\}$ *be the nonlinear semigroup on* $\overline{D(A)}$ *of type* α *generated by* $-A$. *Let* $x_0 \in \overline{D(A)}$. *Then there exists a unique (classical) solution* $x_\lambda(t)$ *of* (1.2.2) *on* $[0, \infty)$ *and*

$$\lim_{\lambda \to 0+} x_\lambda(t) = S(t)x_0,$$

this convergence is uniformly on every bounded interval of $[0, \infty)$.

THEOREM 1.2.5 *Let* $A : X \supset D(A) \to 2^X$ *be a closed operator and* $\alpha \in \boldsymbol{R}$ *such that* $A + \alpha I$ *is accretive in* X *and*

$$R(I + \lambda A) \supset \overline{\mathrm{co}} D(A)$$

for every $\lambda > 0$ *with* $\lambda\alpha < 1$. *Let* $\{S(t) \mid t \geq 0\}$ *be the nonlinear semigroup on* $\overline{D(A)}$ *of the type* α *generated by* $-A$. *Let* $x_0 \in D(A)$. *Then the following (1) and (2) are equivalent:*

(1) A function $x(t)$ *is a unique strong solution on* $[0, \infty)$ *of* (1.2.1)*;*

(2) $x(t) = S(t)x_0$ *for every* $t \geq 0$ *and* $x(t)$ *is differentiable a.e. on* $(0, \infty)$.

Since if X is reflexive an absolutely continuous function $x(t) : [0, \infty) \to X$ is then differentiable a.e. $t \in (0, \infty)$ (by Theorem 1.2.1 and Theorem 1.2.5), we have the following result:

COROLLARY 1.2.1 *Let X be reflexive. Let $A : X \supset D(A) \to 2^X$ be a closed operator and $\alpha \in \boldsymbol{R}$ such that $A + \alpha I$ is accretive in X and*

$$R(I + \lambda A) \supset \overline{\text{co}} D(A)$$

for every $\lambda > 0$ with $\lambda\alpha < 1$. Let $x_0 \in D(A)$. Then a strong solution of (1.2.1) on $[0, \infty)$ exists uniquely.

We have the following result of Crandall which gives the link between strong solutions of (1.2.1) and the nonlinear semigroups generated by $-A$.

THEOREM 1.2.6 (Crandall [22]) *Let $A : X \supset D(A) \to 2^X$ be a closed operator and $\alpha \in \boldsymbol{R}$ such that $A + \alpha I$ is accretive in X and*

$$R(I + \lambda A) \supset \overline{\text{co}} D(A)$$

for every $\lambda > 0$ with $\lambda\alpha < 1$. Let $\{S(t) \mid t \geq 0\}$ be the nonlinear semigroup on $\overline{D(A)}$ of type α generated by $-A$. Let $x_0 \in D(A)$. If (1.2.1) has a strong solution $x(t)$ on $[0, \infty)$ then $x(t) = S(t)x_0$ for every $t \geq 0$.

We consider here linear semigroups

DEFINITION 1.2.3 *Let X be a real Banach space. Let $S(t) : X \to X$ be a one-parameter bounded linear operator for every $t \geq 0$. The family $\{S(t) \mid t \geq 0\}$ is called a linear (C_0)-semigroup on X if:*

(1) $S(0)x = x$ *for every* $x \in X$*;*

(2) $S(t + s)x = S(t)S(s)x$ *for every* $x \in X$ *and* t, $s \geq 0$*;*

(3) $\lim_{t \to 0+} S(t)x = x$ *for every* $x \in X$*;*

A linear (C_0)-semigroup $\{S(t) \mid t \geq 0\}$ is called contraction if

$$\|S(t)\| \leq 1$$

for every $t \geq 0$.

Let $\{S(t) \mid t \geq 0\}$ be a linear (C_0)-semigroup on X. We define an operator $A : D(A) \subset X \to X$ by

$$\begin{aligned}
D(A) &= \left\{ x \in X_0 \;\middle|\; \lim_{t \to 0+} \frac{1}{t}(S(t)x - x) \text{ exists} \right\}, \\
Ax &= \lim_{t \to 0+} \frac{1}{t}(S(t)x - x), \; x \in D(A).
\end{aligned}$$

This operator A is called an infinitesimal generator, or simply a generator, of $\{S(t) \mid t \geq 0\}$.

PROPOSITION 1.2.1 *Let A be the infinitesimal generator of a linear (C_0)-semigroup on X. Then:*

(1) A is closed and $D(A)$ is dense in X;

(2) $(\lambda I - A)^{-1}$ exists, and

$$(\lambda I - A)^{-1}x = \int_0^\infty e^{-\lambda t} S(t)x dt$$

for every $\lambda > 0$ and $x \in X$.

THEOREM 1.2.7 (Lumer–Phillips [102]) *Let $A : D(A) \subset X \to X$ be a linear operator with $\overline{D(A)} = X$. If A is accretive and $R(I + \lambda A) = X$ for some $\lambda > 0$, then A is the infinitesimal generator of a linear (C_0)-semigroup $\{S(t) \mid t \geq 0\}$ on X of contractions.*

1.3. Existence of Solutions

In this section we deal with definitions and existence of solutions of autonomous nonhomogeneous nonlinear evolutions of the type

$$\begin{cases} \dfrac{dx}{dt}(t) + Ax(t) \ni f(t), & t \geq 0 \\ x(0) = x_0, \end{cases} \tag{1.3.1}$$

where $x_0 \in X$ and $f \in L^1(0, \infty; X)$, or

$$\begin{cases} \dfrac{dx}{dt}(t) + Ax(t) \ni f(t), & 0 \leq t \leq T \\ x(0) = x_0, \end{cases} \tag{1.3.2}$$

where $x_0 \in X$ and $f \in L^1(0, T; X)$ for $T > 0$. Let us define several kinds of solutions for (1.3.1) or (1.3.2).

DEFINITION 1.3.1 *1. A continuous function $x(t) : [0, \infty) \to X$ is called a (classical) solution of* (1.3.1) *on* $[0, \infty)$ *if:*

(1) $x(t)$ is continuously differentiable on $[0, \infty)$;

(2) $x(t)$ satisfies (1.3.1) *on* $[0, \infty)$.

2. A continuous function $x(t) : [0, \infty) \to X$ is called a strong solution of (1.3.1) *on* $[0, \infty)$ *if:*

(1) $x(0) = x_0$;

(2) $x(t)$ *is absolutely continuous on every compact interval of* $[0, \infty)$;

(3) $x(t)$ *is differentiable a.e. on* $(0, \infty)$;

(4) $x(t)$ *satisfies*

$$\frac{dx}{dt}(t) + Ax(t) \ni f(t)$$

a.e. on $(0, \infty)$.

3. A continuous function $x(t) : [0, \infty) \to X$ *is called an integral solution of* (1.3.1) *on* $[0, \infty)$ *if:*

(1) $x(0) = x_0$;

(2) $x(t)$ *satisfies*

$$\|x(t) - x\| \leq \|x(s) - x\| + \int_s^t [f(\tau) - y, x(\tau) - x]_+ d\tau.$$

for every $[x, y] \in A$ *and* $0 \leq s \leq t$.

4. A continuous function $x(t) : [0, T] \to X$ *is called a limit solution of* (1.3.2) *on* $[0, T]$ *if:*

(1) $x(0) = x_0$;

(2) $\lim_{n\to\infty} x_n(t) = x(t)$ *uniformly on* $[0, T]$ *and* $\lim_{n\to\infty} f_n = f$ *in* $L^1(0, T; X)$,

satisfying

$$\begin{aligned} 0 &= t_0^n < t_1^n < \cdots < t_{N(n)}^n = T, \\ x_n(0) &= x_0^n, \\ x_n(t) &= x_i^n, \quad t_{i-1}^n < t \leq t_i^n, \\ f_n(t) &= f_i^n, \quad t_{i-1}^n < t \leq t_i^n \end{aligned}$$

and

$$\frac{x_i^n - x_{i-1}^n}{t_i^n - t_{i-1}^n} + Ax_i^n \ni f_i^n$$

for $i = 1, 2, \cdots, N(n)$; $n = 1, 2, \cdots$, *with*

$$\lim_{n\to\infty} \max_{1 \leq i \leq N(n)} (t_i^n - t_{i-1}^n) = 0,$$

$$\{x_i^n \in X \mid i = 0, 1, 2, \cdots, \ N(n); n = 1, 2, \cdots\},$$

$$\{f_i^n \in X \mid i = 1, 2, \cdots, \ N(n); n = 1, 2, \cdots\}.$$

5. A continuous function $x(t) : [0, \infty) \to X$ is called a limit solution of (1.3.1) on $[0, \infty)$ if, for every $T \in (0, \infty)$, $x(t)$ is a limit solution of (1.3.2) on $[0, T]$.

We remark that if $A : X \supset D(A) \to 2^X$ is m-accretive in X, $x_0 \in D(A)$ and $f \in L^1(0, \infty; X)$, then a strong solution of (1.3.1) on $[0, \infty)$ is a limit solution of (1.3.1) on $[0, \infty)$ and a limit solution is an integral solution of (1.3.1) on $[0, \infty)$.

The uniqueness and existence theorems by Bénilan are as follows:

THEOREM 1.3.1 (Bénilan [6], [7]) *Let $A : X \supset D(A) \to 2^X$ be an m-acretive operator of X. Let $f \in L^1(0, \infty; X)$ and let $x_0 \in \overline{D(A)}$. Then (1.3.1) has a unique integral solution on $[0, \infty)$. Moreover, let f, $g \in L^1(0, \infty; X)$ and x_0, $y_0 \in \overline{D(A)}$. Let $x(t)$ and $y(t)$ be integral solutions on $[0, \infty)$ of*

$$\begin{cases} \dfrac{dx}{dt}(t) + Ax(t) \ni f(t), & t \geq 0 \\ x(0) = x_0, \end{cases}$$

and

$$\begin{cases} \dfrac{dy}{dt}(t) + Ay(t) \ni g(t), & t \geq 0 \\ y(0) = y_0, \end{cases}$$

respectively. Then for $0 \leq s \leq t$

$$\|x(t) - y(t)\| \leq \|x(s) - y(s)\| + \int_s^t \|f(\tau) - g(\tau)\| d\tau \qquad (1.3.3)$$

and

$$\|x(t) - y(t)\|^2 \leq \|x(s) - y(s)\|^2 + \int_s^t [f(\tau) - g(\tau), x(\tau) - y(\tau)]_+ d\tau.$$

THEOREM 1.3.2 *Let $A : X \supset D(A) \to 2^X$ and let $f \in L^1(0, \infty; X)$.*

(1) Let X be reflexive and let $x_0 \in D(A)$. Then there exists a unique strong solution $x(t)$ on $[0, \infty)$ of (1.3.2) such that $x \in W^{1,\infty}(0, \infty; X)$ and

$$\begin{aligned} \left\|\frac{dx}{dt}(t)\right\| &= \|(f(t) - Ax(t))^0\| \\ &\leq \|(f(0) - Ax_0)^0\| + \int_0^t \left\|\frac{df}{d\tau}(\tau)\right\| d\tau \end{aligned}$$

a.e. $t \in (0,\infty)$. *Moreover, if* X *and* X^* *are uniformly convex then* $x(t)$ *is differentiable in the right on* $[0,\infty)$ *and*

$$\frac{d^+x}{dt}(t) = (f(t) - Ax(t))^0$$

for every $t \geq 0$;

(2) *Let* $x_0 \in \overline{D(A)}$. *Then there exists a unique integral solution of* (1.3.1) *on* $[0,\infty)$;

(3) *Let* $x_0 \in \overline{D(A)}$. *Then there exists a unique limit solution of* (1.3.2) *on* $[0,T]$.

THEOREM 1.3.3 *Let* $A : X \supset D(A) \to 2^X$ *be an operator and* $\alpha \in \boldsymbol{R}$ *such that* $A + \alpha I$ *is accretive in* X *and*

$$R(I + \lambda A) \supset \overline{D(A)}$$

for every $\lambda > 0$ *with* $\lambda\alpha < 1$. *Let* $\{S(t) \mid t \geq 0\}$ *be the nonlinear semigroup on* $\overline{D(A)}$ *of type* α *generated by* $-A$. *Let* $x_0 \in \overline{D(A)}$. *Then* $x(t) = S(t)x_0$ *is a unique limit solution of* (1.3.1) *on* $[0,\infty)$.

1.4. Generalized Domains

We now define the generalized domain of m-accretive operators.

Let $A : X \supset D(A) \to 2^X$ be an m-accretive operator of X and. If we put

$$|Ax| = \lim_{\lambda \to 0+} \|A_\lambda x\|,$$

then $|Ax| \leq \infty$ for every $x \in X$. So, we define

$$\widehat{D}(A) = \{x \in X \mid |Ax| < \infty\}$$

The set $\widehat{D}(A)$ is called the generalized domain of A.

PROPOSITION 1.4.1 *Let* $A : X \supset D(A) \to 2^X$ *be an* m*-accretive operator of* X. *Then:*

(1) $D(A) \subset \widehat{D}(A) \subset \overline{D(A)}$;

(2) $\|A_\lambda x\| \leq |Ax|$ *for every* $x \in X$ *and* $\lambda > 0$;

(3) $|Ax| \leq \|A^0 x\|$ *for every* $x \in D(A)$;

(4) $|Ax| \leq \underline{\lim}_{n\to\infty} |Ax_n|$ *for every* $x \in X$ *and* $\{x_n\} \subset X$ *with* $x = \lim_{n\to\infty} x_n$. *Moreover, if* $|Ax_n|$ *is bounded then* $x \in \widehat{D}(A)$;

(5) *If* X *is reflexive then* $\widehat{D}(A) = D(A)$.

PROPOSITION 1.4.2 *Let* $A : X \supset D(A) \to 2^X$ *be an operator and* $\alpha \in \boldsymbol{R}$ *such that* $A + \alpha I$ *is accretive in* X *and*

$$R(I + \lambda A) \supset \overline{D(A)}$$

for every $\lambda > 0$ *with* $\lambda\alpha < 1$. *Let* $\{S(t) \mid t \geq 0\}$ *be the nonlinear semigroup on* $\overline{D(A)}$ *of contractions generated by* $-A$. *Then*

$$\widehat{D}(A) = \{x \in \overline{D(A)} \mid S(t)x \text{ is Lipschitz continuous in t} \\ \text{on every bounded subsets of } [0, \infty)\}.$$

Moreover, $S(t) : \widehat{D}(A) \to \widehat{D}(A)$.

Let $T > 0$. We now deal with the generalized domain of a non-autonomous nonlinear operator $A(t)$ for every $t \in [0, T]$. We put the resolvent $J_\lambda(t)$ and the Yosida approximant $A_\lambda(t)$ of $A(t)$ by

$$J_\lambda(t) = (I + \lambda A(t))^{-1} \quad \text{and} \quad A_\lambda(t) = \frac{1}{\lambda}(I - J_\lambda(t))$$

for every $t \in [0, T]$ and $\lambda > 0$. Then

$$D(J_\lambda(t)) = R(I + \lambda A(t)) \quad \text{and} \quad D(A_\lambda(t)) = D(J_\lambda(t))$$

for every $t \in [0, T]$ and $\lambda > 0$.

We consider conditions that the generalized domain $\widehat{D}(A(t))$ of $A(t)$ is independent of $t \in [0, T]$.

PROPOSITION 1.4.3 *Let* $A(t)$ *be m-accretive in* X *for every* $t \in [0, T]$. *If there are a constant* $\lambda_0 > 0$, *a function* $h : [0, T] \to X$ *of bounded variation and a non-decreasing function* $L : [0, \infty) \to [0, \infty)$ *such that for every* $\lambda \in (0, \lambda_0]$, $x \in X$ *and* $t, s \in [0, T]$,

$$\|A_\lambda(t)x - A_\lambda(s)x\| \leq \|h(t) - h(s)\| L(\|x\|)(1 + \|A_\lambda(s)x\|) \quad (1.4.1)$$

holds then the generalized domain $\widehat{D}(A(t))$ *of* $A(t)$ *and the closure* $\overline{D(A(t))}$ *of* $D(A(t))$ *are independent of* $t \in [0, T]$.

Proof. Let $t, s \in [0, T]$. Let $x \in \widehat{D}(A(s))$. Then $|A(s)x| < \infty$. From (1.4.1) we have

$$\|A_\lambda(t)x\| \leq \|A_\lambda(s)x\| + \|h(t) - h(s)\| L(\|x\|)(1 + \|A_\lambda(s)x\|) \quad (1.4.2)$$

and letting $\lambda \to 0+$

$$|A(t)x| \leq |A(s)x| + \|h(t) - h(s)\| L(\|x\|)(1 + |A(s)x|) \quad (1.4.3)$$

Thus $|A(t)x| < \infty$. Hence $x \in \widehat{D}(A(t))$, that is, $\widehat{D}(A(s)) \subset \widehat{D}(A(t))$. Similarly, we have $\widehat{D}(A(t)) \subset \widehat{D}(A(s))$. Therefore, $\widehat{D}(A(t)) = \widehat{D}(A(s))$, and thus $\widehat{D}(A(t))$ is independent of $t \in [0,T]$. By (1) of Proposition 1.4.1 $\overline{D(A(t))} = \overline{\widehat{D}(A(t))}$. Thus $\overline{D(A(t))}$ is independent of $t \in [0,T]$.

PROPOSITION 1.4.4 *Let $A(t) : D \equiv D(A(t)) \subset X \to X$ be an operator for every $t \in [0,T]$ and $\alpha \in \boldsymbol{R}$ such that $A(t) + \alpha I$ is accretive in X and $R(I + \lambda A) = X$ for every $\lambda > 0$ with $\lambda\alpha < 1$. If there exist a function $h : [0,T] \to X$ of bounded variation and a non-decreasing function $L_0 : [0,\infty) \to [0,\infty)$ such that for every $x \in D$ and t, $s \in [0,T]$,*

$$\|A(t)x - A(s)x\| \le \|h(t) - h(s)\| L_0(\|x\|)(1 + \|A(s)x\|) \qquad (1.4.4)$$

holds then (1.4.1) *holds for every $\lambda \in (0,1)$ with $\lambda\alpha < \frac{1}{2}$, $x \in \overline{D}$ and t, $s \in [0,T]$ with $L(\sigma) = 2L_0(K(\sigma+1))$ for some constant $K > 0$.*

On the other hand, if X^* is uniformly convex and (1.4.1) *holds for every sufficiently small $\lambda > 0$, $x \in \overline{D}$ and t, $s \in [0,T]$ with $L = L_0$ then* (1.4.3) *holds for every $x \in D$ and t, $s \in [0,T]$.*

Proof. Let $\lambda \in (0,1)$ with $\lambda\alpha < \frac{1}{2}$, $x \in \overline{D}$ and t, $s \in [0,T]$. From (1.4.4) and the definition of $A_\lambda(t)$

$$\begin{aligned}
&\|A_\lambda(t)x - A_\lambda(s)x\| \\
&\quad = \frac{1}{\lambda}\|J_\lambda(t)x - J_\lambda(s)x\| \\
&\quad = \frac{1}{\lambda}\|J_\lambda(t)(I + \lambda A(s))J_\lambda(s)x - J_\lambda(t)(I + \lambda A(t))J_\lambda(s)x\| \\
&\quad \le \frac{1}{1-\lambda\alpha}\|A(t)J_\lambda(s)x - A(s)J_\lambda(s)x\| \\
&\quad \le 2\|h(t) - h(s)\| L_0(\|J_\lambda(s)x\|)(1 + \|A_\lambda(s)x\|).
\end{aligned}$$

Let $y \in D$. Letting $s = 0$ in (1.4.4) $\|A(t)y\|$ is bounded with respect to $t \in [0,T]$. By

$$\begin{aligned}
&\|J_\lambda(s)x - y\| \\
&\quad = \|J_\lambda(s)x - J_\lambda(s)(I + \lambda A(s))y\| \\
&\quad \le \frac{1}{1-\lambda\alpha}\|x - (y + \lambda A(s)y)\| \\
&\quad \le 2(\|x\| + \|y\| + \|A(s)y\|)
\end{aligned}$$

$\|J_\lambda(s)x\|$ is also bounded with respect to $s \in [0,T]$. Thus there exists a constant $K > 0$ such that

$$\|J_\lambda(s)x\| \le K(\|x\| + 1).$$

Using (1.4.5) we have (1.4.1) with

$$L(\sigma) = 2L_0(K(\sigma + 1))$$

for every $\sigma \in [0, T]$.

On the other hand, let X^* be uniformly convex. Let $x \in D$ and let t, $s \in [0, T]$. Let $\alpha = 0$. From (2) of Propositin 1.1.5

$$A(t)x = w - \lim_{\lambda \to 0+} A_\lambda(t)x$$

for every $x \in D$. Since $|A(s)x| = \|A(s)x\|$, using (1.4.1),

$$\begin{aligned} &\|A(t)x - A(s)x\| \\ &\quad \le \underline{\lim}_{\lambda \to 0+} \|A_\lambda(t)x - A_\lambda(s)x\| \\ &\quad \le \underline{\lim}_{\lambda \to 0+} \|h(t) - h(s)\| L(\|x\|)(1 + \|A_\lambda(s)x\|) \\ &\quad \le \|h(t) - h(s)\| L(\|x\|)(1 + |A(s)x|) \\ &\quad = \|h(t) - h(s)\| L_0(\|x\|)(1 + \|A(s)x\|). \end{aligned}$$

1.5. Evolution Operators

In this section we define nonlinear evolution operators associated with non-autonomous nonlinear operators and consider the problems on generation of evolution operators.

Let $T > 0$ and let

$$\triangle = \{(t, s) \mid 0 \le s \le t \le T\}.$$

DEFINITION 1.5.1 *Let X_0 be a closed subset of X. Let $U(t, s) : X_0 \to X_0$ be a two-parameter nonlinear operator for every $(t, s) \in \triangle$. The family $\{U(t, s) \mid (t, s) \in \triangle\}$ is called an evolution operator on X_0 of type $\alpha \in \boldsymbol{R}$ if:*

(1) $U(t, t)x = x$ for every $x \in X_0$ and $t \in [0, T]$;

(2) $U(t, \sigma)U(\sigma, s)x = U(t, s)x$ for every $x \in X_0$ and (t, σ), $(\sigma, s) \in \triangle$;

(3) $U(t, s)x$ is continuous at every $(t, s) \in \triangle$ for every $x \in X_0$;

(4) For every x, $y \in X_0$, $(t, s) \in \triangle$ and for some $\alpha \in \boldsymbol{R}$,

$$\|U(t, s)x - U(t, s)y\| \le e^{\alpha(t-s)} \|x - y\|.$$

When $\alpha = 0$ $\{U(t, s) \mid (t, s) \in \triangle\}$ is called an evolution operator on X_0 of contractions.

The following is the generation theorem of the evolution operator:

THEOREM 1.5.1 (Crandall–Pazy [29]) *Let $A(t) : X \supset D(A(t)) \equiv D \to 2^X$ be an operator for every $t \in [0,T]$ and $\alpha \in \boldsymbol{R}$ such that $A(t) + \alpha I$ is accretive in X and*

$$R(I + \lambda A(t)) \supset \overline{D}$$

for every $\lambda > 0$ with $\lambda\alpha < 1$. We assume that there exist a continuous function $h : [0,T] \to X$ of bounded variation and a non-decreasing function $L : [0,\infty) \to [0,\infty)$ such that

$$\|A_\lambda(t)x - A_\lambda(s)x\| \le \|h(t) - h(s)\| L(\|x\|)(1 + \|A_\lambda(s)\|) \qquad (1.5.1)$$

for every $\lambda > 0$ with $\lambda\alpha < 1$, $x \in \overline{D}$ and $t,\ s \in [0,T]$. Then the limit

$$\lim_{n\to\infty} \prod_{i=1}^{n} J_{\frac{t-s}{n}}\left(s + i\frac{t-s}{n}\right)x$$

exists for every $x \in \overline{D}$, uniformly in $(t,s) \in \triangle$ with $s < t$.

Let $(t,s) \in \triangle$. We define a two-parameter nonlinear operator $U(t,s) : \overline{D} \to \overline{D}$ by

$$U(t,s)x = \begin{cases} \lim_{n\to\infty} \prod_{i=1}^{n} J_{\frac{t-s}{n}}\left(s + i\frac{t-s}{n}\right)x, \ t > s \\ x, \ t = s \end{cases} \qquad (1.5.2)$$

for every $x \in \overline{D}$.

Then $\{U(t,s) \mid (t,s) \in \triangle\}$ is an evolution operator on $\overline{D}$ of type α

We consider properties of the evolution operator $\{U(t,s) \mid (t,s) \in \triangle\}$ of Theorem 1.5.1.

Let $h(t)$ be a function in (1.5.1). We define a function $\rho : [0,\infty) \to [0,\infty)$ by

$$\rho(\sigma) = \sup\{\|h(t) - h(\tau)\| \mid 0 \le t,\ \tau \le T,\ |t - \tau| \le \sigma\}$$

for every $\sigma \in [0,\infty)$. We say that ρ is the modulus of continuity of h on [0,T]. Clearly, $\rho : [0,\infty) \to [0,\rho(T)]$ is non-decreasing and continuous,

$$\lim_{\sigma\to 0} \rho(\sigma) = \rho(0) = 0,$$

ρ is subadditive and there is a constant $c \ge 0$ such that $\sigma \le c\rho(\sigma)$ unless $\rho \equiv 0$.

THEOREM 1.5.2 *Under the assumptions of Theorem 1.5.1 let $\{U(t,s) \mid (t,s) \in \triangle\}$ be the evolution operator on $\overline{D}$ of type α in* (1.5.2) *and let ρ be the modulus of continuity of $h(t)$ on $[0,T]$. Then:*

(1) *$U(t,s)x$ is continuous on $\triangle$ for every $x \in \overline{D}$;*

(2) *$U(t,s) : \widehat{D} \to \widehat{D}$;*

(3) *For every $x \in \widehat{D}$ there exists a constant $c_1 > 0$ such that*

$$\|U(t,s)x - U(\tau,s)x\| \leq c_1|t-\tau|$$

for every $0 \leq s \leq t,\ \tau \leq T$;

(4) *Let ρ be not identically zero. For every $x \in \widehat{D}$ there is a constant $c_2 > 0$ such that*

$$\|U(t,s)x - U(\tau,s)x\| \leq c_2\rho(|t-\tau|)$$

for every $0 \leq s \leq t,\ \tau \leq T$;

(5) *For every $x \in \widehat{D}$ there exists a constant $c_3 > 0$ such that*

$$\|U(t+\sigma,t)x - U(s+\sigma,s)x\| \leq c_3\rho(|t-s|)$$

for $0 \leq t,\ s,\ s+\sigma \leq T$ with $\sigma \geq 0$;

(6) *For every $x \in \widehat{D}$ there exists a constant $c_4 > 0$ such that*

$$\Big\|U(t,s)x - \prod_{i=1}^{n} J_{\frac{t-s}{n}}\Big(s+i\frac{t-s}{n}\Big)x\Big\| \leq c_4(t-s)\Big(\frac{1}{\sqrt{n}} + \rho\Big(\frac{t-s}{\sqrt{n}}\Big)\Big)$$

for every $(t,s) \in \triangle$.

The approximate result of the evolution operators is given as follows:

THEOREM 1.5.3 *Assume that $A(t) : D \equiv D(A(t)) \subset X \to X$ and $A^{(n)}(t) : D_n \equiv D(A^{(n)}(t)) \subset X \to X$ for $n = 1, 2, \cdots$, satisfy the assumptions of Theorem 1.5.2. Let $\{U(t,s) \mid (t,s) \in \triangle\}$ and $\{U_n(t,s) \mid (t,s) \in \triangle\}$ be evolution operators on $\overline{D}$ and $\overline{D_n}$ corresponding to $A(t)$ and $A^{(n)}(t)$ for every $(t,s) \in \triangle$, respectively. If*

$$(I + \lambda A(t))^{-1}x = \lim_{n\to\infty}\Big(I + \lambda A^{(n)}(t)\Big)^{-1}x$$

uniformly for $0 \leq s \leq t \leq T$, for every $\lambda > 0$ with $\lambda\alpha < 1$ and $x \in \overline{D} \cap \cap_{n=1}^{\infty}\overline{D_n}$ then

$$U(t,s)x = \lim_{n\to\infty} U_n(t,s)x$$

uniformly for $0 \leq s \leq t \leq T$, for every $x \in \overline{D} \cap \cap_{n=1}^{\infty}\overline{D_n}$.

In the next four sections we are concerned with non-autonomous nonlinear evolutions of the type

$$\text{(NNE)} \qquad \begin{cases} \dfrac{dx}{dt}(t) + A(t)x(t) \ni f(t), \quad s \le t \le T \\ x(s) = x_0 \end{cases}$$

in a real Banach space, where $T > 0$, $0 \le s \le T$, $A(t) : X \supset D(A(t)) \to 2^X$ for every $t \in [0,T]$, $f(t) : [0,T] \to X$ and $x_0 \in X$.

Let us define several kinds of solutions for (NNE).

1. A continuous function $x(t) : [s,T] \to X$ is called a (classical) solution of (NNE) on $[s,T]$ if:

(1) $x(t)$ is continuously differentiable on $[s,T]$;
(2) $x(t)$ satisfies (NNE) on $[s,T]$.

2. A continuous function $x(t) : [s,T] \to X$ is called a strong solution of (NNE) on $[s,T]$ if:

(1) $x(s) = x_0$;
(2) $x(t)$ is absolutely continuous on every compact interval of $[s,T]$;
(3) $x(t)$ is differentiable a.e. on (s,T);
(4) $x(t)$ satisfies

$$\frac{dx}{dt}(t) + A(t)x(t) \ni f(t)$$

a.e. on (s,T).

3. A continuous function $x(t) : [s,T] \to X$ is called an integral solution of (NNE) on $[s,T]$ if:

(1) $x(s) = x_0$;
(2) $x(t)$ satisfies

$$\|x(t) - x\| \le \|x(s) - x\| + \int_s^t [f(\tau) - y, x(\tau) - x]_+ d\tau$$

for every $[x,y] \in A(\sigma)$ with $0 \le \sigma \le T$ and $0 \le s \le t$.

4. A continuous function $x(t) : [s,T] \to X$ is called a limit solution of (NNE) on $[s,T]$ if:

(1) $x(s) = x_0$;

associated with an operator $A(t)$ in X for every $t \in [s,T]$, where $T > 0$, $0 \le s \le T$ $A(t) : X \supset D(A(t)) \to 2^X$ is a multi-valued operator for every $t \in [0,T]$ and $x_0 \in X$.

Crandall–Pazy proved that $x(t) = U(t,s)x_0$ is a strong solution on $[s,T]$ of (1.7.1) under the additional assumptions, where $\{U(t,s) \mid (t,s) \in \triangle\}$ is the evolution operator in Section 1.4 for every $x_0 \in \overline{D}$ corresponding to $A(t)$ that $A(t) + \alpha I$ is accretive in X.

PROPOSITION 1.7.1 *Let $A(t) : X \supset D(A(t)) \equiv D \to 2^X$ be an operator of X and $\alpha \in \boldsymbol{R}$ such that $A(t) + \alpha I$ is accretive in X and*

$$R(I + \lambda A(t)) \supset \overline{D}$$

for every $\lambda > 0$ with $\lambda\alpha < 1$ and every $t \in [0,T]$. We assume that there exist a continuous function $h : [0,T] \to X$ of bounded variation and a non-decreasing function $L : [0,\infty) \to [0,\infty)$ such that

$$\|A_\lambda(t)x - A_\lambda(s)x\| \le \|h(t) - h(s)\| L(\|x\|)(1 + \|A_\lambda(s)x\|)$$

for every $\lambda > 0$ with $\lambda\alpha < 1$, $x \in \overline{D}$ and t, $s \in [0,T]$. From Theorem 1.5.1 let $\{U(t,s) \mid (t,s) \in \triangle\}$ be the evolution operator on $\overline{D}$ of type α. If $x(t) : [s,T] \to X$ is a strong solution of (1.7.1) *on $[s,T]$ for every $x_0 \in \widehat{D}$ then $x(t) = U(t,s)x_0$ for every $(t,s) \in \triangle$.*

PROPOSITION 1.7.2 *Let the assumptions of Proposition 1.7.1 be satisfied. Moreover, let $A(t)$ be closed in X for every $t \in [0,T]$. If $U(t,s)x$ is differentiable at $t = t_0 \in (s,T]$ for every $x \in \overline{D}$ then $U(t_0,s)x \in D$ and*

$$\frac{dU}{dt}(t_0,s)x + A(t_0)U(t_0,s)x \ni 0.$$

From Proposition 1.7.1 and Proposition 1.7.2 we have the Crandall–Pazy theorem as follows:

THEOREM 1.7.1 (Crandall–Pazy [29]) *Let X be reflexive. Let the assumptions of Proposition 1.7.1 be satisfied. Moreover, let $A(t)$ be closed in X for every $t \in [0,T]$. Let $x_0 \in \widehat{D}$ and let $0 \le s < T$. Then there exists uniquely a strong solution $x(t)$ of (1.7.1) on $[s,T]$ with*

$$x(t) = U(t,s)x_0 = \lim_{n\to\infty} \prod_{i=1}^{n} J_{\frac{t-s}{n}}\Big(s + i\frac{t-s}{n}\Big)x_0.$$

Moreover, $x(t) \in \widehat{D}$ for every $(t,s) \in \triangle$.

1.8. Evans' Theorem

In this section we deal with Evans' theorem for non-autonomous non-homogeneous nonlinear evolutions of the type

$$\begin{cases} \dfrac{dx}{dt}(t) + A(t)x(t) \ni f(t), \quad 0 \le t \le T \\ x(0) = x_0, \end{cases} \tag{1.8.1}$$

associated with an operator $A(t)$ in X for every $t \in [0,T]$, where $T > 0$, $A(t) : X \supset D(A(t)) \to 2^X$ is a multi-valued operator for every $t \in [0,T]$, $f(t) : [0,T] \to X$ and $x_0 \in X$.

By the definition it is easy to see that if $A(t) : X \supset D(A(t)) \to 2^X$ is m-accretive in X for every $t \in [0,T]$, $x_0 \in D(A)$ and $f \in L^1(0,T;X)$, then a strong solution of (1.8.1) on $[0,T]$ is an integral solution of (1.8.1) on $[0,T]$, and a limit solution of (1.8.1) on $[0,T]$ is also an integral solution of (1.8.1).

The existence and uniqueness theorems are as follows:

THEOREM 1.8.1 (Evans [47]) *Let $A(t) : X \supset D(A(t)) \to 2^X$ be an m-accretive operator of X. We assume that there exist an integrable function $h : [0,T] \to X$, and a non-decreasing continuous function $L : [0,\infty) \to [0,\infty)$ such that*

$$\|A_\lambda(t)x - A_\lambda(s)x\| \le \|h(t) - h(s)\| L(\|x\|)$$

for every sufficiently small $\lambda > 0$, $x \in X$ and a.e. t, $s \in [0,T]$. Let $f(t) : [0,T] \to X$ be an integrable function and let $x_0 \in \overline{D}$. Then there exists a unique limit solution $x(t)$ of (1.8.1) on $[0,T]$.

THEOREM 1.8.2 (Evans [47]) *Let $A(t) : X \supset D(A(t)) \to 2^X$ be an m-accretive operator of X. We assume that there exist a measurable function $h : [0,T] \to X$ of bounded variation, and a non-decreasing continuous function $L : [0,\infty) \to [0,\infty)$ such that*

$$\|A_\lambda(t)x - A_\lambda(s)x\| \le \|h(t) - h(s)\| L(\|x\|)(1 + \|A_\lambda(s)x\|)$$

for every sufficiently small $\lambda > 0$, $x \in X$ and a.e. t, $s \in [0,T]$. Let $f(t) : [0,T] \to X$ be a measurable function of bounded variation and let $x_0 \in \widehat{D}$. Then there exists a unique limit solution $x(t)$ of (1.8.1) on $[0,T]$.

THEOREM 1.8.3 *Let $A(t) : X \supset D(A(t)) \to 2^X$ be an m-accretive operator of X. Let $f(t)$, $g(t) : [0,T] \to X$ be functions in the assumptions* *of Theorem 1.8.1 (or Theorem 1.8.2), and* x_0, $y_0 \in \overline{D}$ *(or* x_0, $y_0 \in \widehat{D}$*). Let* $x(t)$ *and* $y(t)$ *be limit solutions of*

$$\begin{cases} \dfrac{dx}{dt}(t) + A(t)x(t) \ni f(t), & 0 \le t \le T \\ x(0) = x_0, \end{cases}$$

and

$$\begin{cases} \dfrac{dy}{dt}(t) + A(t)y(t) \ni g(t), & 0 \le t \le T \\ y(0) = y_0, \end{cases}$$

on $[0, T]$*, respectively. Then*

$$\|x(t) - y(t)\| \le \|x(s) - y(s)\| + \int_s^t \|f(\tau) - g(\tau)\| d\tau \qquad (1.8.2)$$

for every $(t, s) \in \triangle$.

1.9. Pavel's Theorem

In this section we deal with Pavel's theorem for non-autonomous non-homogeneous nonlinear evolutions of the type

$$\begin{cases} \dfrac{dx}{dt}(t) + A(t)x(t) \ni f(t), & 0 \le t \le T \\ x(0) = x_0, \end{cases} \qquad (1.9.1)$$

associated with an operator $A(t)$ in X for every $t \in [0, T]$, where $T > 0$, $A(t) : X \supset D(A(t)) \to 2^X$ is a multi-valued operator for every $t \in [0, T]$, $f(t) : [0, T] \to X$ and $x_0 \in X$.

We consider existence and uniqueness results of (1.9.1) of Pavel [117].

PROPOSITION 1.9.1 *Let* $A(t) : X \supset D(A(t)) \to 2^X$ *be m-accretive in* X *for every* $t \in [0, T]$*. Assume that there exist two continuous function* $q : [0, T] \to X$ *and* $L : [0, \infty) \to [0, \infty)$ *such that*

$$\langle v_1 - v_2, u_1 - u_2 \rangle_s \ge -\|q(t) - q(s)\| \|u_1 - u_2\| L(\|u_2\|)$$

for every t, $s \in [0, T]$, $[u_1, v_1] \in A(t)$ *and* $[u_2, v_2] \in A(s)$*. Assume that* $u \in \overline{D(A(t))}$ *for* t, $t_n \in [0, T]$ *and* $u_n \in D(A(t_n))$, $n = 1, 2, \cdots$ *with* $\lim_{n \to \infty} t_n = t$ *and* $\lim_{n \to \infty} u_n = u \in X$*. Then* $\overline{D(A(t))}$ *is independent of* $t \in [0, T]$*, and thus so is* $\widehat{D}(A(t))$.

In this case we put $\overline{D} \equiv \overline{D(A(t))}$.

THEOREM 1.9.1 (Pavel [117]) *Under the assumptions of Proposition 1.9.1, let $f(t) : [0, T] \to X$ be continuous. Let $x_0 \in \overline{D}$. Then there exists a unique limit solution $x(t)$ of* (1.9.1) *on* $[-r, T]$.

PROPOSITION 1.9.2 *Under the assumptions of Proposition 1.9.1 let $f(t)$, $g(t) : [0, T] \to X$ be continuous. Let x_0, $y_0 \in \overline{D}$ (or x_0, $y_0 \in \widehat{D}$). Let $x(t)$ and $y(t)$ be limit solutions of*

$$\begin{cases} \dfrac{dx}{dt}(t) + A(t)x(t) \ni f(t), & 0 \le t \le T \\ x(0) = x_0, \end{cases}$$

and

$$\begin{cases} \dfrac{dy}{dt}(t) + A(t)y(t) \ni g(t), & 0 \le t \le T \\ y(0) = y_0, \end{cases}$$

on $[0, T]$, *respectively. Then we have*

$$\|x(t) - y(t)\| \le \|x(s) - y(s)\| + \int_s^t \|f(\tau) - g(\tau)\| d\tau \qquad (1.9.2)$$

for every $(t, s) \in \triangle$.

1.10. Notes for References

The results in this chapter can be easily found in textbooks or monographs on nonlinear functional analysis. For example, see Barbu [5], Brézis [15], Deimling [32], Ha [58], Lakshmikantham–Leela [101], Martin [104], Miyadera [106], Morsanu [108], Pascali–Sburlan [116], Pavel [117], and Vrabie [135]. Thus results are only stated without proof except Proposition 1.4.3 and Proposition 1.4.4 concerning the generalized domain of the non-autonomous nonlinear operator. The proofs can be found in the text books.

Almost of the notions and properties of accretive operators in Section 1.1 and nonlinear semigroups in Section 1.2 are due to Crandall–Liggett [26], while Theorem 1.2.3 for nonlinear compact semigroups is taken from Brézis [16] and Vrabie [135]. Theorem 1.2.6 was first proved by Brézis–Pazy [17] in the case $\alpha = 0$. Theorem 1.2.7 is due to Lumer–Phillips [102] and Pazy [118] for linear semigroups of contraction. Theorem 1.3.1 was formulated and proved by Bénilan [6], [7]. If X is reflexive Theorem 1.3.1

can be strengthened by Theorem 1.3.2. The concept of the generalized domain was from Crandall [24]. Proposition 1.4.3 and Proposition 1.4.4 are due to Kato [95]. Theorems in Section 1.5, Section 1.6, and Section 1.7 are taken from Crandall–Pazy [29]. The results in Section 1.8 owed to Evans [47] and also Kobayashi [97], Kobayashi [98]. Section 1.9 is taken from Pavel [117].

Moreover, one can refer to Davies [31], Dunford–Schwartz [37], Goldstein [53], Hille–Phillips [68], Nagel [109], Pazy [118], and Yosida [140], [141] for linear semigroups in real Banach spaces, and, also Crandall [22], [23], Kato [95] in real Banach spaces, Brézis [14], [15], Komura [99], [100] in real Hilbert spaces, Bénilan [6], [7] in L^1, Bénilan [8] in L^p for $1 \leq p \leq \infty$ and Bénilan–Ha [9], Ha [57] in L^∞ for nonlinear semigroups.

Chapter 2

AUTONOMOUS NONLINEAR FUNCTIONAL EVOLUTIONS

The aim of this chapter is to have results concerning autonomous nonlinear functional evolutions associated with accretive operators in real Banach spaces. In Section 2.1 and Section 2.2 we consider the global existence and uniqueness of solutions with definitions. Section 2.3 and Section 2.4 deal with compactness methods and L^p-space methods for existence of solutions, respectively. Section 2.5 is devoted to general methods for existence of solutions. Section 2.6 contains the stability results of solutions. In Section 2.7 examples and applications are treated. Finally, comments and notes for references are given in Section 2.8.

We consider existence and uniqueness of solutions of autonomous nonlinear functional evolutions of the type

$$\begin{cases} \dfrac{dx}{dt}(t) + Ax(t) \ni G(x_t), & t \geq 0 \\ x(t) = \phi(t), & -r \leq t \leq 0 \end{cases}$$

for every $\phi \in E$, where $E = C([-r,0];X)$ of all continuous functions from $[-r,0]$ into X with the sup-norm $\|\cdot\|_C$ defined by

$$\|\psi\|_C = \sup_{t\in[-r,0]} \|\psi(t)\|$$

for every $\psi \in C([-r,0];X)$, or $E = L^p(-r,0;X)$, $1 \leq p < \infty$, of all mesurable functions $\psi : [-r,0] \to X$ such that $\|\psi(t)\|^p$ is integrable on $[-r,0]$ with the L^p-norm $\|\cdot\|_p$ defined by

$$\|\psi\|_p = \Big(\int_{-r}^{0} \|\psi(t)\|^p dt\Big)^{\frac{1}{p}}$$

for every $\psi \in L^p(-r,0;X)$ for $r>0$, and $x_t : [-r,0] \to X$ defined by $x_t(s) = x(t+s)$ for every $s \in [-r,0]$ and $t \geq 0$.

On the other hand, to study existence and uniqueness of solutions of autonomous nonlinear functional evolutions of the type

$$\begin{cases} \dfrac{dx}{dt}(t) + Ax(t) \ni G(x_t), & t \geq 0 \\ x(t) = \phi(t), & -\infty < t \leq 0 \end{cases}$$

for every $\phi \in E$, E may be chosen to be the weighted sup-norm space of the type

$$E_v = \{\psi \in C((-\infty,0];X) \mid v\psi \in BUC((-\infty,0];X)\}$$

with the norm $\|\cdot\|_{E_v}$ defined by

$$\|\psi\|_{E_v} = \sup_{\tau \in (-\infty,0]} v(\tau)\|\psi(\tau)\|$$

for every $\psi \in E_v$, with the weight function $v : (-\infty,0] \to (0,1]$ having the following properties:

(v1) v is continuous, non-decreasing and $v(0) = 1$;

(v2) There exists a constant $c_v \geq 0$ such that

$$\left|\frac{v(\tau+\sigma)}{v(\tau)} - 1\right| \leq c_v|\sigma|$$

for every τ, $\sigma \leq 0$, where $C((-\infty,0];X)$ is the real Banach space of all continuous functions from $(-\infty,0]$ into X and $BUC((-\infty,0];X)$ is the real Banach space of all bounded uniformly continuous functions from $(-\infty,0]$ into X.

Typical such weight functions are $v(\tau) \equiv 1$, in this case

$$E_v = BUC((-\infty,0];X)$$

with the sup-norm, $v(\tau) = e^{\mu\tau}$, or $v(\tau) = (1+|\tau|)^{-\mu}$, $\mu \geq 0$ (spaces of a 'fading memory type'). E_v are somtimes called UC_g-spaces with $v = 1/g$.

We also can take E as follows:

(a) $C_{v_b}((-\infty,0];X) = \{\psi \in C((-\infty,0];X) \mid v\psi \text{ is bounded}\}$;

(b) $C_{v_0}((-\infty,0];X)$
$= \{\psi \in C((-\infty,0];X) \mid \lim_{\tau\to-\infty} v(\tau)\psi(\tau) = 0\}$;

(c) $E_{v_l} = \{\psi \in E_v \mid \lim_{\tau \to -\infty} v(\tau)\psi(\tau) \text{ exists}\}$;

(d) $E_{v_0} = \{\psi \in E_v \mid \lim_{\tau \to -\infty} v(\tau)\psi(\tau) = 0\}$,

with the weighted sup-norm.

In this chapter we consider the case for $E = C([-r,0];X)$ or $E = L^p(-r,0;X)$, except Section 2.5 in which we treat the case for E with the norm $\|\cdot\|_E$, in general.

2.1. Existence of Solutions of $(dx/dt)(t) = G(x_t)$

In this section we consider nonlinear functional evolutions of the type

$$\begin{cases} \dfrac{dx}{dt}(t) = G(x_t), & t \geq 0 \\ x(t) = \phi(t), & -r \leq t \leq 0. \end{cases} \tag{2.1.1}$$

in a real Banach space X, where $r > 0$, $x(t) : [-r, \infty) \to X$ is an unknown function, $G : C([-r,0];X) \to X$ and $\phi \in C([-r,0];X)$.

We now give the definition for a solution of (2.1.1).

DEFINITION 2.1.1 *A function $x(t) : [-r, \infty) \to X$ is a (classical) solution of* (2.1.1) *on* $[-r, \infty)$ *if:*

(1) $x(t) = \phi(t)$ for every $t \in [-r, 0]$;

(2) $x(t)$ is continuously differentiable on $[0, \infty)$;

(3) $x(t)$ satisfies

$$\frac{dx}{dt}(t) = G(x_t)$$

for every $t \geq 0$.

We deal with existence and uniqueness for solutions of (2.1.1) for every $\phi \in C([-r,0];X)$.

We prove the following existence and uniqueness results for (classical) solutions of (2.1.1) by Picard's iteration method under the assumption of uniform Lipschitz continuity of G, as in the case of ordinary linear evolutions.

THEOREM 2.1.1 *Assume that there exists a constant $\beta > 0$ such that*

$$\|G(\psi_1) - G(\psi_2)\| \leq \beta \|\psi_1 - \psi_2\|_C \tag{2.1.2}$$

for every $\psi_i \in C([-r,0];X)$, $i = 1,2$. *Let* $\phi \in C([-r,0];X)$. *Then* (2.1.1) *has a unique (classical) solution* $x(t) \equiv x(t,\phi)$ *on* $[-r,\infty)$ *in the sense that*

$$x(t) = \begin{cases} \phi(0) + \int_0^t G(x_s)ds, & t \geq 0 \\ \phi(t), & -r \leq t \leq 0. \end{cases} \tag{2.1.3}$$

Proof. First we show existence. Let $\phi \in C([-r,0];X)$. We define a sequence $\{x_n(t) \mid n = 0,1,2,\cdots\}$ such that $x_n(t) = \phi(t)$ for every $t \in [-r,0]$ and $x_n(t)$ is continuous for $t \in [0,\infty)$ into X for $n = 0,1,2,\cdots$, as follows:

We define a function $x_0(t) : [-r,\infty) \to X$ by

$$x_0(t) = \begin{cases} \phi(0), & t \geq 0 \\ \phi(t), & -r \leq t \leq 0. \end{cases} \tag{2.1.4}$$

Then $x_0(t)$ is continuous for $t \in [-r,\infty)$ and thus from (2.1.4)

$$(x_0)_t(\tau) = x_0(t+\tau) = \begin{cases} \phi(0), & t+\tau \geq 0 \\ \phi(t+\tau), & -r \leq t+\tau \leq 0 \end{cases} \tag{2.1.5}$$

for every $t \geq 0$ and $\tau \in [-r,0]$.

We show that $(x_0)_t$ is continuous for $t \in [0,\infty)$ into $C([-r,0];X)$. Let $t_0 \in [-r,\infty)$ and let $h \neq 0$ with $t_0 + h \in [-r,\infty)$. Since $x_0(t)$ is continuous for $t \in [-r,\infty)$ for every $\epsilon > 0$ there exists $\delta = \delta(\epsilon, t_0) > 0$ such that $|h| < \delta$ implies

$$\|x_0(t_0+h) - x_0(t_0)\| < \epsilon.$$

Let $\tau \in [-r,0]$ and let $h \neq 0$ with $\tau + h \in [-r,0]$. Let $t \geq 0$. If $|h| < \delta$ then

$$\|x_0(t+\tau+h) - x_0(t+\tau)\| < \epsilon$$

and thus

$$\|(x_0)_t(\tau+h) - (x_0)_t(\tau)\| < \epsilon.$$

Hence $(x_0)_t \in C([-r,0];X)$ for every $t \geq 0$.

Let $t \geq 0$ and let $h \neq 0$ with $t+h \geq 0$. Let $\tau \in [-r,0]$. If $|h| < \delta$ then

$$\|x_0(t+\tau+h) - x_0(t+\tau)\| < \epsilon$$

and thus by (2.1.5)

$$\|(x_0)_{t+h}(\tau) - (x_0)_t(\tau)\| < \epsilon.$$

Hence

$$\|(x_0)_{t+h} - (x_0)_t\|_C = \sup_{\tau \in [-r,0]} \|(x_0)_{t+h}(\tau) - (x_0)_t(\tau)\| \leq \epsilon.$$

Therefore $(x_0)_t$ is continuous for $t \in [0, \infty)$ into $C([-r, 0]; X)$.

We define a function $x_1(t) : [-r, \infty) \to X$ by

$$x_1(t) = \begin{cases} \phi(0) + \int_0^t G((x_0)_s)ds, & t \geq 0 \\ \phi(t), & -r \leq t \leq 0. \end{cases} \tag{2.1.6}$$

Then $x_1(t)$ is continuous for $t \in [-r, \infty)$ and thus from (2.1.6)

$$\begin{aligned} (x_1)_t(\tau) &= x_1(t+\tau) \\ &= \begin{cases} \phi(0) + \int_0^{t+\tau} G((x_0)_s)ds, & t+\tau \geq 0 \\ \phi(t+\tau), & -r \leq t+\tau \leq 0 \end{cases} \end{aligned} \tag{2.1.7}$$

for every $t \geq 0$ and $\tau \in [-r, 0]$.

As in the case of $(x_0)_t$, from (2.1.6) and (2.1.7), $(x_1)_t \in C([-r, 0]; X)$ for every $t \geq 0$ and $(x_1)_t$ is continuous for $t \in [0, \infty)$ into $C([-r, 0]; X)$.

Now we define a function $x_n(t) : [-r, \infty) \to X$ by

$$x_n(t) = \begin{cases} \phi(0) + \int_0^t G((x_{n-1})_s)ds, & t \geq 0 \\ \phi(t), & -r \leq t \leq 0 \end{cases} \tag{2.1.8}$$

for $n = 2, 3, \cdots$. Then we obtain

$$\begin{aligned} (x_n)_t(\tau) &= x_n(t+\tau) \\ &= \begin{cases} \phi(0) + \int_0^{t+\tau} G((x_{n-1})_s)ds, & t+\tau \geq 0 \\ \phi(t+\tau), & -r \leq t+\tau \leq 0 \end{cases} \end{aligned} \tag{2.1.9}$$

for every $t \geq 0$, $\tau \in [-r, 0]$ and $n = 2, 3, \cdots$.

Assume that $(x_n)_t$ is continuous for $t \in [0, \infty)$ into $C([-r, 0]; X)$ with (2.1.8) and (2.1.9) for $n = 2, 3, \cdots$.

Now we define a function $x_{n+1}(t) : [-r, \infty) \to X$ by

$$x_{n+1}(t) = \begin{cases} \phi(0) + \int_0^t G((x_n)_s)ds, & t \geq 0 \\ \phi(t), & -r \leq t \leq 0 \end{cases} \tag{2.1.10}$$

for $n = 2, 3, \cdots$. Then

$$(x_{n+1})_t(\tau) = x_{n+1}(t+\tau) = \begin{cases} \phi(0) + \int_0^{t+\tau} G((x_n)_s)ds, & t+\tau \geq 0 \\ \phi(t+\tau), & -r \leq t+\tau \leq 0 \end{cases} \tag{2.1.11}$$

for every $t \geq 0$, $\tau \in [-r, 0]$ and $n = 2, 3, \cdots$. From the similar way of the case for $(x_0)_t$, $(x_n)_t$ is continuous for $t \in [0, \infty)$ into $C([-r, 0]; X)$ with (2.1.10) and (2.1.11) for $n = 2, 3, \cdots$.

Hence for $x_n(t)$ defined by (2.1.4), (2.1.6) and (2.1.8), $(x_n)_t$ is continuous for $t \in [0, \infty)$ into $C([-r, 0]; X)$ for $n = 0, 1, 2, \cdots$.

On the other hand by (2.1.2) and (2.1.4)

$$\begin{aligned} \|G((x_0)_s)\| &\leq \|G((x_0)_s) - G(\bar{0})\| + \|G(\bar{0})\| \\ &\leq \beta\|(x_0)_s\|_C + \|G(\bar{0})\| \\ &\leq \beta\|\phi\|_C + \|G(\bar{0})\| \\ &\equiv M \end{aligned}$$

for every $s \geq 0$, where $\bar{0}$ is the zero element of $C([-r, 0]; X)$. Thus it follows from (2.1.4), (2.1.6) and this inequality that

$$\|x_1(t) - x_0(t)\| \leq \int_0^t \|G((x_0)_s)\|ds \leq Mt$$

for every $t \geq 0$ and thus

$$\|(x_1)_t - (x_0)_t\|_C = \sup_{\tau \in [-r,0]} \|x_1(t+\tau) - x_0(t+\tau)\| \leq Mt$$

for every $t \geq 0$. From (2.1.6), (2.1.8) and (2.1.2) we have

$$\begin{aligned} \|(x_2)_t - (x_1)_t\|_C &= \sup_{\tau \in [-r,0]} \|(x_2)_t(\tau) - (x_1)_t(\tau)\| \\ &= \sup_{\tau \in [-r,0]} \|x_2(t+\tau) - x_1(t+\tau)\| \\ &\leq \sup_{\tau \in [-r,0]} \int_0^{t+\tau} \|G((x_1)_s) - G((x_0)_s)\|_C ds \\ &\leq \beta \int_0^t \|(x_1)_s - (x_0)_s\|_C ds \\ &\leq \frac{M\beta}{2} t^2 \end{aligned}$$

for every $t \geq 0$. We assume that

$$\|(x_n)_t - (x_{n-1})_t\|_C \leq \frac{M\beta^{n-1}}{n!}t^n$$

for every $t \geq 0$ and $n = 3, 4, \cdots$. Then by (2.1.9) and (2.1.2)

$$\begin{aligned}\|(x_{n+1})_t - (x_n)_t\|_C &\leq \beta \int_0^t \|(x_n)_s - (x_{n-1})_s\|_C ds \\ &\leq \frac{M\beta^n}{(n+1)!}t^{n+1}\end{aligned}$$

for every $t \geq 0$ and $n = 3, 4, \cdots$. Using mathematical induction

$$\|(x_n)_t - (x_{n-1})_t\|_C \leq \frac{M\beta^{n-1}}{n!}t^n \tag{2.1.12}$$

for every $t \geq 0$ and $n = 1, 2, \cdots$ and then

$$\begin{aligned}\|x_{n+1}(t) - x_n(t)\| &\leq \beta \int_0^t \|(x_n)_s - (x_{n-1})_s\|_C ds \\ &\leq \frac{M\beta^n}{(n+1)!}t^{n+1} \qquad (2.1.13)\end{aligned}$$

for every $t \geq 0$ and $n = 0, 1, 2, \cdots$. Thus (2.1.13) implies that the series

$$\sum_{i-0}^{\infty}(x_{i+1}(t) - x_i(t))$$

converges uniformly on every bounded interval of $[0, \infty)$. Since

$$x_n(t) = x_0(t) + \sum_{i=1}^{n}(x_i(t) - x_{i-1}(t))$$

for every $t \geq 0$ and $n = 1, 2, \cdots$ the sequence $\{x_n(t)\}$ converges uniformly on every bounded interval of $[0, \infty)$. Similarly from (2.1.12) the sequence $\{(x_n)_s\}$ converges uniformly on every bounded interval of $[0, \infty)$ and thus from (2.1.2) so does the sequence $\{G((x_n)_s)\}$.

We define a function $x(t) : [-r, \infty) \to X$ by

$$x(t) = \lim_{n\to\infty} x_n(t)$$

for every $t \in [-r, \infty)$. Then $x(t) = \phi(t)$ for every $t \in [-r, 0]$ and $x(t)$ is continuous for $t \in [-r, \infty)$. Thus $x_t \in C([-r, 0]; X)$. Letting $n \to \infty$ in (2.1.8) we obtain (2.1.3). Thus $x(t)$ is differentiable on $[0, \infty)$,

$$\frac{dx}{dt}(t) = G(x_t)$$

for every $t \geq 0$ and hence $x(t)$ is differentially continuous on $[0, \infty)$. Therefore $x(t)$ is a (classical) solution of (2.1.1) on $[-r, \infty)$.

Next, we now consider the uniqueness. Let $x(t)$, $y(t)$ be two (classical) solutions of (2.1.1) on $[-r, \infty)$. Then from (2.1.2) and (2.1.3)

$$\|x_t - y_t\|_C \leq \beta \int_0^t \|x_s - y_s\|_C ds$$

for every $t \geq 0$. By Gronwall's inequality $x_t = y_t$ in $C([-r, 0]; X)$ for every $t \geq 0$ and thus $x(t) = y(t)$ in X on $[-r, \infty)$.

We are concerned with the representation theorem for solutions of (2.1.1) which is obtained as nonlinear semigroups in $C([-r, 0]; X)$.

We define a nonlinear operator $B : D(B) \subset C([-r, 0]; X) \to C([-r, 0]; X)$ by

$$\begin{aligned} D(B) &= \left\{ \psi \in C([-r, 0]; X) \;\middle|\; \frac{d\psi}{dt} \in C([-r, 0]; X),\ \frac{d\psi}{dt}(0) = G(\psi) \right\}, \\ B\psi &= -\frac{d\psi}{dt}, \quad \psi \in D(B). \end{aligned} \tag{2.1.14}$$

PROPOSITION 2.1.1 *Under the assumptions of Theorem 2.1.1 let B be the operator as in* (2.1.14). *Let $\gamma \geq \beta$. Then:*

(1) $B + \gamma I$ is accretive in $C([-r, 0]; X)$;
(2) $R(I + \lambda B) = C([-r, 0]; X)$ for every $\lambda > 0$ with $\lambda\gamma < 1$;
(3) $\overline{D(B)} = C([-r, 0]; X)$.

Proof (1) Let $\lambda > 0$. Let

$$\psi_i = (I + \lambda(B + \gamma I))^{-1} \phi_i$$

for every $\phi_i \in D((I + \lambda(B + \gamma I))^{-1})$, $i = 1, 2$. Then $\psi_i \in D(B)$ and $(I + \lambda(B + \gamma I))\psi_i = \phi_i$, $i = 1, 2$. From the definition of B in (2.1.15), for $i = 1, 2$

$$(1 + \lambda\gamma)\psi_i(t) - \lambda \frac{d\psi_i}{dt}(t) = \phi_i(t) \tag{2.1.15}$$

for every $t \in [-r, 0]$, and

$$(1 + \lambda\gamma)\psi_i(0) - \lambda G(\psi_i) = \phi_i(0). \tag{2.1.16}$$

By (2.1.15) we have

$$\psi_i(t) = e^{(\gamma + \frac{1}{\lambda})t} \psi_i(0) + \frac{1}{\lambda} \int_t^0 e^{(\gamma + \frac{1}{\lambda})(t - \sigma)} \phi_i(\sigma) d\sigma$$

for every $t \in [-r, 0]$ and $i = 1, 2$. Thus

$$\begin{aligned}\|\psi_1(t) - \psi_2(t)\| &\leq e^{(\gamma+\frac{1}{\lambda})t}\|\psi_1(0) - \psi_2(0)\| \\ &\quad + \frac{1}{1+\lambda\gamma}(1 - e^{(\gamma+\frac{1}{\lambda})t})\|\phi_1 - \phi_2\|_C\end{aligned}$$

for every $t \in [-r, 0]$ and then

$$\begin{aligned}&\|\psi_1 - \psi_2\|_C \\ &\quad \leq \max\Big\{\|\psi_1(0) - \psi_2(0)\|, \frac{1}{1+\lambda\gamma}\|\phi_1 - \phi_2\|_C\Big\}. \qquad (2.1.17)\end{aligned}$$

Since using (2.1.16)

$$\begin{aligned}&\|\psi_1(0) - \psi_2(0)\| \\ &\quad \leq \frac{1}{1+\lambda\gamma}\|\phi_1(0) - \phi_2(0)\| + \frac{\lambda\beta}{1+\lambda\gamma}\|\psi_1 - \psi_2\|_C \\ &\quad \leq \frac{1}{1+\lambda\gamma}\|\phi_1 - \phi_2\|_C + \frac{\lambda\beta}{1+\lambda\gamma}\|\psi_1 - \psi_2\|_C\end{aligned}$$

(2.1.17) implies

$$\|\psi_1 - \psi_2\|_C \leq \frac{1}{1+\lambda\gamma}\|\phi_1 - \phi_2\|_C + \frac{\lambda\beta}{1+\lambda\gamma}\|\psi_1 - \psi_2\|_C$$

and thus

$$\|\psi_1 - \psi_2\|_C \leq \|\phi_1 - \phi_2\|_C.$$

Therefore $B + \gamma I$ is accretive in $C([-r, 0] : X)$ for every $\gamma \geq \beta$.

(2) Let $\phi \in C([-r, 0]; X)$. We define a function $\psi(t) : [-r, 0] \to X$ by

$$\psi(t) = e^{\frac{t}{\lambda}}\psi(0) + \frac{1}{\lambda}\int_t^0 e^{\frac{t-\sigma}{\lambda}}\phi(\sigma)d\sigma \qquad (2.1.18)$$

for every $t \in [-r, 0]$. Then $\psi \in C([-r, 0]; X)$, and since

$$\frac{d\psi}{dt}(t) = \frac{1}{\lambda}(\psi(t) - \phi(t)) \qquad (2.1.19)$$

for every $t \in [-r, 0]$ $\dfrac{d\psi}{dt} \in C([-r, 0]; X)$.

On the other hand, we define an operator $Q : X \to X$ by

$$Qx = \phi(0) + \lambda G\Big(xe^{\frac{t}{\lambda}} + \frac{1}{\lambda}\int_t^0 e^{\frac{t-\sigma}{\lambda}}\phi(\sigma)d\sigma\Big) \qquad (2.1.20)$$

for every $x \in X$ and $t \in [-r, 0]$. Then

$$\|Qx_1 - Qx_2\| \leq \lambda\beta\|x_1 - x_2\|$$

for every $x_i \in X$, $i = 1, 2$. Since $0 < \lambda\beta \leq \lambda\gamma < 1$ from Banach's contraction principle there exists $x \in X$ uniquely such that $Qx = x$.

We put $\psi(0) = x$, by (2.1.18) and (2.1.20)

$$\psi(0) = \phi(0) + \lambda G(\psi),$$

and thus from (2.1.19)

$$G(\psi) = \frac{1}{\lambda}\{\psi(0) - \phi(0)\} = \frac{d\psi}{dt}(0).$$

Thus $\psi \in D(B)$ and from (2.1.19)

$$(I + \lambda B)\psi = \psi - \lambda\frac{d\psi}{dt} = \phi.$$

Hence $R(I + \lambda B) = C([-r, 0]; X)$ for every $\lambda > 0$ with $\lambda\gamma < 1$.

(3) Let $\phi \in C([-r, 0]; X)$. Let $0 < \alpha \leq \dfrac{1}{\gamma}$ and $0 < \lambda < \alpha$. It follows from (2) that there exists $\psi_\lambda \in D(B)$ with

$$\psi_\lambda + \lambda B\psi_\lambda = \phi.$$

By (2.1.19)

$$\psi_\lambda(0) - \lambda G(\psi_\lambda) = \phi(0), \tag{2.1.21}$$

$$\psi_\lambda(t) - \lambda\frac{d\psi_\lambda}{dt}(t) = \phi(t) \tag{2.1.22}$$

for every $t \in [-r, 0]$. From (2.1.22)

$$\psi_\lambda(t) = \psi_\lambda(0)e^{\frac{t}{\lambda}} + \frac{1}{\lambda}\int_t^0 e^{\frac{t-\sigma}{\lambda}}\phi(\sigma)d\sigma \tag{2.1.23}$$

for every $t \in [-r, 0]$ and

$$\begin{aligned}\|\psi_\lambda(t)\| &\leq \|\psi_\lambda(0)\|e^{\frac{t}{\lambda}} + \|\phi\|_C(1 - e^{\frac{t}{\lambda}}) \\ &\leq \|\psi_\lambda(0)\| + \|\phi\|_C\end{aligned}$$

for every $t \in [-r, 0]$. Thus

$$\|\psi_\lambda\|_C \leq \|\psi_\lambda(0)\| + \|\phi\|_C.$$

On the other hand from (2.1.21)

$$\begin{aligned} \|\psi_\lambda(0)-\phi(0)\| &= \lambda\|G(\psi_\lambda)\| \\ &\le \lambda(\beta\|\psi_\lambda\|_C + \|G(\bar{0})\|) \\ &\le \lambda\beta\|\psi_\lambda(0)-\phi(0)\| \\ &\quad +\lambda(\beta\|\phi(0)\| + \|\phi\|_C) + \|G(\bar{0})\|) \end{aligned}$$

and since $\lambda\beta < 1$

$$\|\psi_\lambda(0)-\phi(0)\| \le \frac{\lambda}{1-\lambda\beta}(\beta(\|\phi(0)\| + \|\phi\|_C) + \|G(\bar{0})\|).$$

Thus $\lim_{\lambda\to 0+}\psi_\lambda(0) = \phi(0)$. Using (2.1.23)

$$\begin{aligned} &\psi_\lambda(t)-\phi(t) \\ &\quad = e^{\frac{t}{\lambda}}(\psi_\lambda(0)-\phi(t)) + \frac{1}{\lambda}\int_t^0 e^{\frac{t-\sigma}{\lambda}}(\phi(\sigma)-\phi(t))d\sigma. \end{aligned} \tag{2.1.24}$$

Let $\delta > 0$. Let

$$\epsilon(\delta) = \sup_{|\sigma-t|<\delta}\|\phi(\sigma)-\phi(t)\|.$$

Since $\phi(t)$ is uniformly continuous for $t\in[-r,0]$ $\lim_{\delta\to 0+}\epsilon(\delta) = 0$.

Let $-\delta \le t \le 0$. By (2.1.24)

$$\|\psi_\lambda(t)-\phi(t)\| \le \|\psi_\lambda(0)-\phi(0)\| + 2\epsilon(\delta). \tag{2.1.25}$$

Let $-r \le t \le -\delta$. Again, (2.1.24) implies that

$$\begin{aligned} &\|\psi_\lambda(t)-\phi(t)\| \\ &\quad \le e^{-\frac{\delta}{\lambda}}(\|\psi_\lambda(0)\| + \|\phi\|_C) \\ &\qquad +\frac{1}{\lambda}\int_t^{t+\delta} e^{\frac{t-\sigma}{\lambda}}\|\phi(\sigma)-\phi(t)\|d\sigma \\ &\qquad +\frac{1}{\lambda}\int_{t+\delta}^{0} e^{\frac{t-\sigma}{\lambda}}\|\phi(\sigma)-\phi(t)\|d\sigma \\ &\quad \le e^{-\frac{\delta}{\lambda}}(\|\psi_\lambda(0)\| + \|\phi\|_C) + \epsilon(\delta) + 2e^{-\frac{\delta}{\lambda}}\|\phi\|_C. \end{aligned} \tag{2.1.26}$$

From (2.1.25) and (2.1.26)

$$\lim_{\lambda\to 0+}\|\psi_\lambda-\phi\|_C = 0.$$

Since $\psi_\lambda \in D(B)$ $\phi \in \overline{D(B)}$. Hence $\overline{D(B)} = C([-r,0]:X)$.

From Theorem 1.2.1 let $\{S(t) \mid t \geq 0\}$ be the nonlinear semigroup on $\overline{D(B)}$ of type γ generated by $-B$. Let $x(t)$ be the unique (classical) solution of (2.1.1) on $[-r, \infty)$. We show that $S(t)\phi = x_t$ for every $t \geq 0$ and $\phi \in \overline{D(B)}$.

Proposition 2.1.2 *Under the assumptions of Theorem 2.1.1 let* $x(t) \equiv x(t, \phi)$ *be the unique solution of* (2.1.1) *on* $[-r, \infty)$ *for every* $\phi \in \overline{D(B)}$. *Define a function* $y(t) : [0, \infty) \to C([-r, 0]; X)$ *by* $y(t) = x_t$ *for every* $t \in [0, \infty)$. *Then*

(1) $y(t)$ *is a (classical) solution of*

$$\begin{cases} \dfrac{dy}{dt}(t) + By(t) = 0, \quad t \geq 0 \\ y(0) = \phi \end{cases} \tag{2.1.27}$$

on $[0, \infty)$ *for every* $\phi \in D(B)$;

(2) $x_t \equiv x_t(\phi) : D(B) \subset C([-r, 0]; X) \to C([-r, 0]; X)$ *is continuous in* $\phi \in D(B)$ *for every* $t \geq 0$.

Proof. (1) Let $\phi \in D(B)$ and let $x(t) \equiv x(t, \phi)$ be the unique solution of (2.1.1) on $[-r, \infty)$ with the form (2.1.3). Then $x_t \in C([-r, 0]; X)$ for every $t \geq 0$ and $x_t : [0, \infty) \to C([-r, 0]; X)$ is continuous, as in the proof of Theorem 2.1.1. By (2.1.3)

$$x_t(\tau) = \begin{cases} \phi(0) + \displaystyle\int_0^{t+\tau} G(x_s)ds, \quad -t \leq \tau \leq 0, \\ \phi(t + \tau), \quad -r \leq \tau \leq -t. \end{cases} \tag{2.1.28}$$

Now we show that $x_t \in D(B)$ for every $t \geq 0$. Since $\phi \in D(B)$

$$\frac{d\phi}{dt} \in C([-r, 0]; X),$$
$$\frac{d\phi}{dt}(0) = G(\phi)$$

and thus

$$\phi(\tau) = \phi(0) + \int_0^{\tau} \frac{d\phi}{d\sigma}(\sigma)d\sigma$$

for every $\tau \in [-r, 0]$. Since from (2.1.3)

$$\phi(0) = x(t) - \int_0^t G(x_s)ds$$

for every $t \geq 0$ by (2.1.28)

$$x_t(\tau) = \begin{cases} x(t) + \int_t^{t+\tau} G(x_s)ds, & -t \leq \tau \leq 0, \\ x(t) - \int_0^t G(x_s)ds + \int_0^{t+\tau} \frac{d\phi}{ds}(s)ds, & -r \leq \tau \leq -t. \end{cases} \tag{2.1.29}$$

We define a function $u(s) : [-r, \infty) \to X$ by

$$u(s) = \begin{cases} G(x_s), & s \geq 0, \\ \frac{d\phi}{ds}(s), & -r \leq s \leq 0. \end{cases} \tag{2.1.30}$$

Then $u_s \in C([-r, 0]; X)$ for every $s \geq 0$ and u_s is continuous for $s \geq 0$. Using (2.1.29) and (2.1.30) we obtain

$$\begin{aligned} x_t(\tau) &= x(t) + \int_t^{t+\tau} u(s)ds \\ &= x_t(0) + \int_0^{\tau} u_t(s)ds \end{aligned} \tag{2.1.31}$$

for every $t \geq 0$ and $\tau \in [-r, 0]$. Thus

$$\frac{dx_t}{d\tau}(\tau) = u_t(\tau) \in C([-r, 0]; X)$$

and

$$\frac{dx_t}{d\tau}(0) = u_t(0) = u(t) = G(x_t)$$

for every $t \geq 0$. Hence $x_t \in D(B)$ and

$$Bx_t = -\frac{dx_t}{d\tau} = -u_t \tag{2.1.32}$$

for every $t \geq 0$.

We show that x_t is continuously differentiable for every $t \geq 0$. x_t is continuous for every $t \geq 0$. Let $t \geq 0$ and $s \geq t$. From (2.1.31)

$$x(s + \tau) - x(s) = \int_0^{\tau} u(s + \sigma)d\sigma \tag{2.1.33}$$

for every $s \geq 0$ and $\tau \in [-r, 0]$. We put $\tau = t - s + \theta$ for $\theta \in [-r, 0]$ in (2.1.33),

$$\begin{aligned} x(t + \theta) - x(s) &= \int_0^{t-s+\theta} u(s + \sigma)d\sigma \\ &= \int_s^{t+\theta} u(\sigma)d\sigma. \end{aligned} \tag{2.1.34}$$

By (2.1.33) and (2.1.34)

$$\begin{aligned} x_t(\theta) - x_s(\theta) &= x(t+\theta) - x(s+\theta) \\ &= \int_s^{t+\theta} u(\sigma)d\sigma - \int_s^{s+\theta} u(\sigma)d\sigma \\ &= \int_{s+\theta}^{t+\theta} u(\sigma)d\sigma = \int_s^t u(\sigma+\theta)d\sigma \\ &= \int_s^t u_\sigma(\theta)d\sigma \end{aligned} \tag{2.1.35}$$

for every $\theta \in [-r, 0]$. Thus from (2.1.35)

$$x_t - x_s = \int_s^t u_\sigma d\sigma$$

for every t, $s \geq 0$ and

$$\frac{dx_t}{dt} = u_t \tag{2.1.36}$$

for every $t \geq 0$.

Hence x_t is continuously differentiable for every $t \geq 0$. (2.1.32) and (2.1.36) imply

$$\frac{dx_t}{dt} + Bx_t = 0$$

for every $t \geq 0$. Therefore $y(t) = x_t$ is a (classical) solution of (2.1.27) on $[0, \infty)$ with $y(0) = x_0 = \phi$. Since, from (1) of Proposition 2.1.1, $B + \gamma I$ is accretive for $\gamma \geq \beta$, the uniqueness of (2.1.27) follows.

(2) Let ϕ, $\psi \in D(B)$. Let $x(t) \equiv x(t, \phi)$ and $y(t) \equiv y(t, \psi)$ be two (classical) solutions of (2.1.1) on $[-r, \infty)$. From (2.1.2) and (2.1.3)

$$\begin{aligned} &\|x(s,\phi) - y(s,\psi)\| \\ &\quad \leq \|\phi(0) - \psi(0)\| + \beta \int_0^s \|x_\sigma(\phi) - y_\sigma(\psi)\|_C d\sigma \end{aligned} \tag{2.1.37}$$

for every $s \geq 0$. From definition of the sup-norm

$$\begin{aligned} &\|x_t(\phi) - y_t(\psi)\|_C \\ &\quad = \sup_{-r \leq s \leq 0} \|x_t(\phi)(s) - y_t(\psi)(s)\| \\ &\quad = \sup_{-r \leq s \leq 0} \|x(t+s,\phi) - y(t+s,\psi)\| \\ &\quad \leq \sup_{-r \leq s \leq t} \|x(s,\phi) - y(s,\psi)\| \leq \end{aligned}$$

$$\leq \sup_{-r\leq s\leq 0} \|x(s,\phi) - y(s,\psi)\| + \sup_{0\leq s\leq t} \|x(s,\phi) - y(s,\psi)\|$$
$$= \|\phi - \psi\|_C + \sup_{0\leq s\leq t} \|x(s,\phi) - y(s,\psi)\| \qquad (2.1.38)$$

for every $t \geq 0$. Thus by (2.1.37) and (2.1.38)

$$\|x_t(\phi) - y_t(\psi)\|_C$$
$$\leq 2\|\phi - \psi\|_C + \beta \int_0^t \|x_\sigma(\phi) - y_\sigma(\psi)\|_C d\sigma$$

for every $t \geq 0$. From Gronwall's inequality

$$\|x_t(\phi) - y_t(\psi)\|_C \leq 2e^{\beta t}\|\phi - \psi\|_C$$

for every $t \geq 0$.

COROLLARY 2.1.1 *Under the assumptions of Theorem 2.1.1, let* $\{S(t) \mid t \geq 0\}$ *be the nonlinear semigroup on* $\overline{D(B)}$ *of type* γ *generated by* $-B$ *and let* $x(t)$ *be the unique (classical) solution of* (2.1.1) *on* $[-r, \infty)$. *Then* $S(t)\phi = x_t$ *for every* $\phi \in D(B)$ *and* $t \geq 0$.

Proof. Then from Proposition 2.1.2 x_t is the unique (classical) solution of (2.1.27) on $[0, \infty)$. Thus it follows from Theorem 1.2.6 that $S(t)\phi = x_t$ for every $\phi \in D(B)$ and $t \geq 0$.

We have the representation theorem for (classical) solutions of (2.1.1) on $[-r, \infty)$ as follows:

THEOREM 2.1.2 *Assume that* (2.1.2) *is satisfied and let* B *be the operator defined as in* (2.1.15). *Let* $\{S(t) \mid t \geq 0\}$ *be a nonlinear semigroup on* $C([-r, 0]; X)$, *of type* $\gamma \geq \beta$, *generated by* $-B$. *Let* $x(t) \equiv x(t, \phi)$ *be the solution of* (2.1.1) *on* $[-r, \infty)$ *for every* $\phi \in D(B)$. *Then* $S(t)\phi = x_t(\phi)$ *for every* $\phi \in C([-r, 0]; X)$ *and* $t \geq 0$.

Proof. From Corollary 2.1.1 $S(t)\phi = x_t(\phi)$ for every $\phi \in D(B)$ and $t \geq 0$. Since $\{S(t) \mid t \geq 0\}$ is of type γ on $\overline{D(B)}$, from (3) of Proposition 2.1.1, and (2) of Proposition 2.1.2 $S(t)\phi = x_t(\phi)$ for every $\phi \in C([-r, 0]; X) = \overline{D(B)}$ and $t \geq 0$.

Now we construct the unique (classical) solution of (2.1.1) on $[-r, \infty)$ from the nonlinear semigroup generated by $-B$. First of all, we give a basic result and definition of translation.

LEMMA 2.1.1 *Let* I *and* J *be closed intervals. Let a function* $f : I \times J \to X$ *be continuous. Assume that*

$$\int_\tau^t (f(s,\theta) - f(\sigma,\theta))d\theta = \int_\sigma^s (f(\zeta,t) - f(\zeta,\tau))d\zeta \qquad (2.1.39)$$

holds. Then

$$f(\zeta_1, \theta_1) = f(\zeta_2, \theta_2)$$

for every $(\zeta_i, \theta_i) \in I \times J$, $i = 1, 2$ *with* $\zeta_1 + \theta_1 = \zeta_2 + \theta_2$.

Proof. We define a function $g : I \times I \times J \times J \to X$ by

$$g(\sigma, s, \tau, t) = \int_\tau^t \int_\sigma^s f(\zeta, \theta) d\zeta d\theta$$

for every $(\sigma, s, \tau, t) \in I \times I \times J \times J$. Then g is continuously partially differentiable on $I \times I \times J \times J$. From (2.1.39)

$$\frac{\partial g}{\partial \sigma} + \frac{\partial g}{\partial s} = \frac{\partial g}{\partial \tau} + \frac{\partial g}{\partial t}. \tag{2.1.40}$$

We define a function $h : I \times I \times J \times J \to X$ by

$$h(\alpha) = g(\sigma - \alpha, s - \alpha, \tau + \alpha, t + \alpha)$$

for every $\alpha \in \mathbf{R}$ with

$$(\sigma - \alpha, s - \alpha, \tau + \alpha, t + \alpha) \in I \times I \times J \times J.$$

By (2.1.40) $\dfrac{dh}{d\alpha} = 0$ and thus $h(\alpha) = h(0)$. The proof is completed.

DEFINITION 2.1.2 *Let I and J be closed intervals with the right endpoint 0 and the left endpoint 0, respectively. A function $U : J \to C(I; X)$ is called a translation if $U(t) = x_t$ for every $t \in J$, where $x : I \cup J \to X$ is defined by*

$$x(s) = \begin{cases} U(0)(s), & s \in I \\ U(s)(0), & s \in J. \end{cases}$$

PROPOSITION 2.1.3 *Assume that B is an operator and $\gamma \in \boldsymbol{R}$ such that $B + \gamma I$ is accretive in $C([-r, 0]; X)$ and $R(I + \lambda B) = C([-r, 0]; X)$ for every $\lambda > 0$ with $\lambda\gamma < 1$. Let $\{S(t) \mid t \geq 0\}$ be a nonlinear semigroup on $\overline{D(B)}$, of type γ, generated by $-B$. We define the function $U : [0, \infty) \to C([-r, 0]; X)$ by $U(t) = S(t)\phi$ for every $t \geq 0$ and $\phi \in \overline{D(B)}$. Then U is a translation.*

Proof. We define a function $f : [-r, 0] \times [0, \infty) \to X$ by

$$f(\sigma, t) = (S(t)\phi)(\sigma)$$

for every $(\sigma, t) \in [-r, 0] \times [0, \infty)$.

Then f is continuous on $[-r, 0] \times [0, \infty)$. From Theorem 1.2.1,

$$S(t)\phi = \lim_{n \to \infty,\ \frac{m}{n} \to t} \Big(I + \frac{1}{n}B\Big)^{-m} \phi$$

for every $t \geq 0$ and $\phi \in \overline{D(B)}$. A function $f_n : [-r, 0] \times [0, \infty) \to X$ by

$$f_n(\sigma, t) = \sum_i \chi_{(\frac{i-1}{n}, \frac{i}{n}]}(t) \Big(\Big(I + \frac{1}{n}B\Big)^{-1} \phi\Big)(\sigma)$$

for every $(\sigma, t) \in [-r, 0] \times [0, \infty)$ and $n = 1, 2, \cdots$. Then $\lim_{n \to \infty} f_n = f$ uniformly on every compact subset of $[-r, 0] \times [0, \infty)$. We put $x_i = f_n\Big(\cdot, \frac{i}{n}\Big)$ for $i = 1, 2, \cdots$. Then

$$\begin{cases} n(x_i - x_{i-1}) + Bx_i = 0 \\ x_0 = \phi \end{cases} \tag{2.1.41}$$

and

$$Bx_i = -\frac{\partial x_i}{\partial \zeta}.$$

Thus from (2.1.41)

$$\begin{aligned} f_n\Big(s, \frac{i}{n}\Big) &- f_n\Big(\sigma, \frac{i}{n}\Big) \\ &= \int_\sigma^s \frac{\partial}{\partial \zeta} f_n\Big(\zeta, \frac{i}{n}\Big) d\zeta \\ &= n \int_\sigma^s \Big(f_n\Big(\zeta, \frac{i}{n}\Big) - f_n\Big(\zeta, \frac{i-1}{n}\Big)\Big) d\zeta. \end{aligned} \tag{2.1.42}$$

Dividing by n and summing (2.1.42) for $i = j + 1, \cdots, k$

$$\int_{\frac{j}{n}}^{\frac{k}{n}} (f_n(s, \theta) - f_n(\sigma, \theta)) d\theta = \int_\sigma^s \Big(f_n\Big(\zeta, \frac{k}{n}\Big) - f_n\Big(\zeta, \frac{j}{n}\Big)\Big) d\zeta.$$

Letting $n \to \infty$, $\frac{j}{n} \to \tau$ and $\frac{k}{n} \to t$, we have

$$\int_\tau^t (f(s, \theta) - f(\sigma, \theta)) d\theta = \int_\sigma^s (f(\zeta, t) - f(\zeta, \tau)) d\zeta$$

for every σ, $\tau \in (-\infty, 0]$ and t, $s \in [0, \infty)$. Let $(\sigma, t) \in (-\infty, 0] \times [0, \infty]$. Using Lemma 2.1.1

$$f(\sigma, t) = \begin{cases} f(0, \sigma + t), & t + \sigma \geq 0 \\ f(t + \sigma, 0), & t + \sigma \leq 0. \end{cases} \tag{2.1.43}$$

for every $\phi \in C([-r,0];X)$ and $t \geq 0$. Thus $x(t)$ is differentiable on $[0,\infty)$, and

$$\frac{dx}{dt}(t) = G(x_t)$$

for every $t \geq 0$. Hence $x(t)$ is continuously continuous on $[0,\infty)$. From (2.1.45) $x(t) = \phi(t)$ for every $t \in [-r,0]$. Therefore $x(t)$ is a (classical) solution of (2.1.1) on $[-r,\infty)$. The uniqueness is trivial from its representation.

2.2. Existence of Solutions of $(dx/dt)(t) + Ax(t) \ni G(x_t)$

In this section we deal with autonomous nonlinear functional evolutions of the type

$$\begin{cases} \dfrac{dx}{dt}(t) + Ax(t) \ni G(x_t), & t \geq 0 \\ x(t) = \phi(t), & -r \leq t \leq 0. \end{cases} \tag{2.2.1}$$

associated with an operator A in X, where $x(t) : [-r,\infty) \to X$ is an unknown function, $G : C([-r,0];X) \to X$ and $\phi \in C([-r,0];X)$.

We give definitions of several solutions for (2.2.1).

DEFINITION 2.2.1 *1. A function $x(t) : [-r,\infty) \to X$ is a (classical) solution of* (2.2.1) *on $[-r,\infty)$ if:*

(1) $x(t) = \phi(t)$ for every $t \in [-r,0]$;

(2) $x(t)$ is continuously differentiable on $[0,\infty)$;

(3) $x(t)$ satisfies

$$\frac{dx}{dt}(t) + Ax(t) \ni G(x_t)$$

for every $t \geq 0$.

2. A function $x(t) : [-r,\infty) \to X$ is called a strong solution of (2.2.1) *on $[-r,\infty)$ if:*

(1) $x(t) = \phi(t)$ for every $t \in [-r,0]$;

(2) $x(t)$ is continuous on $[0,\infty)$;

(3) $x(t)$ is absolutely continuous on every compact interval of $[0,\infty)$;

(4) $x(t)$ is differentiable a.e. on $(0,\infty)$;

(5) $x(t)$ satisfies

$$\frac{dx}{dt}(t) + Ax(t) \ni G(x_t)$$

a.e. on $(0, \infty)$.

3. A function $x(t) : [-r, \infty) \to X$ *is called an integral solution of* (2.2.1) *on* $[-r, \infty)$ *if:*

(1) $x(t) = \phi(t)$ *for every* $t \in [-r, 0]$;

(2) $x(t)$ *is continuous on* $[0, \infty)$;

(3) $x(t)$ *satisfies*

$$\|x(t) - x\| \le \|x(s) - x\| + \int_s^t [G(x_\tau) - y, x(\tau) - x]_+ d\tau.$$

for every $[x, y] \in A$ *and* $t,\ s \in [0, \infty)$.

4. A function $x(t); [-r, \infty) \to X$ *is called a limit solution of* (2.2.1) *on* $[-r, \infty)$ *if:*

(1) $x(t) = \phi(t)$ *for every* $t \in [-r, 0]$;

(2) $x(t)$ *is continuous on* $[0, \infty)$;

(3) $x(t)$ *is a limit solution of*

$$\begin{cases} \dfrac{dx}{dt}(t) + Ax(t) \ni f(t), \ t \ge 0 \\ x(0) = \phi(0) \end{cases}$$

on $[0, \infty)$ *with* $f(t) = G(x_t)$ *for every* $t \ge 0$.

For existence and uniqueness of solutions of (2.2.1), as in Section 2.1, we define a nonlinear operator $B : D(B) \subset C([-r, 0]; X) \to C([-r, 0; X)$ by

$$\begin{aligned} D(B) &= \Big\{\psi \in C([-r, 0]; X) \ \Big| \ \frac{d\psi}{dt} \in C([-r, 0]; X), \ \psi(0) \in D(A), \\ &\qquad\qquad \frac{d\psi}{dt}(0) + A\psi(0) \ni G(\psi)\Big\}, \\ B\psi &= -\frac{d\psi}{dt}, \quad \psi \in D(B). \end{aligned} \tag{2.2.2}$$

PROPOSITION 2.2.1 *Let* $A : X \supset D(A) \to 2^X$ *be an operator and* $\alpha \in \boldsymbol{R}$ *such that* $A + \alpha I$ *is accretive in* X *and* $R(I + \lambda A) = X$ *for every* $\lambda > 0$ *with* $\lambda\alpha < 1$. *Assume that* $G : C([-r, 0]; X) \to X$ *satisfies*

$$\|G(\psi_1) - G(\psi_2)\| \le \beta\|\psi_1 - \psi_2\|_C \tag{2.2.3}$$

for every $\psi_i \in C([-r,0];X)$, $i = 1,2$, *and for some* $\beta > 0$. *Let* B *be the operator defined in* (2.2.2). *We put* $\gamma = \max\{0, \alpha + \beta\}$. *Then:*

(1) $B + \gamma I$ *is accretive in* $C([-r,0];X)$;

(2) $R(I + \lambda B) = C([-r,0];X)$ *for every* $\lambda > 0$ *with* $\lambda\gamma < 1$;

(3) $\overline{D(B)} = \{\phi \in C([-r,0];X) \mid \phi(0) \in \overline{D(A)}\}$.

Proof. We first prove (2). Let $\phi \in C([-r,0];X)$. We define a function $\zeta : [-r,0] \to X$ by

$$\zeta(t) = \int_t^0 e^{\frac{t-\sigma}{\lambda}} \phi(\sigma) d\sigma \tag{2.2.4}$$

for every $t \in [-r,0]$. Then $\zeta \in C([-r,0];X)$. We define an operator $Q : X \to X$ by

$$Qx = J_\lambda^A\Big(\phi(0) + \lambda G\Big(e^{\frac{t}{\lambda}} x + \frac{1}{\lambda}\zeta\Big)\Big) \tag{2.2.5}$$

for every $x \in X$ and $t \in [-r,0]$. Then by (2.2.3)

$$\begin{aligned}
&\|Qx_1 - Qx_2\| \\
&\le \frac{\lambda}{1-\lambda\alpha}\Big\|G\Big(e^{\frac{t}{\lambda}}x_1 + \frac{1}{\lambda}\zeta\Big) - G\Big(e^{\frac{t}{\lambda}}x_2 + \frac{1}{\lambda}\zeta\Big)\Big\| \\
&\le \frac{\lambda\beta}{1-\lambda\alpha}\|e^{\frac{t}{\lambda}}x_1 - e^{\frac{t}{\lambda}}x_2\|_C \\
&\le \frac{\lambda\beta}{1-\lambda\alpha}\sup_{-r\le t\le 0} e^{\frac{t}{\lambda}}\|x_1 - x_2\| \\
&\le \frac{\lambda\beta}{1-\lambda\alpha}\|x_1 - x_2\|
\end{aligned}$$

for every $x_i \in X$, $i = 1,2$ and $t \in [-r,0]$. Since $0 < \dfrac{\lambda\beta}{1-\lambda\alpha} < 1$ Q is a strict contraction. From Banach's contraction principle there exists $x_0 \in X$ uniquely such that $Qx_0 = x_0$. From (2.2.4) and (2.2.5)

$$x_0 = J_\lambda^A\Big(\phi(0) + \lambda G\Big(e^{\frac{t}{\lambda}}x_0 + \frac{1}{\lambda}\int_t^0 e^{-\frac{t-\sigma}{\lambda}}\phi(\sigma)d\sigma\Big)\Big) \tag{2.2.6}$$

for every $t \in [-r,0]$ and thus $x_0 \in D(A)$. We put

$$\psi(t) = e^{\frac{t}{\lambda}}x_0 + \frac{1}{\lambda}\zeta(t) = e^{\frac{t}{\lambda}}x_0 + \frac{1}{\lambda}\int_t^0 e^{-\frac{t-\sigma}{\lambda}}\phi(\sigma)d\sigma \tag{2.2.7}$$

for every $t \in [-r, 0]$. Then $\psi \in C([-r, 0]; X)$. By (2.2.7) $\psi(0) = x_0$ and $\frac{d\psi}{dt} \in C([-r, 0]; X)$. Using (2.2.6)

$$\psi(0) = J_\lambda^A(\phi(0) + \lambda G(\psi)), \tag{2.2.8}$$

$$\frac{1}{\lambda}(\psi(0) - \phi(0)) \in G(\psi) - A\psi(0). \tag{2.2.9}$$

But then since from (2.2.7) for every $t \in [-r, 0]$

$$\begin{aligned} \frac{d\psi}{dt}(t) &= \frac{1}{\lambda}(\psi(t) - \phi(t)), \\ \frac{d\psi}{dt}(0) &= \frac{1}{\lambda}(\psi(0) - \phi(0)), \end{aligned} \tag{2.2.10}$$

it follows from (2.2.9) and (2.2.10) that we have

$$\frac{d\psi}{dt}(0) \in G(\psi) - A\psi(0).$$

Thus $\psi \in D(B)$. From

$$\psi(t) - \lambda\frac{d\psi}{dt}(t) = \phi(t)$$

for every $t \in [-r, 0]$

$$(I + \lambda B)\psi = \psi - \lambda\frac{d\psi}{dt} = \phi$$

and then

$$R(I + \lambda B) = C([-r, 0]; X)$$

for every $\lambda > 0$ with $\lambda\gamma < 1$.

(1) Let $\phi_i \in C([-r, 0]; X)$, $i = 1, 2$. From (2) there exists $\psi_i \in D(B)$ such that

$$R(I + \lambda B)\psi_i = \phi_i$$

for $i = 1, 2$. Thus (2.2.3), (2.2.7) and (2.2.8) imply

$$\begin{aligned} &\|\psi_1(t) - \psi_2(t)\| \\ &\quad\le e^{\frac{t}{\lambda}}\|\psi_1(0) - \psi_2(0)\| + \frac{1}{\lambda}\int_t^0 e^{\frac{t-\sigma}{\lambda}}\|\phi_1(\sigma) - \phi_2(\sigma)\|d\sigma \\ &\quad\le \frac{e^{\frac{t}{\lambda}}}{1 - \lambda\alpha}(\|\phi_1(0) - \phi_2(0)\| + \lambda\beta\|\psi_1 - \psi_2\|_C) \\ &\qquad+(1 - e^{\frac{t}{\lambda}})\|\phi_1 - \phi_2\|_C \\ &\quad\le \frac{1 - \lambda\alpha + \lambda\alpha e^{\frac{t}{\lambda}}}{1 - \lambda\alpha}\|\phi_1 - \phi_2\|_C + \frac{\lambda\beta e^{\frac{t}{\lambda}}}{1 - \lambda\alpha}\|\psi_1 - \psi_2\|_C. \end{aligned}$$

for every $t \in [-r, 0]$. Thus

$$\|\psi_1 - \psi_2\|_C \leq \frac{1 - \lambda\alpha + \lambda\alpha e^{\frac{t}{\lambda}}}{1 - \lambda\alpha - \lambda\beta e^{\frac{t}{\lambda}}} \|\phi_1 - \phi_2\|_C \tag{2.2.11}$$

for every $t \in [-r, 0]$.

If $\alpha + \beta \geq 0$

$$\frac{1 - \lambda\alpha + \lambda\alpha e^{\frac{t}{\lambda}}}{1 - \lambda\alpha - \lambda\beta e^{\frac{t}{\lambda}}} \leq \frac{1}{1 - \lambda(\alpha + \beta)} = \frac{1}{1 - \lambda\gamma}$$

and if $\alpha + \beta \geq 0$

$$\frac{1 - \lambda\alpha + \lambda\alpha e^{\frac{t}{\lambda}}}{1 - \lambda\alpha - \lambda\beta e^{\frac{t}{\lambda}}} \leq 1 < \frac{1}{1 - \lambda\gamma}.$$

From (2.2.11)

$$\|\psi_1 - \psi_2\|_C \leq \frac{1}{1 - \lambda\gamma} \|\phi_1 - \phi_2\|_C. \tag{2.2.12}$$

Thus $(I + \lambda B)^{-1}$ exists for every $\lambda > 0$ with $\lambda\gamma < 1$, and by (2.2.12)

$$\|(I + \lambda B)^{-1}\phi_1 - (I + \lambda B)^{-1}\phi_2\|_C \leq \frac{1}{1 - \lambda\gamma} \|\phi_1 - \phi_2\|_C$$

Hence $B + \gamma I$ is accretive in $C([-r, 0]; X)$ for every $\lambda > 0$ with $\lambda\gamma < 1$.

(3) From (1) and (2) above, and (3) of Theorem 1.1.4

$$\overline{D(B)} = \{\phi \in C([-r, 0]; X) \mid \lim_{\lambda \to 0+} J_\lambda^B \phi = \phi\}.$$

We put

$$D = \{\phi \in C([-r, 0]; X) \mid \phi(0) \in \overline{D(A)}\},$$
$$D_0 = \{\phi \in C([-r, 0]; X) \mid \lim_{\lambda \to 0+} J_\lambda^B \phi = \phi\}$$

and we prove $D = D_0$.

Indeed, let $\phi \in D$. Then $\phi \in C([-r, 0]; X)$ and $\phi(0) \in \overline{D(A)}$. By (2.2.7)

$$\begin{aligned} &\|\psi(t) - \phi(t)\| \\ &= \left\| e^{\frac{t}{\lambda}} \psi(0) + \frac{1}{\lambda} \int_0^t e^{\frac{t-\sigma}{\lambda}} \phi(\sigma) d\sigma - \phi(t) \right\| \\ &\leq e^{\frac{t}{\lambda}} \|\psi(0) - \phi(0)\| \\ &\quad + \left\| \frac{1}{\lambda} \int_t^0 e^{\frac{t-\sigma}{\lambda}} (\phi(\sigma) - \phi(0)) d\sigma - (\phi(t) - \phi(0)) \right\| \end{aligned} \tag{2.2.13}$$

for every $t \in [-r, 0]$. Since, using (2.2.8) and (2.2.3)

$$\begin{aligned}
&\|\psi(0) - \phi(0)\| \\
&\quad = \|J_\lambda^A(\phi(0) + \lambda G(\psi)) - \phi(0)\| \\
&\quad \le \|J_\lambda^A(\phi(0) + \lambda G(\psi)) - J_\lambda^A \phi(0)\| + \|J_\lambda^A \phi(0) - \phi(0)\| \\
&\quad \le \frac{\lambda}{1-\lambda\alpha}\|G(\psi)\| + \|J_\lambda^A \phi(0) - \phi(0)\| \\
&\quad \le \frac{\lambda}{1-\lambda\alpha}(\beta\|\psi - \phi\|_C + \|G(\phi)\|) + \|J_\lambda^A \phi(0) - \phi(0)\|,
\end{aligned}$$

(2.2.13) implies

$$\begin{aligned}
&\|\psi - \phi\|_C \\
&\quad \le \frac{1-\lambda}{1-\lambda(\alpha+\beta)}\Big(\Big\|\frac{1}{\lambda}\int_t^0 e^{\frac{t-\sigma}{\lambda}}(\phi(\sigma) - \phi(0))d\sigma - (\phi(t) - \phi(0))\Big\| \\
&\qquad\qquad + \frac{\lambda}{1-\lambda\alpha}\|G(\phi)\| + \|J_\lambda^A \phi(0) - \phi(0)\|\Big).
\end{aligned}$$

Hence $\lim_{\lambda\to 0+} J_\lambda^B \phi = \phi$ and thus $\phi \in D_0$.

Let $\phi \in D_0$. Then $\phi \in C([-r, 0]; X)$ and $\lim_{\lambda\to 0+} J_\lambda^B \phi = \phi$. From (2.2.8)

$$\begin{aligned}
&\|J_\lambda^A \psi(0) - \phi(0)\| \\
&\quad = \|J_\lambda^A \psi(0) - J_\lambda^A(\psi(0) + \lambda G(\phi))\| \\
&\quad \le \frac{\lambda}{1-\lambda\alpha}\|G(\psi)\| \\
&\quad \le \frac{\lambda}{1-\lambda\alpha}(\beta\|\phi - J_\lambda^B \phi\|_C + \|G(\psi)\|).
\end{aligned}$$

Hence $\lim_{\lambda\to 0+} J_\lambda^A \psi(0) = \phi(0)$ and thus $\phi(0) \in \overline{D(A)}$. That is, $\phi \in D$. Therefore $D = D_0$.

In the following theorems we obtain existence and uniqueness results for solutions of (2.2.1).

THEOREM 2.2.1 *Let $A : D(A) = X \to X$ be a continuous m-accretive operator of X. Under the assumptions of Proposition 2.2.1 let B be the operator defined in (2.2.2). Then there exists a unique (classical) solution $x(t) \equiv x(t, \phi)$ of (2.2.1) on $[-r, \infty)$ for every $\phi \in \widehat{D}(B)$.*

Proof. From (3) of Proposition 2.2.1 $\overline{D(B)} = C([-r, 0]; X)$. Let $\{S(t) \mid t \ge 0\}$ be the nonlinear semigroup on $C([-r, 0]; X)$ of type γ generated by $-B$. We define $x(t) \equiv x(t, \phi) : [-r, \infty) \to X$ by

$$x(t) = \begin{cases} \phi(t), & -r \le t \le 0 \\ (S(t)\phi)(0), & t \ge 0 \end{cases} \tag{2.2.14}$$

for every $\phi \in C([-r, 0]; X)$. By Proposition 2.1.3 and (2.2.14)

$$S(t)\phi = x_t(\phi) \tag{2.2.15}$$

for every $t \ge 0$ and $\phi \in C([-r, 0]; X)$. Let $\lambda\gamma < 1$. Using Theorem 1.2.4 there exists the unique (classical) solution $x_\lambda(t)$ of

$$\begin{cases} \dfrac{dx_\lambda}{dt}(t) + B_\lambda x_\lambda(t) = 0, \quad t \ge 0 \\ x_\lambda(0) = \phi \end{cases} \tag{2.2.16}$$

on $[0, \infty)$ and

$$\lim_{\lambda \to 0+} x_\lambda(t) = S(t)\phi \tag{2.2.17}$$

in $C([-r, 0]; X)$ uniformly on every bounded interval of $[0, \infty)$. (2.2.16) and the definition of B in (2.2.2) imply

$$\begin{aligned} (x_\lambda(t))(0) &= \phi(0) + \int_0^t (-BJ_\lambda^B x_\lambda(s))(0)ds \\ &= \phi(0) + \int_0^t \frac{d(J_\lambda x_\lambda(s))}{d\sigma}(0)ds \\ &= \phi(0) \\ &\quad + \int_0^t (-(AJ_\lambda^B x_\lambda(s)(0) + G(J_\lambda^B x_\lambda(s)))ds \end{aligned} \tag{2.2.18}$$

for every $t \ge 0$. From (2.2.17)

$$\lim_{\lambda \to 0+} J_\lambda^B x_\lambda(s) = S(s)\phi.$$

Since A and G are continuous

$$\lim_{\lambda \to 0+} \Big((-AJ_\lambda^B x_\lambda(s))(0) + G(J_\lambda^B x_\lambda(s))\Big) = -(AS(s)\phi)(0) + G(S(s)\phi)$$

for every $s \ge 0$. Moreover,

$$\begin{aligned} &\|AJ_\lambda^B x_\lambda(s)(0) - G(J_\lambda^B x_\lambda(s))\| \\ &\qquad = \|(BJ_\lambda^B x_\lambda(s))(0)\| \\ &\qquad \le \|J_\lambda^B x_\lambda(s)\|_C \\ &\qquad \le \|BJ_\lambda^B \phi\|_C \end{aligned}$$

for every $\phi \in C([-r,0];X)$, $s \in [0,t]$ and $\lambda > 0$, and thus

$$\begin{aligned}\lim_{\lambda\to 0+} \|AJ_\lambda^B x_\lambda(s)(0) - G(J_\lambda^B x_\lambda(s))\| \\ &\leq \lim_{\lambda\to 0+} \|BJ_\lambda^B \phi\|_C \\ &= |B\phi| < \infty\end{aligned}$$

for every $\phi \in \widehat{D}(B)$ and $s \in [0,t]$. Letting $\lambda \to 0+$ in (2.2.18), by Lebesgue's bounded convergence theorem we have

$$(S(t)\phi)(0) = \phi(0) + \int_0^t (-AS(s)\phi)(0) + G(S(s)\phi)))ds$$

for every $t \geq 0$ and thus from (2.2.14) and (2.2.15)

$$x(t) = \phi(0) + \int_0^t (-Ax(s) + G(x_s))ds$$

for every $t \geq 0$ and from (2.2.14) $x(t) = \phi(t)$ for every $t \in [-r,0]$. Therefore $x(t)$ is a (classical) solution of (2.2.1) on $[-r,\infty)$.

We show uniqueness for solutions. Let $x_i(t) \equiv x_i(t;\phi_i)$ be a (classical) solution of (2.2.1) for $\phi_i \in \widehat{D}(B)$, $i = 1,2$. Since A is accretive

$$\Big(\frac{dx_1}{dt}(t) - G((x_1)_t) - \frac{dx_2}{dt}(t) + G((x_2)_t), x_1(t) - x_2(t)\Big)_+ \leq 0$$

for every $t \geq 0$. This implies that

$$\begin{aligned}\|x_1(t) - x_2(t)\| \frac{d}{dt}\|x_1(t) - x_2(t)\| \\ &= \Big[\frac{dx_1}{dt}(t) - \frac{dx_2}{dt}(t), x_1(t) - x_2(t)\Big]_- \\ &\leq [G((x_1)_t) - G((x_2)_t), x_1(t) \quad x_2(t)]_- \\ &\leq \|G((x_1)_t) - G((x_2)_t)\| \|x_1(t) - x_2(t)\| \\ &\leq \beta \|(x_1)_t - (x_2)_t\|_C \|x_1(t) - x_2(t)\|\end{aligned}$$

for every $t \geq 0$ and thus

$$\frac{d}{dt}\|x_1(t) - x_2(t)\| \leq \beta \|(x_1)_t - (x_2)_t\|_C$$

for every $t \geq 0$. Integrating this on $[0,t]$,

$$\|x_1(t) - x_2(t)\| \leq \|\phi_1(0) - \phi_2(0)\| + \beta \int_0^t \|(x_1)_s - (x_2)_s\| ds$$

for every $t \geq 0$. Thus

$$\begin{aligned}
&\|(x_1)_t - (x_2)_t\|_C \\
&\quad = \sup_{\tau\in[-r,0]} \|(x_1)_t(\tau) - (x_2)_t(\tau)\| \\
&\quad = \sup_{\tau\in[-r,0]} \|x_1(t+\tau) - x_2(t+\tau)\| \\
&\quad \leq \|\phi_1(0) - \phi_2(0)\| + \beta \sup_{\tau\in[-r,0]} \int_0^{t+\tau} \|(x_1)_s - (x_2)_s\|_C ds \\
&\quad \leq \|\phi_1(0) - \phi_2(0)\| + \beta \int_0^t \|(x_1)_s - (x_2)_s\|_C ds
\end{aligned}$$

for every $t \geq 0$. From Gronwall's inequality we have

$$\|(x_1)_t - (x_2)_t\|_C \leq \|\phi_1(0) - \phi_2(0)\| e^{\beta t}$$

for every $t \geq 0$. This implies uniqueness.

THEOREM 2.2.2 *Let the dual space X^* of X be uniformly convex. Under the assumptions of Proposition 2.2.1 with $\alpha = 0$ let B be the operator defined as in* (2.2.2). *Then there exists the unique strong solution $x(t) \equiv x(t,\phi)$ of* (2.2.1) *on $[-r,\infty)$ for every $\phi \in \widehat{D}(B)$.*

Proof. Let $\lambda > 0$ with $\lambda\beta < 1$. As in the proof of Theorem 2.2.1, we have

$$(x_\lambda(t))(0) = \phi(0) + \int_0^t (-y_\lambda(s))(0) + G(J_\lambda^B x_\lambda(s)))ds \qquad (2.2.19)$$

for every $s \in [0,t]$, and $\lim_{\lambda\to 0+} x_\lambda(s) = S(s)\phi$ for every $s \in [0,t]$ and $\phi \in \overline{D(B)}$. Here, $y_\lambda(s) \in AJ_\lambda^B x_\lambda(s)$ for every $s \in [0,t]$. Since $J_\lambda^B x_\lambda(s)$ is bounded in λ for every $s \in [0,t]$ from (2.2.3) so is $G(J_\lambda^B x_\lambda(s))$. By

$$\begin{aligned}
&\|(y_\lambda(s))(0) - G(J_\lambda^B x_\lambda(s))\| \\
&\qquad = \|(BJ_\lambda^B x_\lambda(s))(0)\| \\
&\qquad \leq \|BJ_\lambda^B x_\lambda(s)\|_C \\
&\qquad \leq \|BJ_\lambda^B \phi\|_C
\end{aligned}$$

and thus

$$\lim_{\lambda\to 0+} \|(y_\lambda(s))(0) - G(J_\lambda^B x_\lambda(s))\| \leq \lim_{\lambda\to 0+} \|BJ_\lambda^B x_\lambda(s)\|_C = |B\phi|$$

for every $s \in [0,t]$ $(AJ_\lambda^B x_\lambda(s))(0)$ is also bounded as $\lambda \to 0+$ for every $\phi \in \widehat{D}(B)$. Since X is reflexive there exist $y \in X$ and $\lambda_n > 0$ as $n \to \infty$

with $\lambda_n \to 0+$ such that $w - \lim_{n\to\infty}(y_\lambda(s))(0) = y$ for every $s \in [0,t]$. Since by (3) of Proposition 1.1.2 A is demiclosed, $(S(s)\phi)(0) \in D(A)$ and $y \in (AS(s)\phi)(0)$ for every $s \in [0,t]$. Using (2.2.19)

$$\begin{aligned}((x_{\lambda_n}(t)(0), x^*) \quad\quad (2.2.20)\\ = (\phi(0), x^*) + \int_0^t ((-y_{\lambda_n}(s))(0) + G(J^B_{\lambda_n} x_{\lambda_n}(s))), x^*)ds\end{aligned}$$

for every $t \geq 0$ and $x^* \in X^*$. Letting $n \to \infty$ in (2.2.20) by Lebesgue's dominated convergence theorem

$$\begin{aligned}((S(t)\phi)(0), x^*)\\ = (\phi(0), x^*) + \int_0^t ((-y(s))(0) + G(S(s)\phi)(s))), x^*)ds\end{aligned}$$

for every $t \geq 0$ and $x^* \in X^*$. Hence $x(t) = (S(t)\phi)(0)$ is weakly differentiable a.e. $t \in (0,\infty)$. Since $x(t)$ is absolutely continuous in t $x(t)$ is differentiable a.e. $t \in (0,\infty)$. Hence

$$\frac{dx}{dt}(t) + y(t) = G(x_t)$$

a.e. $t \in (0,\infty)$ and thus

$$\frac{dx}{dt}(t) + Ax(t) \ni G(x_t)$$

a.e. $t \in (0,\infty)$. Therefore $x(t)$ is a strong solution of (2.2.1) on $[-r,\infty)$.

We have uniqueness for strong solutions by the same way in the proof of Theorem 2.2.1.

THEOREM 2.2.3 *Under the assumptions of Proposition 2.2.1 let B be the operator defined as in (2.2.2). We define*

$$x(t) = \begin{cases} \phi(t), & -r \leq t \leq 0 \\ (S(t)\phi)(0), & t \geq 0. \end{cases} \quad\quad (2.2.21)$$

Then $x(t) \equiv x(t,\phi)$ is a limit solution of (2.2.1) on $[-r,\infty)$ for every $\phi \in \overline{D(B)}$. In particular, it is an integral solution of (2.2.1) on $[-r,\infty)$, that is,

$$\begin{aligned}\|x(t) - x\| - \|x(s) - x\|\\ \leq \alpha \int_s^t \|x(\tau) - x\|d\tau + \int_s^t [G(x_\tau) - y, x(\tau) - x]_+ d\tau \quad (2.2.22)\end{aligned}$$

for every $0 \le s \le t$ *and* $[x, y] \in A$.

Finally, let $x(t) \equiv x(t, \phi)$ *be an integral solution of* (2.2.1) *on* $[-r, \infty)$ *for* $\phi \in \overline{D(B)}$ *and let* $y(t) \equiv y(t, \psi)$ *be a limit solution of* (2.2.1) *on* $[-r, \infty)$ *for* $\psi \in \overline{D(B)}$. *Then*

$$e^{-\alpha t}\|x(t) - y(t)\| - e^{-\alpha s}\|x(s) - y(s)\|$$
$$\le \int_s^t e^{-\alpha\sigma}[G(S(\sigma)\phi) - G(S(\sigma)\psi), x((\sigma) - y(\sigma)]_+ d\sigma \quad (2.2.23)$$

for every $0 \le s \le t$.

Proof. From Proposition 2.2.1 let $\{S(t) \mid t \ge 0\}$ be the nonlinear semigroup on $\overline{D(B)}$ of type γ, generated by $-B$. Then by Theorem 1.2.1

$$S(t)\phi = \lim_{\lambda \to 0+} (J_\lambda^B)^{[\frac{t}{\lambda}]}\phi \tag{2.2.24}$$

for every $t \ge 0$ and $\phi \in \overline{D(B)}$, where $J_\lambda^B = (I + \lambda B)^{-1}$ for $\lambda > 0$. This convergence is uniform on every bounded interval of $[0, \infty)$ and $[t/\lambda]$ denotes the largest integer $\le t/\lambda$ for $0 < \lambda \le t$. For arbitrarily fixed $T > 0$ choose $\lambda > 0$ with $\lambda\gamma < 1$. Si:nce $B + \gamma I$ is accretive

$$(J_\lambda^B)^i\phi(0) - (J_\lambda^B)^{i-1}\phi(0) = -\lambda B(J_\lambda^B)^i\phi(0).$$

Let $N(n)$ be the smallest integer and $\{\lambda_n\}$, $\lambda_n > 0$, $\lambda_n \to 0+$ such that $\lambda_n N(n) \ge T$ for $n = 1, 2, \cdots$. We put

$$x_0^n = \phi(0), \; x_i^n = ((J_{\lambda_n}^B)^i\phi)(0), \; t_i^n = i\lambda_n \tag{2.2.25}$$

for $i = 1, 2, \cdots, N(n)$ and $n = 1, 2, \cdots$. Using (2.2.25) and the definition of B

$$\frac{((J_{\lambda_n}^B)^i\phi)(0) - ((J_{\lambda_n}^B)^{i-1}\phi)(0)}{\lambda_n}$$
$$= -(B(J_{\lambda_n}^B)^i\phi)(0) \in G((J_{\lambda_n}^B)^i\phi) - A((J_{\lambda_n}^B)^i\phi)(0).$$

Thus

$$\frac{x_i^n - x_{i-1}^n}{t_i^n - t_{i-1}^n} + Ax_i^n \in G((J_{\lambda_n}^B)^i\phi). \tag{2.2.26}$$

for $i = 1, 2, \cdots, N(n)$ and $n = 1, 2, \cdots$. We define

$$\overline{x_i^n}(t) : [-r, \lambda_n N(n)] \to X$$

by

$$\overline{x_i^n}(t) = \begin{cases} \phi(t), \ t \in [-r, 0] \\ x_j^n, \ t \in (t_{j-1}^n, t_j^n], \ j = 1, 2, \cdots, i-1 \\ ((J_{\lambda_n}^B)^i \phi)(0), \ t \in (t_{i-1}^n, \lambda_n N(n)] \end{cases}$$

for $i = 1, 2, \cdots, N(n)$ and $n = 1, 2, \cdots$. We put $x_n(t) = (\overline{x_{N(n)}^n})(t)$ for every $t \in [-r, \lambda_n N(n)]$. Then using (2.2.24)

$$(S(t)\phi)(0) = \lim_{n \to \infty} x_n(t) \tag{2.2.27}$$

uniformly on every bounded interval of $[0, \infty)$. From (2.2.3) and (2.2.24)

$$\begin{aligned} &\sum_{i=1}^{N(n)} \int_{(i-1)\lambda_n}^{i\lambda_n} \|G(S(\sigma)\phi) - G((J_{\lambda_n}^B)^i \phi)\| d\sigma \\ &\leq \int_0^{T+1} \|G(S(\sigma)\phi) - G((J_{\lambda_n}^B)^{[\frac{\sigma}{\lambda_n}]+1} \phi)\| d\sigma \\ &\leq \beta \int_0^{T+1} \|S(\sigma)\phi - (J_{\lambda_n}^B)^{[\frac{\sigma}{\lambda_n}]+1} \phi\| d\sigma \to 0, \end{aligned}$$

and by Proposition 2.1.3

$$f(t) = G(S(t)\phi) = G(x_t)$$

for every $t \in [0, \infty)$. Hence $x(t)$ is a limit solution of (2.2.1) on $[-r, \infty)$.

We now show that (2.2.22) holds. Let $x(t)$ be a limit solution of (2.2.1) on $[-r, \infty)$ and let $[x, y] \in A$. From (2.2.27) $x(t) = \lim_{n \to \infty} x_n(t)$. Using (2.2.26) there exists $y_i^n \in Ax_i^n$ such that

$$\begin{aligned} \frac{x_i^n - x_{i-1}^n}{\lambda_n} &= G((J_{\lambda_n}^B)^i \phi) - y_i^n, \\ (x_i^n - x) - (x_{i-1}^n - x) &= \lambda_n (G((J_{\lambda_n}^B)^i \phi) - y_i^n). \end{aligned}$$

Thus using Lemma 1.1.1 and accretivity of $A + \alpha I$

$$\begin{aligned} &\|x_i^n - x\| - \|x_{i-1}^n - x\| \\ &= \|x_i^n - x\| - \|(x_i^n - x) - \lambda_n (G((J_{\lambda_n}^B)^i \phi) - y_i^n)\| \\ &\leq \lambda_n [G((J_{\lambda_n}^B)^i \phi) - y_i^n, x_i^n - x]_- \\ &\leq \lambda_n [y - y_i^n, x_i^n - x]_- + \lambda_n [G((J_{\lambda_n}^B)^i \phi) - y, x_i^n - x]_+ \\ &\leq \lambda_n \alpha \|x_i^n - x\| + \lambda_n [G((J_{\lambda_n}^B)^i \phi) - y, x_i^n - x]_+ \end{aligned}$$

for $i = 1, 2, \cdots, N(n)$ and $n = 1, 2, \cdots$. (2.2.25) and this inequality imply

$$\begin{aligned}&\|((J^B_{\lambda_n})^i\phi)(0) - x\| - \|((J^B_{\lambda_n})^{i-1}\phi)(0) - x\| \\ &\quad \leq \lambda_n \alpha \|((J^B_{\lambda_n})^i\phi)(0) - x\| \\ &\qquad + \lambda_n [G((J^B_{\lambda_n})^i\phi) - y, ((J^B_{\lambda_n})^i\phi)(0) - x]_+\end{aligned}$$

for $i = 1, 2, \cdots, N(n)$ and $n = 1, 2, \cdots$. From this inequality and

$$((J^B_{\lambda_n})^{[\frac{\tau}{\lambda_n}]})(0) = ((J^B_{\lambda_n})^i\phi)(0)$$

for $i\lambda_n \leq \tau < (i+1)\lambda_n$ we have

$$\begin{aligned}&\|((J^B_{\lambda_n})^i\phi)(0) - x\| - \|((J^B_{\lambda_n})^{i-1}\phi)(0) - x\| \\ &\quad \leq \alpha \int_{i\lambda_n}^{(i+1)\lambda_n} \|((J^B_{\lambda_n})^{[\frac{\tau}{\lambda_n}]}\phi)(0) - x\| d\tau \\ &\qquad + \int_{i\lambda_n}^{(i+1)\lambda_n} [G((J^B_{\lambda_n})^{[\frac{\tau}{\lambda_n}]}\phi) - y, ((J^B_{\lambda_n})^{[\frac{\tau}{\lambda_n}]}\phi)(0) - x]_+ d\tau\end{aligned} \tag{2.2.28}$$

for $i = 1, 2, \cdots, N(n)$ and $n = 1, 2, \cdots$. Let $\lambda_n \leq s < t$ and adding (2.2.28) for $i = [s/\lambda_n] + 1, \cdots, [t/\lambda_n]$, we have

$$\begin{aligned}&\|((J^B_{\lambda_n})^{[\frac{t}{\lambda_n}]}\phi)(0) - x\| - \|((J^B_{\lambda_n})^{[\frac{s}{\lambda_n}]}\phi)(0) - x\| \\ &\quad \leq \alpha \int_{([\frac{s}{\lambda_n}]+1)\lambda_n}^{([\frac{t}{\lambda_n}]+1)\lambda_n} \|((J^B_{\lambda_n})^{[\frac{\tau}{\lambda_n}]}\phi)(0) - x\| d\tau \\ &\qquad + \int_{([\frac{s}{\lambda_n}]+1)\lambda_n}^{([\frac{t}{\lambda_n}]+1)\lambda_n} [G((J^B_{\lambda_n})^{[\frac{\tau}{\lambda_n}]}\phi) - y, ((J^B_{\lambda_n})^{[\frac{\tau}{\lambda_n}]}\phi)(0) - x]_+ d\tau\end{aligned} \tag{2.2.29}$$

for $n = 1, 2, \cdots$. Letting $n \to \infty$ in (2.2.29), by Lebesgue's dominated convergence theorem and (2.2.27) we obtain

$$\begin{aligned}&\|(S(t)\phi)(0) - x\| - \|(S(s)\phi)(0) - x\| \\ &\quad \leq \alpha \int_s^t \|(S(\tau)\phi)(0) - x\| d\tau \\ &\qquad + \int_s^t [G(S(\tau)\phi) - y, (S(\tau)\phi)(0) - x]_+ d\tau\end{aligned}$$

for $t \geq s \geq 0$. Thus from (2.2.27) and Proposition 2.1.3

$$\begin{aligned}&\|x(t) - x\| - \|x(s) - x\| \\ &\quad \leq \alpha \int_s^t \|x(\tau) - x\| d\tau + \int_s^t [G(x_\tau) - y, x(\tau) - x]_+ d\tau\end{aligned}$$

for $0 \le s \le t$, which is (2.2.22).

We next prove that (2.2.23) holds. For simplicity let $\alpha = 0$. To show this we need the following lemma.

LEMMA 2.2.1 *Let $u(t)$, $v(t) : [0,T] \to X$ be continuous. Let f, $g \in L^1(0,T;X)$. Assume that*

$$\int_q^p (\|u(t) - v(\tau)\| - \|u(s) - v(\tau)\|)d\tau$$
$$+ \int_s^t (\|u(\sigma) - v(p)\| - \|u(\tau) - v(q)\|)d\sigma$$
$$\le \int_q^p \Big(\int_s^t [f(\sigma) - g(\tau), u(\sigma) - v(\tau)]_+ d\sigma\Big)d\tau$$

for $0 \le q \le p \le T$ and $0 \le s \le t \le T$. Then

$$\|u(t) - v(t)\| - \|u(s) - v(s)\| \le \int_s^t [f(\sigma) - g(\sigma), u(\sigma) - v(\sigma)]_+ d\sigma$$

for $0 \le s \le t \le T$.

Now let $x(t) \equiv x(t,\phi)$ be an integral solution of (2.2.1) on $[-r,\infty)$ for $\phi \in \overline{D(B)}$ and let $y(t) \equiv y(t,\psi)$ be a limit solution of (2.2.1) on $[-r,\infty)$ for $\psi \in \overline{D(B)}$. By the definition of a limit solution

(1) $y(t) = \phi(t)$ for every $t \in [-r,0]$,

(2) $y(t)$ is continuous on $[0,\infty)$,

(3) $y(t) = \lim_{n\to\infty} y_n(t)$ uniformly on $[0,T]$ for every $T \in (0,\infty)$ and $G(y_t) = \lim_{n\to\infty} G_n(t)$ in $L^1(0,T;X)$

satisfying

$$\begin{aligned}
0 &= t_0^n < t_1^n < \cdots < t_{N(n)}^n = T \text{ for every } T > 0,\\
y_n(0) &= y_0^n = \phi(0),\\
y_n(t) &= y_l^n,\ t_l^n < t \le t_l^n,\\
G_n(t) &= G_l^n,\ t_l^n < t \le t_l^n
\end{aligned}$$

and

$$\frac{y_l^n - y_{l-1}^n}{t_l^n - t_{l-1}^n} + Ay_l^n \ni G_l^n$$

for $l = 1, 2, \cdots, N(n)$; $n = 1, 2, \cdots$, with

$$\lim_{n\to\infty} \max_{1\le l\le N(n)}(t_l^n - t_{l-1}^n) = 0,$$
$$\{y_l^n \in X \mid l = 0, 1, 2, \cdots, N(n); n = 1, 2, \cdots\},$$
$$\{G_l^n \in X \mid l = 0, 1, 2, \cdots, N(n); n = 1, 2, \cdots\}.$$

We put $h_l^n = t_l^n - t_{l-1}^n$ for $l = 1, 2, \cdots, N(n)$. Let $0 \le s \le t \le T$. Since $y_l^n \in D(B)$ and

$$G_l^n - \frac{y_l^n - y_{l-1}^n}{h_l^n} \in Ay_l^n$$

(2.2.22) implies

$$\begin{aligned}
&\|x(t) - y_l^n\| - \|x(s) - y_l^n\| \\
&\quad \le \int_s^t \left[G(x_\sigma) - G_l^n + \frac{y_l^n - y_{l-1}^n}{t_l^n - t_{l-1}^n}, x(\sigma) - y_l^n\right]_+ d\sigma \\
&\quad \le \int_s^t [G(x_\sigma) - G_l^n, x(\sigma) - y_l^n]_+ d\sigma \\
&\qquad + \int_s^t \left[\frac{y_l^n - y_{l-1}^n}{h_l^n}, x(\sigma) - y_l^n\right]_+ d\sigma
\end{aligned}$$

for every $0 \le s \le t$. By the property of $[\cdot, \cdot]_+$,

$$\begin{aligned}
&\left[\frac{y_l^n - y_{l-1}^n}{h_l^n}, x(\sigma) - y_l^n\right]_+ \\
&\quad \le \frac{1}{h_l^n}(\|(x(\sigma) - y_l^n) + (y_l^n - y_{l-1}^n)\| - \|x(\sigma) - y_l^n\|) \\
&\quad = \frac{1}{h_l^n}(\|x(\sigma) - y_{l-1}^n\| - \|x(\sigma) - y_l^n\|).
\end{aligned}$$

Thus we have

$$\begin{aligned}
&h_l^n(\|x(t) - y_l^n\| - \|x(s) - y_l^n\|) \\
&\qquad + \int_s^t (\|x(\sigma) - y_l^n\| - \|x(\sigma) - y_{l-1}^n\|) d\sigma \\
&\quad \le h_l^n \int_s^t [G(x_\sigma) - G_l^n, x(\sigma) - y_l^n]_+ d\sigma.
\end{aligned}$$

Adding these inequalities for $l = j, j + 1, \cdots, i$, we obtain

$$\sum_{l=j+1}^{i} h_l^n(\|x(t) - y_l^n\| - \|x(s) - y_l^n\|)+$$

$$+\int_s^t (\|x(\sigma) - y_i^n\| - \|x(\sigma) - y_j^n\|)d\sigma$$
$$\leq \sum_{l=j+1}^{i} \Big(\int_s^t [G(x_\sigma) - G_l^n, x(\sigma) - y_l^n]_+ d\sigma\Big) h_l^n.$$

Hence

$$\int_{t_{j+1}^n}^{t_i^n} (\|x(t) - y_n(\tau)\| - \|x(s) - y_n(\tau)\|)d\tau$$
$$+\int_s^t (\|x(\sigma) - y_n(t_i^n)\| - \|x(\sigma) - y_n(t_j^n)\|)d\sigma$$
$$\leq \int_{t_{j+1}^n}^{t_i^n} \Big(\int_s^t [G(x_\sigma) - G_n(\tau), x(\sigma) - y_n(\tau)]_+ d\sigma\Big) d\tau$$
$$\leq \int_{t_{j+1}^n}^{t_i^n} \Big(\int_s^t [G(x_\sigma) - G(y_\tau), x(\sigma) - y_n(\tau)]_+ d\sigma\Big) d\tau$$
$$+\int_{t_{j+1}^n}^{t_i^n} \Big(\int_s^t \|G_n(\tau) - G(y_\tau)\| d\sigma\Big) d\tau$$
$$\leq \int_{t_{j+1}^n}^{t_i^n} \Big(\int_s^t [G(x_\sigma) - G(y_\tau), x(\sigma) - y_n(\tau)]_+ d\tau\Big) d\sigma$$
$$+T\int_0^T \|G_n(\tau) - G(y_\tau)\| d\tau$$

We suppose that $t_i^n \to p$ and $t_j^n \to q$ as $n \to \infty$. Then by Lebesgue's dominated convergence theorem and the upper continuity of $[\cdot,\cdot]_+$ we obtain

$$\int_q^p (\|x(t) - y(\tau)\| - \|x(s) - y(\tau)\|)d\tau$$
$$+\int_s^t (\|x(\sigma) - y(p)\| - \|x(\sigma) - y(q)\|)d\sigma$$
$$\leq \int_q^p \int_s^t \overline{\lim}_{n\to\infty} [G(x_\sigma) - G(y_\tau), x(\sigma) - y_n(\tau)]_+ d\sigma d\tau$$
$$\leq \int_q^p \int_s^t [G(x_\sigma) - G(y_\tau), x(\sigma) - y(\tau)]_+ d\sigma d\tau.$$

Therefore from Lemma 2.2.1 we have (2.2.22) with $\alpha = 0$.

COROLLARY 2.2.1 *Let X be reflexive. Under the assumptions of Theorem 2.2.3 let $A : X \supset D(A) \to 2^X$ be m-accretive. Then $x(t) \equiv x(t;\phi)$ is a unique strong solution of (2.2.1) on $[-r,\infty)$.*

$T \in (0, \min\{T_0, r\}]$ such that

$$\|\phi(t+\tau) - \phi(\tau)\| \leq \frac{\epsilon}{3} \tag{2.3.5}$$

for every $\tau \in [-r, 0]$ and $t \in [0, T]$ with $t + \tau \leq 0$, and

$$\|S(t)\phi(0) - \phi(0)\| + TM \leq \frac{2\epsilon}{3} \tag{2.3.6}$$

for every $t \in [0, T]$. We define

$$K_0 = \{u \in C([0, T]; X) \mid u(0) = \phi(0)\}.$$

Then K_0 is a nonempty closed subset of $C([0, T]; X)$. Let $u \in C([0, T]; X)$. We define a function $\overline{u} : [-r, T] \to X$ by

$$\overline{u}(t) = \begin{cases} \phi(t), & -r \leq t \leq 0 \\ u(t), & 0 \leq t \leq T. \end{cases} \tag{2.3.7}$$

We set

$$K = \{u \in K_0 \mid (\overline{u})_t \in B(\phi, \epsilon) \text{ for every } t \in [0, T]\}.$$

Then K is a nonempty bounded closed and convex subset of $C([0, T]; X)$.

Let $u \in K$. Then by (2.3.3), (2.3.4) and the definition of K there exists the unique integral solution $x(t)$ for nonlinear evolutions of the type

$$\begin{cases} \frac{dx}{dt}(t) + Ax(t) \ni G(t, (\overline{u})_t), & 0 \leq t \leq T \\ x(0) = \phi(0) \end{cases} \tag{2.3.8}$$

on $[-r, T]$. So we define the operator $Q : K \to K_0$ by

$$(Qu)(t) = x(t)$$

for every $u \in K$ and $t \in [0, T]$, where $x(t)$ is the unique integral solution of (2.3.8) on $[-r, T]$.

Now we show that $Q(K) \subset K$. To this aim observe that $(Qu)_t \in B(\phi, \epsilon)$ if and only if

$$\|x(t+\tau) - \phi(\tau)\| \leq \epsilon, \; x = Qu$$

for every $t \in [0, T]$ and $\tau \in [-r, 0]$.

If $t+\tau > 0$ then we have

$$\begin{aligned}&\|x(t+\tau)-\phi(\tau)\|\\&\quad\leq \|x(t+\tau)-S(t+\tau)\phi(0)\|\\&\qquad+\|S(t+\tau)\phi(0)-\phi(0)\|+\|\phi(0)-\phi(\tau)\|.\end{aligned}$$

Since $S(t)\phi(0)$ is an integral solution of (2.3.8) on $[0,\infty)$ with $G \equiv 0$ using (1.3.3) and (2.3.4)–(2.3.6) we obtain

$$\begin{aligned}&\|x(t+\tau)-\phi(\tau)\|\\&\quad\leq \int_0^{t+\tau}\|G(\sigma,(\overline{u})_\sigma\|d\sigma+\|S(t+\tau)\phi(0)-\phi(0)\|+\|\phi(0)-\phi(\tau)\|\\&\quad\leq \int_0^{T}\|G(\sigma,(\overline{u})_\sigma\|d\sigma+\|S(t+\tau)\phi(0)-\phi(0)\|+\|\phi(0)-\phi(\tau)\|\\&\quad\leq TM+\|x(t+\tau)\phi(0)-\phi(0)\|+\|\phi(0)-\phi(\tau)\|\\&\quad\leq \frac{2\epsilon}{3}+\frac{\epsilon}{3}=\epsilon.\end{aligned}$$

If $t+\tau \leq 0$ then we have

$$\|x(t+\tau)-\phi(\tau)\| = \|\phi(t+\tau)-\phi(\tau)\| \leq \frac{\epsilon}{3} < \epsilon,$$

and therefore $Q(K) \subset K$.

Let $x_i(t)$ be integral solutions of (2.3.8) on $[0,T]$ corresponding to $u_i \in K$ for $i = 1,2$. Then

$$\begin{aligned}\|Qu_1-Qu_2\|_{C([0,T];X)} &= \|x_1-x_2\|_{C([0,T];X)}\\&\leq \sup_{t\in[0,T]}\|x_1(t)-x_2(t)\|\\&\leq \sup_{t\in[0,T]}\int_0^t\|G(\sigma,(u_1)_\sigma)-G(\sigma,(u_2)_\sigma)\|d\sigma\\&\leq \int_0^T\|G(\sigma,(u_1)_\sigma)-G(\sigma,(u_2)_\sigma)\|d\sigma.\end{aligned}$$

Applying (1.3.3), (2.3.4), the continuity of G, and Lebesgue's dominated convergence theorem we conclude that $Q : K \to K$ is continuous in the induced norm topology on both domain and range.

Finally, from (2.3.4) and the definition of K the set

$$V = \{f_u \in L^1(0,T;X) \mid f_u(t) = G(t,(\overline{u})_t) \text{ } a.e. \text{ } t \in (0,T), \text{ } u \in K\}$$

is bounded in $L^\infty(0,T;X)$. Hence from (1) of Remark 2.3.1 V is uniformly integrable in $L^1(0,T;X)$. Thus from Proposition 2.3.1 $Q(K)$ is relatively compact in $C([0,T];X)$. From Schauder's fixed point theorem there exists $u \in K$ such that $Qu = u$. Thus $x(t) = u(t)$ for every $t \in [0,T]$. By (2.3.7)

$$\bar{x}(t) = \begin{cases} \phi(t), & -r \leq t \leq 0 \\ x(t), & 0 \leq t \leq T. \end{cases}$$

Then $\bar{x}(t)$ is a local integral solution of (2.3.1) on $[-r,\infty)$.

Next we show the existence of a local strong solution of (2.3.1) on $[-r,\infty)$ with the additional assumptions of G owing to its specific form as an integral operator and the compactness of J_λ^A.

We set

$$C_A([-r,0];X) = \{\psi \in C([-r,0];X) \mid \psi(\tau) \in D(A) \text{ a.e. } \tau \in [-r,0]\}$$

and we consider nonlinear functional evolutions of the type

$$\begin{cases} \dfrac{dx}{dt}(t) + Ax(t) \ni \displaystyle\int_{t-r}^{t} k(t-\sigma)g(x(\sigma))d\sigma, & t \geq 0 \\ x(t) = \phi(t), \quad -r \leq t \leq 0, \end{cases} \tag{2.3.9}$$

where k is an operator kernel and $g : D(A) \to X$ is a mapping.

If we define $G : [0,\infty) \times C_A([-r,0];X) \to X$ by

$$G(t,\psi) = \int_{-r}^{0} k(-\tau)g(\psi(\tau))d\tau. \tag{2.3.10}$$

for every $t \in [0,\infty)$ and $\psi \in C_A([-r,0];X)$ then (2.3.9) may be rewritten in the form (2.3.1) because we have by a simple change of variable

$$\begin{aligned} G(t,x_t) &= \int_{-r}^{0} k(-\tau)g(x_t(\tau))d\tau \\ &= \int_{-r}^{0} k(-\tau)g(x(t+\tau))d\tau \\ &= \int_{t-r}^{t} k(t-\sigma)g(x(\sigma))d\sigma \end{aligned}$$

for every $t \in [0,\infty)$.

But then $C_A([-r,0];X)$ is not open in $C([-r,0];X)$ in general, and thus Theorem 2.3.1 can not be applied in the case $U = C_A([-r,0];X)$.

We begin with the definition of A-demiclosedness.

DEFINITION 2.3.4 *A mapping $g : D(A) \to X$ is called A-demiclosed if for every $T > 0$, the following conditions (1) and (2) are satisfied:*

(1) For every $u \in C_A([0,T];X)$ such that there exists $v \in L^2(0,T;X)$ with $[u(t),v(t)] \in A$ a.e. $t \in (0,T)$, $g(u(t))$ is strongly mesurable on $[0,T]$ into X;

(2) If $\{u_n\} \subset C_A([0,T];X)$, $u \in C_A([0,T];X)$, $[u_n(t),v_n(t)] \in A$ a.e. $t \in (0,T)$ and $n = 1,2,\cdots$, $[u(t),v(t)] \in A$ a.e. $t \in (0,T)$, and if, in addition,

$$\begin{aligned}\lim_{n\to\infty} u_n &= u \text{ strongly in } C([0,T];X),\\ \lim_{n\to\infty} v_n &= v \text{ weakly in } L^2(0,T;X),\\ \lim_{n\to\infty} g(u_n) &= \overline{g} \text{ weakly in } L^2(0,T;X),\end{aligned}$$

then $\overline{g}(t) = g(u(t))$ a.e. $t \in (0,T)$;

(3) There exists a monotone non-decreasing function $f : [0,\infty) \to [0,\infty)$ such that

$$\|g(x)\| \leq f(\|A^0 x\|)$$

for every $x \in D(A)$.

We then have our result.

THEOREM 2.3.2 *Let X^* be uniformly convex. Let $A : X \supset D(A) \to 2^X$ be an operator and $\alpha \in \boldsymbol{R}$ such that $A + \alpha I$ is accretive in X and $R(I + \lambda A) = X$ for every $\lambda > 0$ with $\lambda\alpha < 1$. Let $J_\lambda = (I + \lambda A)^{-1}$ be compact for every λ with $\lambda\alpha < 1$. Let $k : [0,r] \to L(X)$ be a mapping of class C^1. Let $g : D(A) \to X$ be an A-demiclosed mapping. Then for every $\phi \in C_A([0,-r];X)$ with $\phi(0) \in D(A)$ and $g(\phi) \in L^2(-r,0;X)$, (2.3.9) has at least one local strong solution on $[-r,\infty)$.*

Proof. Let $T_0 \in (0,\infty)$. Let $\epsilon > 0$ and $T \in (0,\min\{T_0,r\}]$. We define

$$K_\epsilon^T = \{u \in W^{1,2}(0,T;X) \mid \|u\|_{1,2,T} \leq \epsilon,\ f(0) = G(0,\phi)\},$$

where $\|\cdot\|_{1,2,T}$ is the usual norm on $W^{1,2}(0,T;X)$ defined by

$$\|u\|_{1,2,T} = \left(\int_0^T \left(\|u(t)\|^2 + \left\|\frac{du}{dt}(t)\right\|^2\right)dt\right)^{\frac{1}{2}}$$

for every $u \in W^{1,2}(0,T;X)$ and G is given in (2.3.10).

By virtue of (1) of Theorem 1.3.2 let $x(t)$ be the unique strong solution of the nonlinear evolution

$$\begin{cases} \dfrac{dx}{dt}(t) + Ax(t) \ni u(t), \quad 0 \leq t \leq T \\ x(0) = \phi(0) \end{cases} \tag{2.3.11}$$

for every $u \in K_\epsilon^T$, with

$$\left\| \frac{dx}{dt}(t) \right\| \leq \|(A\phi(0) - u(0))^0\| + \int_0^t \left\| \frac{du}{d\sigma}(\sigma) \right\| d\sigma$$

a.e. $t \in (0, T)$. Thus since $u(0) = G(0, \phi)$ we have

$$\begin{aligned} \|A^0 x(t)\| &\leq \left\| \frac{dx}{dt}(t) \right\| + \|u(t)\| \\ &\leq \|(A\phi(0) - u(0))^0\| \\ &\quad + \int_0^t \left\| \frac{du}{d\sigma}(\sigma) \right\| d\sigma + \|G(0, \phi)\| + \int_0^t \left\| \frac{du}{d\sigma}(\sigma) \right\| d\sigma \\ &\leq \|(A\phi(0) - u(0))^0\| + \|G(0, \phi)\| + 2 \int_0^t \left\| \frac{du}{d\sigma}(\sigma) \right\| d\sigma \\ &\leq \|(A\phi(0) - u(0))^0\| + \|G(0, \phi)\| \\ &\quad + 2 \Big(\int_0^t \left\| \frac{du}{d\sigma}(\sigma) \right\| d\sigma \Big)^{\frac{1}{2}} \Big(\int_0^T d\sigma \Big)^{\frac{1}{2}} \\ &= \|(A\phi(0) - u(0))^0\| + \|G(0, \phi)\| + 2\epsilon T^{\frac{1}{2}} \\ &= c(T, \epsilon) \end{aligned} \tag{2.3.12}$$

for every $u \in K_\epsilon^T$ and a.e. $t \in (0, T)$. Since k is a mapping of class C^1, there exists $c_0 > 0$ such that

$$\|k(\sigma)\|_{L(X)} \leq c_0, \quad \left\| \frac{dk}{d\sigma}(\sigma) \right\|_{L(X)} \leq c_0 \tag{2.3.13}$$

for every $\sigma \in [0, T]$. We put

$$c_1 = \Big(\int_{-r}^0 \|g(\phi(\tau)\|^2 d\tau \Big)^{\frac{1}{2}}.$$

Then $0 \leq c_1 < \infty$ from the assumption. By (2.3.12), (2.3.13), and the assumptions

$$\left\| \int_{t-r}^t k(t - \sigma) g(\overline{x}(\sigma)) d\sigma \right\| \leq$$

$$
\begin{aligned}
&\le c_0 \int_{t-r}^{0} \|g(\phi(\sigma))\| d\sigma + c_0 \int_0^t \|g(x(\sigma))\| d\sigma \\
&\le c_0 \Big(\int_{-r}^{0} \|g(\phi(\sigma))\|^2 d\sigma \Big)^{\frac{1}{2}} \Big(\int_{-r}^{0} d\sigma \Big)^{\frac{1}{2}} + c_0 \int_0^t f(\|A^0 x(\sigma)\|) d\sigma \\
&\le c_0 c_1 r^{\frac{1}{2}} + c_0 f(c(T,\epsilon)) T
\end{aligned}
$$

for every $\epsilon > 0$, $T \in (0, \min\{T_0, r\}]$, $u \in K_\epsilon^T$ and $t \in [0, T]$. Thus we have

$$
\begin{aligned}
&\int_0^T \Big\| \int_{t-r}^{t} k(t-\sigma) g(\overline{x}(\sigma)) d\sigma \Big\|^2 dt \qquad (2.3.14) \\
&\qquad \le T(c_0 c_1 r^{\frac{1}{2}} + c_0 f(c(T,\epsilon))T)^2
\end{aligned}
$$

for every $\epsilon > 0$, $T \in (0, \min\{T_0, r\}]$ and $u \in K_\epsilon^T$. Using ways similar to the above we also have

$$
\begin{aligned}
&\int_0^T \Big\| \int_{t-r}^{t} \frac{dk}{d\sigma}(t-\sigma) g(\overline{x}(\sigma)) d\sigma \Big\|^2 dt \qquad (2.3.15) \\
&\qquad \le T(c_0 c_1 r^{\frac{1}{2}} + c_0 f(c(T,\epsilon))T)^2
\end{aligned}
$$

for every $\epsilon > 0$, $T \in (0, \min\{T_0, r\}]$ and $u \in K_\epsilon^T$.

We define the operator $Q : D(Q) \subset K_\epsilon^T \to W^{1,2}(0, T; X)$ by

$$
(Qu)(t) = G(t, (\overline{x})_t) = \int_{t-r}^{t} k(t-\sigma) g(\overline{x}(\sigma)) d\sigma
$$

for every $u \in D(Q)$ and $t \in [0, T]$. Here, $x(t) \equiv x(t; u)$ is the unique strong solution of (2.3.11) and $\overline{x}(t)$ is defined by

$$
\overline{x}(t) = \begin{cases} \phi(t), & -r \le t \le 0 \\ x(t), & t \ge 0 \end{cases}
$$

and hence

$$
D(Q) = \{u \in K_\epsilon^T \mid t \to G(t, (\overline{x})_t) \text{ belongs to } W^{1,2}(0, T; X)\}.
$$

We show first that $D(Q) = K_\epsilon^T$ and $QK_\epsilon^T \subset K_\epsilon^T$ for every $\epsilon > 0$ and $T \in (0, \min\{T_0, r\}]$.

In fact, if $T \in (0, \min\{T_0, r\}]$, then $g(\overline{x}(t-r)) = g(\phi(t-r))$ for every $t \in [0, T]$ and thus

$$
\begin{aligned}
\frac{d(Qu)}{dt}(t) &= k(0) g(x(t)) - k(r) g(\phi(t-r)) \\
&\quad + \int_{t-r}^{t} \frac{dk}{d\sigma}(t-\sigma) g(\overline{x})(\sigma)) d\sigma
\end{aligned}
$$

for every $u \in K_\epsilon^T$ and a.e. $t \in (0,T)$. Since from (2.3.12)–(2.3.15) and (3) of Definition 2.3.4

$$\begin{aligned}
&\int_0^T \left\| \frac{d(Qu)}{dt}(t) \right\|^2 dt \\
&\quad = \int_0^T \Big\| k(0)g(x(t)) - k(r)g(\phi(t-r)) \\
&\qquad + \int_{t-r}^t \frac{dk}{d\sigma}(t-\sigma)g(\overline{x}(\sigma))d\sigma \Big\|^2 dt \\
&\quad \le \int_0^T \Big(\|k(0)g(x(t)) - k(r)g(\phi(t-r))\| \\
&\qquad + \Big\| \int_{t-r}^t \frac{dk}{d\sigma}(t-\sigma)g(\overline{x}(\sigma))d\sigma \Big\| \Big)^2 dt \\
&\quad \le \int_0^T \|k(0)g(x(t)) - k(r)g(\phi(t-r))\|^2 dt \\
&\qquad +2 \int_0^T \Big(\|k(0)g(x(t)) - k(r)g(\phi(t-r))\| \\
&\qquad\qquad \times \Big\| \int_{t-r}^t \frac{dk}{d\sigma}(t-\sigma)g(\overline{x}(\sigma))d\sigma \Big\| \Big) dt \\
&\qquad + \int_0^T \Big\| \int_{t-r}^t \frac{dk}{d\sigma}(t-\sigma)g(\overline{x}(\sigma))d\sigma \Big\|^2 dt \\
&\quad \le c_0^2 c_1^2 + \int_0^T (2\|k(0)g(x(t))\| \|k(r)g(\phi(t-r))\| \\
&\qquad\qquad + \|k(r)g(\phi(t-r))\|^2) dt \\
&\qquad +2 \int_0^T \Big(\|k(0)g(x(t)) - k(r)g(\phi(t-r))\| \\
&\qquad\qquad \times \Big\| \int_{t-r}^t \frac{dk}{d\sigma}(t-\sigma)g(\overline{x}(\sigma))d\sigma \Big\|^2 \Big) dt \\
&\qquad + \int_0^T \Big\| \int_{t-r}^t \frac{dk}{d\sigma}(t-\sigma)g(\overline{x}(\sigma))d\sigma \Big\|^2 dt \le \\
&\quad \le c_0^2 c_1^2 + h(T,\epsilon),
\end{aligned}$$

where $\lim_{T \to 0+} h(T,\epsilon) = 0$ for every $\epsilon > 0$

$$\begin{aligned}
\|Qu\|_{1,2,T}^2 &= \int_0^T \left(\|(Qu)(t)\|^2 + \left\| \frac{d(Qu)}{dt}(t) \right\|^2 \right) dt \\
&\le c_2^2 T + c_0^2 c_1^2 + h(T,\epsilon) =
\end{aligned}$$

$$= c_0^2 c_1^2 + \overline{h}(T, \epsilon)$$

for every $\epsilon > 0$, $T \in (0, \min\{T_0, r\}]$ and $u \in K_\epsilon^T$. Here, $\lim_{T\to 0+} \overline{h}(T, \epsilon) = 0$ for every $\epsilon > 0$. Thus by this inequality for every $\epsilon > c_0 c_1$ there exists $T \in (0, \min\{T_0, r\}]$ such that $c_0^2 c_1^2 + \overline{h}(T, \epsilon) \leq \epsilon^2$. Hence $\|Qu\|_{1,2,T} \leq \epsilon$ and thus $QK_\epsilon^T \subset K_\epsilon^T$ for $\epsilon > 0$ and $T \in (0, \min\{T_0, r\}]$ as above.

Since K_ϵ^T is bounded, closed, and convex in $W^{1,2}(0, T; X)$ and X is reflexive, it is weakly compact in $W^{1,2}(0, T; X)$. We show that the graph $gr(Q)$ is weakly sequentially closed in $K_\epsilon^T \times K_\epsilon^T$. Indeed, let $\{[u_n, v_n]\} \subset gr(Q)$ such that $\lim_{n\to\infty} u_n = u$ and $\lim_{n\to\infty} v_n = v$ weakly in $W^{1,2}(0, T; X)$. Let $x_n(t) \equiv x(t; u_n)$. Then from Proposition 2.3.3

$$\lim_{n\to\infty} x_n = x \text{ strongly in } C([0, T]; X).$$

Let $\{y_n\} \subset L^2(0, T; X)$ defined by

$$y_n(t) = -\frac{dx_n}{dt}(t) + u_n(t)$$

for $n = 1, 2, \cdots$ and a.e. $t \in (0, T)$. Since $y_n(t) \in Ax_n(t)$ for $n = 1, 2, \cdots$ and a.e. $t \in (0, T)$ from (2.3.12) $\{y_n\}$ is bounded in $L^\infty(0, T; X)$ and thus bounded in $L^2(0, T; X)$. Hence we may assume without loss of generality that

$$\lim_{n\to\infty} y_n = y \text{ weakly in } L^2(0, T; X).$$

We now define the operator $\widetilde{A} : L^2(0, T; X) \supset D(\widetilde{A}) \to 2^{L^2(0,T;X)}$ by

$$\begin{aligned}
D(\widetilde{A}) &= \{x \in L^2(0, T; X) \mid x(t) \in D(A) \text{ a.e. } t \in (0, T) \text{ and} \\
&\qquad \exists\, y \in L^2(0, T; X) \text{ such that} \\
&\qquad y(t) \in Ax(t) \text{ a.e. } t \in (0, T)\}, \\
\widetilde{A}x &= \{y \in L^2(0, T; X) \mid y(t) \in Ax(t) \text{ a.e. } t \in (0, T)\}, \quad x \in D(\widetilde{A}).
\end{aligned}$$

Then $\widetilde{A} + \alpha I$ is accretive in $L^2(0, T; X)$ and $R(I + \lambda\widetilde{A}) = L^2(0, T; X)$ for every $\lambda > 0$ with $\lambda\alpha < 1$. Since X^* is uniformly convex, $(L^2(0, T; X))^* = L^2(0, T; X^*)$ is uniformly convex. Thus $\widetilde{A}$ is demiclosed and then $y \in \widetilde{A}x$ or equivalently, $y(t) \in \widetilde{A}x(t)$ a.e. $t \in (0, T)$. From (2.3.12) and (3) of Definition 2.3.4 we may assume without loss of generality that

$$\lim_{n\to\infty} g(x_n) = g_0 \text{ weakly in } L^2(0, T; X).$$

From (2) of Definition 2.3.4 $g_0(t) = g(x(t))$ a.e. $t \in (0, T)$.

Since

$$v_n(t) = \int_{t-r}^{t} k(t - \sigma) g((\overline{x}_n)(\sigma)) d\sigma$$

for $n = 1, 2, \cdots$ and $t \in [0, T]$ we easily conclude that

$$v(t) = \int_{t-r}^{t} k(t-\sigma)g(\overline{x}(\sigma))d\sigma = (Qu)(t)$$

for every $t \in [0, T]$. Thus $gr(Q)$ is weakly sequentially closed in $K_\epsilon^T \times K_\epsilon^T$. Since K_ϵ^T is weakly compact Q is weakly sequentially continuous from K_ϵ^T into K_ϵ^T. From Lemma 2.3.1 Q has at least one fixed point $u \in K_\epsilon^T$ which by means of $u \to \overline{x}(t) \equiv x(t; u)$ defines a local strong solution of (2.3.9) on $[-r, \infty)$ thereby completing the proof.

2.4. L^p-Space Methods

Let X be a real Banach space with norm $\|\cdot\|$. Let $r > 0$ and let $1 \leq p < \infty$. Let $L^p(-r, 0; X)$ be the usual space of all measurable functions $\psi : [-r, 0] \to X$ such that $\|\psi(t)\|^p$ is integrable on $[-r, 0]$ with norm $\|\cdot\|_p$ defined by

$$\|\psi\|_p = \Big(\int_{-r}^{0} \|\psi(\tau)\|^p d\tau\Big)^{\frac{1}{p}}$$

for every $\psi \in L^p(-r, 0; X)$.

Since the evaluation of $\phi \in L^p(-r, 0; X)$ at points in $[-r, 0]$ makes no sense, the initial value $x(0) = h \in X$ has to be considered as an additional datum. Thus in this section we deal with autonomous nonlinear functional evolutions of the type

$$\begin{cases} \dfrac{dx}{dt}(t) + Ax(t) \ni G(x_t), \ t \geq 0 \\ x_0 = \phi \\ x(0) = h \end{cases} \tag{2.4.1}$$

associated with an operator $A : X \supset D(A) \to 2^X$ instead of (2.2.1), where $r > 0$, $x(t) : [-r, \infty) \to X$ is an unknown function, $G : L^p(-r, 0; X) \to X$, $\phi \in L^p(-r, 0; X)$, $1 \leq p < \infty$, and $h \in X$.

DEFINITION 2.4.1 *1. A function $x(t) : [-r, \infty) \to X$ is called a strong solution of (2.4.1) on $[-r, \infty)$ if:*

(1) $x_0 = \phi$;

(2) $x(0) = h$;

(3) $x(t)$ is continuous on $[0, \infty)$;

(4) $x(t)$ is absolutely continuous on every compact interval of $[0, \infty)$;

(5) $x_t \in L^p(-r, 0; X)$ *for every* $t \in [0, \infty)$*;*

(6) the function $f : [0, \infty) \to X$ *defined by* $f(t) = G(x_t)$ *a.e.* $t \in (0, \infty)$ *belongs to* $L^1(0, \infty; X)$*;*

(7) $x(t)$ *is differentiable a.e. on* $(0, \infty)$*;*

(8) $x(t)$ *satisfies*

$$\frac{dx}{dt}(t) + Ax(t) \ni G(x_t)$$

a.e. $t \in (0, \infty)$.

2. A function $x(t) : [-r, \infty) \to X$ *is called an integral solution of* (2.4.1) *on* $[-r, \infty)$ *if*

(1) $x = \phi$*;*

(2) $x(0) = h$*;*

(3) $x(t)$ *is continuous on* $[0, \infty)$*;*

(4) $x_t \in L^p(-r, 0; X)$ *for every* $t \in [0, \infty)$*;*

(5) the function $f : [0, \infty) \to X$ *defined by* $f(t) = G(x_t)$ *a.e.* $t \in (0, \infty)$ *belongs to* $L^1(0, \infty; X)$*;*

(6) $x(t)$ *satisfies*

$$\begin{aligned} &\|x(t) - x\| - \|x(s) - x\| \\ &\quad \le \alpha \int_s^t \|x(\tau) - x\| d\tau + \int_s^t [G(x_\tau) - y, x(\tau) - x]_+ d\tau. \end{aligned}$$

for every $[x, y] \in A$ *and* $t,\ s \in [0, \infty)$.

Let $Y = L^p(-r, 0; X) \times X$. We define a norm $\|\cdot\|_Y$ on Y by

$$\|\{\phi, h\}\|_Y = \left(\int_{-r}^0 \|\phi(\tau)\|^p d\tau + \|h\|^p \right)^{\frac{1}{p}} \tag{2.4.2}$$

for every $\{\phi, h\} \in Y$. Then $(Y, \|\cdot\|_Y)$ is a real Banach space and

$$\|\{\phi, h\}\|_Y \ge \|\phi\|_p, \ \|\{\phi, h\}\|_Y \ge \|h\|. \tag{2.4.3}$$

We define an operator $B : Y \supset D(B) \to 2^Y$ by

$$\begin{aligned} D(B) &= \Big\{ \{\phi, h\} \in Y \ \Big| \ \phi : [-r, 0] \to X \text{ is absolutely continuous,} \\ &\qquad \frac{d\phi}{dt} \in L^p(-r, 0; X), \ h = \phi(0) \in D(A) \Big\}, \\ B\{\phi, h\} &= \Big\{ -\frac{d\phi}{dt}, A\phi(0) - G(\phi) \Big\}, \quad \{\phi, h\} \in D(B). \end{aligned} \tag{2.4.4}$$

To show the accretivity of B in (2.4.4), at first, we are concerned with the duality mappings F_Y on Y, F on X and F_{L^p} on $L^p(-r,0;X)$.

LEMMA 2.4.1 *Let $\{\phi, h\} \in Y$. Let $\phi^* \in F_p(\phi)$ and $h^* \in F(h)$. We define $y^* : Y \to \boldsymbol{R}$ by*

$$(\{\psi, k\}, y^*) \tag{2.4.5}$$
$$= \|\{\phi, h\}\|_Y^{2-p}\Big(\int_{-r}^{0}(\psi(\tau), \phi^*(\tau))\|\phi(\tau)\|^{p-2}d\tau + (k, h^*)\|h\|^{p-2}\Big)$$

for every $\{\psi, k\} \in Y$. Then $y^* \in F_Y(\{\phi, h\})$.

Proof. It is easy to see from (2.4.5) that $y^* \in Y^*$.

When $p = 1$ since

$$\Big|\int_{-r}^{0}(\psi(\tau), \phi^*(\tau))\|\phi(\tau)\|^{-1}d\tau\Big| \leq \int_{-r}^{0}\|\psi(\tau)\|d\tau < \infty$$

(2.4.5) makes sense. We put $\{\psi, k\} = \{\phi, h\}$ in (2.4.5). Then we obtain

$$\begin{aligned}(\{\phi, h\}, y^*) &= \|\{\phi, h\}\|_Y\Big(\int_{-r}^{0}\|\phi(\tau)\|d\tau + \|h\|\Big)\\ &= \|\{\phi, h\}\|_Y^2\end{aligned}$$

and thus $\|y^*\| \geq \|\{\phi, h\}\|_Y$. By (2.4.5) we have

$$\begin{aligned}(\{\psi, k\}, y^*) &\leq \|\{\phi, h\}\|_Y\Big(\int_{-r}^{0}\|\psi(\tau)\|d\tau + \|k\|\Big)\\ &= \|\{\phi, h\}\|_Y\|\{\psi, k\}\|_Y\end{aligned}$$

and thus $\|y^*\| \leq \|\{\phi, h\}\|_Y$. Hence $\|y^*\| = \|\{\phi, h\}\|_Y$. Therefore $y^* \in F_{L^1}(\{\phi, h\})$.

When $p > 1$ let $q > 1$ with $\dfrac{1}{p} + \dfrac{1}{q} = 1$. Since using Cauchy–Schwartz inequality

$$\begin{aligned}&\Big|\int_{-r}^{0}(\psi(\tau), \phi^*(\tau))\|\phi(\tau)\|^{p-2}d\tau\Big| \leq\\ &\leq \int_{-r}^{0}\|\psi(\tau)\|\|\phi(\tau)\|^{p-1}d\tau\\ &\leq \Big(\int_{-r}^{0}\|\psi(\tau)\|^p d\tau\Big)^{\frac{1}{p}}\Big(\int_{-r}^{0}\|\phi(\tau)\|^{(p-1)q}d\tau\Big)^{\frac{1}{q}}\\ &= \Big(\int_{-r}^{0}\|\psi(\tau)\|^p d\tau\Big)^{\frac{1}{p}}\Big(\int_{-r}^{0}\|\phi(\tau)\|^p d\tau\Big)^{\frac{1}{q}} < \infty\end{aligned}$$

(2.4.5) also makes sense. We put $\{\psi, k\} = \{\phi, h\}$ in (2.4.5). Then from (2.4.2)

$$\begin{aligned} (\{\phi, h\}, y^*) &= \|\{\phi, h\}\|_Y^{2-p}\Big(\int_{-r}^{0} \|\phi(\tau)\|^p d\tau + \|h\|^p\Big) \\ &= \|\{\phi, h\}\|_Y^{2-p}\|\{\phi, h\}\|_Y^p = \|\{\phi, h\}\|_Y^2 \end{aligned}$$

and thus $\|y^*\| \geq \|\{\phi, h\|\|_Y^2$. From (2.4.5) and Cauchy–Schwartz inequality

$$\begin{aligned} &(\{\psi, k\}, y^*) \\ &\leq \|\{\phi, h\}\|_Y^{2-p}\Big(\int_{-r}^{0} \|\psi(\tau)\|\|\phi(\tau)\|^{p-1} d\tau + \|k\|\|h\|^{p-1}\Big) \\ &\leq \|\{\phi, h\}\|_Y^{2-p}\Big(\Big(\int_{-r}^{0} \|\psi(\tau)\|^p d\tau\Big)^{\frac{1}{p}} \\ &\qquad \times\Big(\int_{-r}^{0} \|\phi(\tau)\|^{(p-1)q} d\tau\Big)^{\frac{1}{q}} + \|k\|\|h\|^{p-1}\Big) \\ &\leq \|\{\phi, h\}\|_Y^{2-p}\|\{\phi, h\}\|_Y^{\frac{p}{q}}\|\{\psi, k\}\|_Y \\ &= \|\{\phi, h\}\|_Y\|\{\psi, k\}\|_Y \end{aligned}$$

for every $\{\psi, k\} \in Y$. Here we have used the inequality

$$a^{\frac{1}{p}}b^{\frac{1}{q}} + cd^{p-1} \leq (a + d^p)^{\frac{1}{p}}(b + c^p)^{\frac{1}{q}}, \quad (a,\ b,\ c,\ d > 0)$$

for

$$a = \int_{-r}^{0} \|\psi(\tau)\|^p d\tau, \quad b = \int_{-r}^{0} \|\phi(\tau)\|^p d\tau, \quad c = \|h\| \text{ and } d = \|k\|.$$

Thus $\|y^*\| \leq \|\{\phi, h\}\|_Y$. Hence $\|y^*\| = \|\{\phi, h\}\|_Y$. Therefore $y^* \in F_{L^p}(\{\phi, h\})$ for $1 \leq p < \infty$.

Proposition 2.4.1 *Let $A : X \supset D(A) \to 2^X$ be an operator and $\alpha \in \boldsymbol{R}$ such that $A + \alpha I$ is accretive in X and $R(I + \lambda A) = X$ for every $\lambda > 0$ with $\lambda\alpha < 1$. Let $1 \leq p < \infty$. Assume that $G : L^p(-r, 0; X) \to X$ satisfies*

$$\|G(\psi_1) - G(\psi_2)\| \leq \beta\|\psi_1 - \psi_2\|_{L^p} \qquad (2.4.6)$$

for every $\psi_i \in L^p(-r, 0; X)$, $i = 1, 2$, and for some $\beta > 0$. Let B be the operator defined in (2.4.4). *We put $\gamma = \max\left\{0, \dfrac{1}{p} + \alpha\right\} + \beta$. Then:*

To give solutions of (2.4.1) associated with $S(t)\{\phi, h\}$ for $\{\phi, h\} \in \overline{D(B)}$, let $\pi_1 : Y \to L^p(-r, 0; X)$ and $\pi_2 : Y \to X$ be projections satisfying $\pi_1\{\phi, h\} = \phi$ and $\pi_2\{\phi, h\} = h$ for every $\{\phi, h\} \in Y$, and we define

$$x(t) \equiv x(t; \{\phi, h\}) : [-r, \infty) \to X$$

by

$$x(t) = \begin{cases} \phi(t), & -r \le t \le 0 \\ \pi_2 S(t)\{\phi, h\}, & t \ge 0. \end{cases} \tag{2.4.10}$$

Then from Theorem 1.2.1 $x(t)$ is continuous in $t \ge 0$. In order to show that $x(t)$ in (2.4.10) is a solution of (2.4.1) on $[-r, \infty)$, consider a relationship between $\{S(t) \mid t \ge 0\}$ and a nonlinear semigroup associated with (2.4.1) in $C([-r, 0]; X)$.

We now define an operator $Q : D(Q) \subset Y \to C([-r, 0]; X)$ by

$$\begin{aligned} D(Q) &= \{\{\phi, h\} \in Y \mid \phi = \psi \text{ a.e. on } (-r, 0) \text{ for some} \\ &\qquad\qquad \psi \in C([-r, 0]; X) \text{ and } \psi(0) = h\}, \\ Q\{\phi, h\} &= \psi, \quad \{\phi, h\} \in D(Q). \end{aligned} \tag{2.4.11}$$

Without confusion we will sometimes write $Q\{\phi, h\} = \phi$, where ϕ means the continuous function identified with $\phi \in L^p(-r, 0; X)$.

LEMMA 2.4.2 *Let $A : X \supset D(A) \to 2^X$ be an operator and $\alpha \in \boldsymbol{R}$ such that $A + \alpha I$ is accretive in X and $R(I + \lambda A) = X$ for every $\lambda > 0$ with $\lambda\gamma < 1$. Let G satisfy* (2.4.6). *Let B and Q be operators defined in* (2.4.4) *and* (2.4.11), *respectively. Then:*

(1) $D(B) \subset D(Q)$;

(2) Q is one to one and $Q^{-1} : C([-r, 0]; X) \to Y$ is continuous.

Proof. (1) It is trivial from definitions of B and Q in (2.4.4) and (2.4.11), respectively.

(2) Let $\{\phi_i, h_i\} \in D(Q)$, $i = 1, 2$. Then $Q\{\phi_i, h_i\} = \phi_i$ and $\phi_i(0) = h_i$, $i = 1, 2$. Since by (2.4.2)

$$\begin{aligned} &\|\{\phi_1, h_1\} - \{\phi_2, h_2\}\|_Y \\ &\quad = \|\{\phi_1 - \phi_2, h_1 - h_2\}\|_Y \\ &\quad = \Big(\int_{-r}^0 \|\phi_1(\tau) - \phi_2(\tau)\|^p d\tau + \|h_1 - h_2\|^p\Big)^{\frac{1}{p}} \\ &\quad = \Big(\int_{-r}^0 \|\phi_1(\tau) - \phi_2(\tau)\|^p d\tau + \|\phi_1(0) - \phi_2(0)\|^p\Big)^{\frac{1}{p}} \le \end{aligned}$$

$$\leq \Big(\int_{-r}^{0} d\tau + 1\Big)^{\frac{1}{p}} \|\phi_1 - \phi_2\|_C$$
$$= (1+r)^{\frac{1}{p}} \|Q\{\phi_1, h_1\} - Q\{\phi_2, h_2\}\|_C, \tag{2.4.12}$$

there exists $Q^{-1} : C([-r,0];X) \to Y$ and Q^{-1} is continuous. This completes the proof of (2).

We define also an operator $B_C : C([-r,0];X) \to C([-r,0];X)$ by

$$\begin{aligned} D(B_C) &= \Big\{\phi \in C([-r,0];X) \;\Big|\; \frac{d\phi}{d\tau} \in C([-r,0];X),\ \phi(0) \in D(A), \\ &\qquad\qquad \frac{d\phi}{d\tau} \in G(\phi) - A\phi(0)\Big\}, \\ B_C\phi &= -\frac{d\phi}{d\tau}, \quad \phi \in D(B_C). \end{aligned} \tag{2.4.13}$$

Since

$$\|G(\phi_1) - G(\phi_2)\| \leq \beta r^{\frac{1}{p}} \|\phi_1 - \phi_2\|_C$$

for every $\phi_i \in C([-r,0];X)$, $i = 1, 2$ if we put $\beta_C = \beta r^{\frac{1}{p}}$ and $\gamma_C = \max\{0, \alpha + \beta_C\}$ then from Proposition 2.2.1 $B_C + \gamma_C I$ is accretive in $C([-r,0];X)$ and $R(I + \lambda B_C) = C([-r,0];X)$ for every $\lambda > 0$ with $\lambda\gamma_C < 1$. Thus by Theorem 1.2.1 the nonlinear semigroup $\{S_C(t) \mid t \geq 0\}$ is generated by $-B_C$.

PROPOSITION 2.4.2 *Let $A : X \supset D(A) \to 2^X$ be an operator and $\alpha \in \boldsymbol{R}$ such that $A + \alpha I$ is accretive in X and $R(I + \lambda A) = X$ for every $\lambda > 0$ with $\lambda\gamma < 1$. Let G satisfy* (2.4.6). *Let B, Q and B_C be operators defined in* (2.4.4), (2.4.11) *and* (2.4.13), *respectively. Let $\{S(t) \mid t \geq 0\}$ and $\{S_C(t) \mid t \geq 0\}$ be nonlinear semigroups generated by $-B$ and $-B_C$, respectively. Then:*

(1) If $Q\{\phi, h\} \in D(B_C)$ then $QB\{\phi, h\} = B_C Q\{\phi, h\}$;

(2) If $\{\phi, h\} \in D(Q)$ then $QS(t)\{\phi, h\} = S_C(t)Q\{\phi, h\}$ for every $t \geq 0$.

Proof. (1) Let $Q\{\phi, h\} \in D(B_C)$. Then by (2.4.13)

$$B_C Q\{\phi, h\} = B_C\phi = -\frac{d\phi}{d\tau}$$

and since using (2.4.4)

$$QB\{\phi, h\} = Q\Big\{-\frac{d\phi}{d\tau}, A\phi(0) - G(\phi)\Big\} = -\frac{d\phi}{d\tau}$$

the proof of (1) is completed.

(2) Let $\phi \in C([-r,0];X)$ and $\lambda > 0$ with $\lambda\gamma_C < 1$. Then by $R(I+\lambda B_C) = C([-r,0];X)$ there exists $\{\psi,k\} \in D(B)$ such that $Q\{\psi,k\} \in D(B_C)$ and $(I+\lambda B_C)Q\{\psi,k\} = \phi$. From (1) we have

$$Q(I+\lambda B)\{\psi,k\} = (I+\lambda B_C)Q\{\psi,k\} = \phi$$

and thus

$$J_\lambda^{B_C}\phi = QJ_\lambda^B Q^{-1}\phi$$

for every $\phi \in \overline{D(B_C)}$. Hence

$$\left(J_{\frac{t}{n}}^{B_C}\right)^n \phi = Q\left(J_{\frac{t}{n}}^{B}\right)^n Q^{-1}\phi$$

for every $t > 0$ and $n = 1,2,\cdots$. Thus from (2.4.12)

$$\begin{aligned}
&\|S(t)Q^{-1}\phi - Q^{-1}S_C(t)\phi\|_Y \\
&= \lim_{n\to\infty}\left\|\left(J_{\frac{t}{n}}^{B}\right)^n Q^{-1}\phi - Q^{-1}S_C(t)\phi\right\|_Y \\
&\le (1+r)^{\frac{1}{p}} \lim_{n\to\infty}\left\|Q\left(J_{\frac{t}{n}}^{B}\right)^n Q^{-1}\phi - S_C(t)\phi\right\|_C \\
&= (1+r)^{\frac{1}{p}} \lim_{n\to\infty}\left\|\left(J_{\frac{t}{n}}^{B_C}\right)^n \phi - S_C(t)\phi\right\|_C \\
&= 0.
\end{aligned}$$

Hence

$$\begin{aligned}
QS(t)\{\phi,h\} &= QS(t)Q^{-1}Q\{\phi,h\} \\
&= QQ^{-1}S_C(t)Q\{\phi,h\} \\
&= S_C(t)Q\{\phi,h\}
\end{aligned}$$

for every $t \ge 0$ and $\{\phi,h\} \in \overline{D(B)}$. This completes the proof of (2).

As in (2.1.46), we define a function $y(t) \equiv y(t;\phi) : [-r,\infty) \to X$ by

$$y(t) = \begin{cases} \phi(t), & -r \le t \le 0 \\ (S_C(t)\phi)(0), & t \ge 0. \end{cases} \tag{2.4.14}$$

Then from Proposition 2.1.3

$$S_C(t)\phi = y_t(\phi) \tag{2.4.15}$$

for every $t \ge 0$.

PROPOSITION 2.4.3 *Let $A : X \supset D(A) \to 2^X$ be an operator and $\alpha \in$ **R** such that $A + \alpha I$ is accretive in X and $R(I+\lambda A) = X$ for every*

$\lambda > 0$ with $\lambda\gamma < 1$. Let G satisfy (2.4.6). Let Q be operators defined in (2.4.11). Let $x(t)$ and $y(t)$ be functions defined in (2.4.10) and (2.4.14), respectively. Then:

(1) $x(t) = y(Q\{\phi, h\})(t)$ for every $t \geq -r$ and $\{\phi, h\} \in D(Q)$;

(2) $x_t = \pi_1 S(t)\{\phi, h\}$ for every $t \geq 0$ and $\{\phi, h\} \in \overline{D(B)}$.

Proof. (1) Let $\{\phi, h\} \in D(Q)$. Then, from (2.4.10), (2.4.14) and (1) of Proposition 2.4.2

$$\begin{aligned} x(t; \{\phi, h\}) &= \pi_2 S(t)\{\phi, h\} = \pi_2 Q^{-1} Q S(t)\{\phi, h\} \\ &= \pi_2 Q^{-1} S_C(t) Q\{\phi, h\} = (S_C(t) Q\{\phi, h\})(0) \\ &= y(t; Q\{\phi, h\}) \end{aligned}$$

for every $t \geq 0$ and

$$x(t; \{\phi, h\}) = \phi(t) = y(t; \phi) = y(t; Q\{\phi, h\})$$

for every $t \in [-r, 0]$. Hence (1) holds for every $t \geq -r$.

(2) Let, at first, $\{\phi, h\} \in D(B)$. Since $D(B) \subset D(Q)$, by (1), (2.4.15) and (2) of Proposition 2.4.2

$$\begin{aligned} x_t(\{\phi, h\}) &= y_t(Q\{\phi, h\}) = S_C(t) Q\{\phi, h\} \\ &= \pi_1 Q^{-1} S_C(t) Q\{\phi, h\} = \pi_1 S(t)\{\phi, h\} \end{aligned} \quad (2.4.16)$$

for every $t \geq 0$. Let, next, $\{\phi, h\} \in \overline{D(B)}$. Then there exists $\{\{\phi_n, h_n\}\} \in D(B)$ such that $\lim_{n\to\infty}\{\phi_n, h_n\} = \{\phi, h\}$ in Y. Thus

$$\begin{aligned} &\|x_t(\{\phi, h\}) - \pi_1 S(t)\{\phi, h\}\|_p \\ &\leq \|x_t(\{\phi, h\}) - x_t(\{\phi_n, h_n\})\|_p \\ &\quad + \|x_t(\{\phi_n, h_n\}) - \pi_1 S(t)\{\phi_n, h_n\}\|_p \\ &\quad + \|\pi_1 S(t)\{\phi_n, h_n\} - \pi_1 S(t)\{\phi, h\}\|_p \\ &= (1)_n + (2)_n + (3)_n \end{aligned}$$

for every $t \geq 0$ and $n = 1, 2, \cdots$.

Using (2.4.16) $(2)_n = 0$ for $n = 1, 2, \cdots$. From Definition 1.2.1

$$(3)_n \leq e^{\gamma t}\|S(t)\{\phi_n, h_n\} - S(t)\{\phi, h\}\|_p \to 0$$

as $n \to \infty$. Finally we estimate $(1)_n$. We obtain

$$((1)_n)^p = \int_{-r}^{0} \|x_t(\{\phi, h\})(\tau) - x_t(\{\phi_n, h_n\})(\tau)\|^p d\tau =$$

$$
\begin{aligned}
&= \int_{-r}^{0} \|x(t+\tau;\{\phi,h\}) - x(t+\tau;\{\phi_n,h_n\})\|^p d\tau \\
&= \int_{-r}^{-t} \|x(t+\tau;\{\phi,h\}) - x(t+\tau;\{\phi_n,h_n\})\|^p d\tau \\
&\quad + \int_{-t}^{0} \|x(t+\tau;\{\phi,h\}) - x(t+\tau;\{\phi_n,h_n\})\|^p d\tau \\
&= \int_{-r}^{-t} \|\phi(t+\tau) - \phi_n(t+\tau)\|^p d\tau \\
&\quad + \int_{-t}^{0} \|\pi_2 S(t+\tau)\{\phi_n,h_n\} - \pi_2 S(t+\tau)\{\phi,h\}\|^p d\tau \\
&\leq \int_{-r}^{-t} \|\phi(t+\tau) - \phi_n(t+\tau)\|^p d\tau \\
&\quad + e^{\gamma t}\|\{\phi_n,h_n\} - \{\phi,h\}\|_Y d\tau \to 0
\end{aligned}
$$

as $n \to \infty$ for every $t \geq 0$. Hence

$$\|x_t(\{\phi,h\}) - \pi_1 S(t)\{\phi,h\}\|_p = 0$$

for every $t \geq 0$ and $\{\phi,h\} \in \overline{D(B)}$. Therefore the proof of (2) is completed.

We now consider the existence and the uniqueness for solutions of (2.4.1).

THEOREM 2.4.1 *Let $A : X \supset D(A) \to 2^X$ be an operator and $\alpha \in \boldsymbol{R}$ such that $A + \alpha I$ is accretive in X and $R(I + \lambda A) = X$ for every $\lambda > 0$ with $\lambda\gamma < 1$. Let G satisfy (2.4.6). Let B be the operator in (2.4.4) and let $1 \leq p < \infty$. Then:*

(1) $x(t) \equiv x(t;\{\phi,h\})$ in (2.4.10) is an integral solution of (2.4.1) on $[-r,\infty)$ for every $\{\phi,h\} \in \overline{D(B)}$;

(2) If X is reflexive then $x(t)$ of (2.4.10) is a strong solution of (2.4.1) on $[-r,\infty)$ for every $\{\phi,h\} \in D(B)$.

Proof. (1) Let $\{\phi,h\} \in \overline{D(B)}$. From Theorem 1.3.3 $u(t) \equiv S(t)\{\phi,h\}$ is the unique limit solution of the nonlinear evolution

$$
\begin{cases}
\dfrac{du}{dt}(t) + Bu(t) \ni 0, \quad t \geq 0 \\
u(0) = \{\phi,h\}
\end{cases}
\tag{2.4.17}
$$

in Y. Here,

$$u(t) = \lim_{n\to\infty} u_n(t)$$

uniformly on every bounded interval of $[0,\infty)$,

$$\begin{aligned} u_n(0) &= u_0^n \\ u^n(t) &= u_i^n, \quad t \in (t_{j-1}^n, t_j^n] \end{aligned}$$

and

$$\frac{u_i^n - u_{i-1}^n}{t_i^n - t_{i-1}^n} + Bu_i^n \ni 0 \tag{2.4.18}$$

for $i = 1, 2, \cdots, N(n)$; $n = 1, 2, \cdots$, with

$$\begin{aligned} &0 = t_0^n < t_1^n < \cdots < t_{N(n)}^n = T, \\ &\lim_{n\to\infty} \max_{1\le i\le N(n)}(t_i^n - t_{i-1}^n) = 0, \\ &\{u_i^n \in D(B) \mid i = 0, 1, 2, \cdots, N(n);\ n = 1, 2, \cdots\}, \\ &u_0^n = \{\phi, h\} \end{aligned}$$

for every $T > 0$. We put $u_i^n = \{\phi_i^n, h_i^n\} \in D(B)$ for $i = 1, 2, \cdots, N(n)$; $n = 1, 2, \cdots$. If we project (2.4.18) into X from (2.4.4) we have

$$\frac{h_i^n - h_{i-1}^n}{t_i^n - t_{i-1}^n} + Ah_i^n \ni G(\phi_i^n)$$

for $i = 1, 2, \cdots, N(n)$; $n = 1, 2, \cdots$.

From (2.4.10)

$$x(0) = \pi_2 S(0)\{\phi, h\} = \pi_2\{\phi, h\} = h,$$

and from (2) of Proposition 2.4.3

$$x_0 = \pi_1 S(0)\{\phi, h\} = \pi_1\{\phi, h\} = \phi.$$

Let $x \in D(A)$. Since for $x^* \in F(h_i^n - x)$

$$\begin{aligned} &\|h_i^n - x\|^2 \\ &\quad = (h_i^n - x, x^*) \\ &\quad = (h_i^n - h_{i-1}^n, x^*) + (h_{i-1}^n - x, x^*) \\ &\quad \le (h_i^n - h_{i-1}^n, x^*) + \|h_{i-1}^n - x\|\|h_i^n - x\|, \end{aligned}$$

we obtain

$$\begin{aligned}
&\|h_i^n - x\|^2 - \|h_{i-1}^n - x\|\|h_i^n - x\| \\
&\qquad \le (h_i^n - h_{i-1}^n, x^*) \\
&\qquad = (t_i^n - t_{i-1}^n)(-y_i^n + G(\phi_i^n), x^*) \\
&\qquad = -(t_i^n - t_{i-1}^n)(y_i^n - y, x^*) + (t_i^n - t_{i-1}^n)(-y + G(\phi_i^n), x^*) \\
&\qquad \le (t_i^n - t_{i-1}^n)\alpha\|h_i^n - x\|^2 \\
&\qquad\quad +(t_i^n - t_{i-1}^n)(-y + G(\phi_i^n), h_i^n - x)_+ \\
&\qquad = (t_i^n - t_{i-1}^n)\alpha\|h_i^n - x\|^2 \\
&\qquad\quad +(t_i^n - t_{i-1}^n)[-y + G(\phi_n^i), h_i^n - x]_+\|h_i^n - x\|
\end{aligned}$$

for every $[h_i^n, y_i^n] \in A$ and $[x, y] \in A$. Hence

$$\begin{aligned}
&\|h_i^n - x\| - \|h_{i-1}^n - x\| \\
&\qquad \le (t_i^n - t_{i-1}^n)\alpha\|h_i^n - x\| + (t_i^n - t_{i-1}^n)[-y + G(\phi_i^n), h_i^n - x]_+
\end{aligned}$$

for every $[x, y] \in A$. Iterating for $i = j+1, \cdots, k, (j+1 < k)$ we obtain

$$\begin{aligned}
&\|h_k^n - x\| - \|h_j^n - x\| \\
&\qquad \le \sum_{i=j+1}^{k} (\alpha\|h_i^n - x\| + [-y + G(\phi_i^n), h_i^n - x]_+)
\end{aligned}$$

for every $[x, y] \in A$. Let $k = k_n$ and $j = j_n$ be such that $t \in (t_{k_n-1}^n, k_{k_n}^n]$ and $s \in (t_{j_n-1}^n, t_{j_n}^n]$. Then

$$\begin{aligned}
&\|\pi_2(u_n(t)) - x\| - \|\pi_2(u_n(s)) - x\| \\
&\qquad \le \int_{t_{jn}^n}^{t_{kn}^n} (\alpha\|\pi_2(u_n(\tau)) - x\| \\
&\qquad\quad +[-y + G(\pi_1(u_n(\tau)), \pi_2(u_n(\tau)) - x]_+)d\tau
\end{aligned}$$

for every $[x, y] \in A$. Passing to the limit as $n \to \infty$ we have

$$\begin{aligned}
&\|\pi_2 u(t) - x\| - \|\pi_2 u(s) - x\| \\
&\qquad \le \int_s^t (\alpha\|\pi_2 u(\tau) - x\| + [-y + G(\pi_1 u(\tau)), \pi_2 u(\tau) - x]_+)d\tau
\end{aligned}$$

for every $[x, y] \in A$. From (2.4.10) and (2) of Proposition 2.4.3 $\pi_1 u(t) = x_t$ and $\pi_2 u(t) = x(t)$ for every $t \ge 0$. Therefore

$$\begin{aligned}
&\|x(t) - x\| - \|x(s) - x\| \\
&\qquad \le \int_s^t (\alpha\|x(\tau) - x\| + [-y + G(x_\tau), x(\tau) - x]_+)d\tau
\end{aligned}$$

for every $[x, y] \in A$. It implies that $x(t)$ is an integral solution of (2.4.1) on $[-r, \infty)$.

(2) As in the Proof of Corollary 2.2.1, the proof is completed.

THEOREM 2.4.2 *Let X be a real Hilbert space. Let $A : X \supset D(A) \to 2^X$ be an operator and $\alpha \in \boldsymbol{R}$ such that $A + \alpha I$ is monotone in X and $R(I + \lambda A) = X$ for every $\lambda > 0$ with $\lambda\gamma < 1$. Let G satisfy* (2.4.6). *Let B be the operator in* (2.4.4) *and let $1 < p < \infty$. Then $x(t) \equiv x(t; \{\phi, h\})$ in* (2.4.10) *is a unique strong solution of* (2.4.1) *on $[-r, \infty)$ for every $\{\phi, h\} \in D(B)$.*

Proof. Let $1 < p < \infty$. Then Y is reflexive. Let $\{S(t) \mid t \geq 0\}$ be the nonlinear semigroup of type γ on $\overline{D(B)}$, generated by $-B$. From Corollary 1.2.1 and Theorem 1.2.6

$$\frac{d}{dt}S(t)\{\phi, h\} + BS(t)\{\phi, h\} \ni 0$$

a.e. $t \in (0, \infty)$ for every $\{\phi, h\} \in D(B)$. From (2.4.4), (2.4.10) and (2) of Proposition 2.4.3

$$\begin{aligned} &BS(t)\{\phi, h\} \\ &\quad = B\{\pi_1 S(t)\{\phi, h\}, \pi_2 S(t)\{\phi, h\}\} \\ &\quad = B\{x_t(\{\phi, h\}), x(\{\phi, h\})(t)\} \\ &\quad = B\{x_t(\{\phi, h\}), x_t(\{\phi, h\})(0)\} \\ &\quad = \Big\{-\frac{dx_t}{dt}(\{\phi, h\}), Ax_t(\{\phi, h\})(0) - G(x_t(\{\phi, h\}))\Big\} \\ &\quad = \Big\{-\frac{dx_t}{dt}(\{\phi, h\}), Ax(\{\phi, h\})(t) - G(x_t(\{\phi, h\}))\Big\} \end{aligned}$$

for every $t \geq 0$. Thus by (2.4.18) we have

$$\frac{d}{dt}S(t)\{\phi, h\} + \Big\{\frac{dx_t}{dt}(\{\phi, h\}), -Ax(\{\phi, h\})(t) + G(x_t(\{\phi, h\}))\Big\} \ni 0$$

a.e. $t \in (0, \infty)$.

Since from (2.4.10)

$$\begin{aligned} \frac{d}{dt}x(t; \{\phi, h\}) &= \frac{d}{dt}\pi_2 S(t)\{\phi, h\} = \pi_2 \frac{d}{dt}S(t)\{\phi, h\} \\ &\in -Ax(t; \{\phi, h\}) + G(x_t(\{\phi, h\})) \end{aligned}$$

a.e. $t \in (0, \infty)$ and thus,

$$\frac{d}{dt}x(t; \{\phi, h\}) + Ax(t; \{\phi, h\}) \ni G(x_t(\{\phi, h\})),$$

a.e. $t \in (0, \infty)$,

$$x(0) = \pi_2 S(0)\{\phi, h\} = \pi_2\{\phi, h\} = h$$

and

$$x_0 = \pi_1\{\phi, h\} = \phi.$$

Hence $x(t)$ in (2.4.10) is a strong solution of (2.4.1) on $[-r, \infty)$. The uniqueness is as in Proof of Theorem 2.2.1.

2.5. General Methods of Ruess–Summer

In this section we consider nonlinear functional evolutions of the type

$$\begin{cases} \dfrac{dx}{dt}(t) + \alpha x(t) + Ax(t) \ni G(x_t), & t \geq 0 \\ x(t) = \phi(t), & t \in I \end{cases} \tag{2.5.1}$$

in a real Banach space X, where $\alpha \in \mathbf{R}$, $A : X \supset D(A) \to 2^X$, $G : D(G) \subset E \to X$ with an appropriate real Banach space E of functions from $I \to X$, where $I = [-r, 0]$, $r > 0$, or $I = (-\infty, 0]$, and $\phi \in E$.

We are dealing with (2.5.1) under general assumptions including as wide a range of applications as possible.

We assume that E is a real Banach space of continuous functions $\phi : I \to X$ with norm $\| \cdot \|_E$ satisfying:

(E1) (1) For every $\phi \in E$
  (a) $\|\phi(0)\| \leq \|\phi\|_E$,
  (b) Define $\tilde{\phi} : I \to X$ by $\tilde{\phi}(\tau) = \phi(0)$ for every $\tau \in I$. Then $\tilde{\phi} \in E$;
 (2) Let ϕ, $\phi_n \in E$, $n = 1, 2, \cdots$. If $\|\phi_n - \phi\|_E \to 0$ then

$$\|\phi_n(\tau) - \phi(\tau)\| \to 0$$

 for every $\tau \in I$.

(E2) Let $\lambda > 0$, $x \in X$ and $\phi \in E$. If $\psi \in C^1(I; X)$ is the solution of

$$\psi - \lambda \frac{d\psi}{d\tau} = \phi, \quad \psi(0) = x,$$

 then $\psi \in E$ and $\|\psi\|_E \leq \max\{\|x\|, \|\phi\|_E\}$.

(E3) (1) If $x(t) : I \cup [0, \infty) \to X$ is continuous and $x_{|_I} \in E$ then
  (a) $x_t \in E$ for every $t \geq 0$,
  (b) the map $t \to x_t$ is continuous from $[0, \infty)$ into E,

where $x_{|_I}$ is the restriction of x on I;

(2) There exist a constant $M_0 \geq 1$ and a locally bounded function $M_1 : [0, \infty) \to [0, \infty)$ such that, given $x(t) : I \cup [0, \infty) \to X$ as in (a) above,

$$\|x_t\|_E \leq M_0 \|x_0\|_E + M_1(t) \max_{0 \leq s \leq t} \|x(s)\|$$

for every $t \geq 0$.

We also give the following assumptions associated with the conditions of E above.

(A1) (1) $\widetilde{X}$ is a closed subset of X;

(2) $\widetilde{E}$ is a closed convex subset of E;

(3) $A : X \supset D(A) \to 2^X$ is an accretive operator;

(4) For every $\psi_i \in \widetilde{E}$, $i = 1, 2$,

$$\|G(\psi_1) - G(\psi_2)\| \leq \beta \|\psi_1 - \psi_2\|_E$$

for some $\beta > 0$;

(5) $\alpha \in \mathbf{R}$ and $\gamma = \max\{0, \beta - \alpha\}$.

(A2) If $x \in \widetilde{X}$, $\phi \in \widetilde{E}$, $\lambda > 0$ with $\lambda\gamma < 1$, and ψ is the solution of

$$\psi - \lambda \frac{d\psi}{d\tau} = \phi, \quad \psi(0) = x,$$

then $\psi \in \widetilde{E}$.

(A3) If $\phi \in \widetilde{E}$ and $\lambda > 0$ with $\lambda\gamma < 1$ then

$$\frac{1}{1 + \lambda\alpha}(\phi(0) + \lambda G(\psi)) \in \Big(I + \frac{\lambda}{1 + \lambda\alpha} A\Big)(D(A) \cap \widetilde{X})$$

for every $x \in \widetilde{X}$.

To obtain principal results on semigroup solutions for (2.5.1) we define a nonlinear operator $B : D(B) \subset E \to E$ as similar in (2.2.2) by

$$\begin{aligned} D(B) &= \Big\{\phi \in \widetilde{E} \;\Big|\; \frac{d\phi}{dt} \in E, \phi(0) \in D(A), \\ &\qquad \frac{d\phi}{dt}(0) \in G(\phi) - \alpha\phi(0) - A\phi(0)\Big\}, \\ B\phi &= -\frac{d\phi}{dt}, \quad \phi \in D(B). \end{aligned} \tag{2.5.2}$$

PROPOSITION 2.5.1 *Assume that (E1)–(E3) and (A1)–(A3) are satisfied. Let B be the operator defined as in* (2.5.2)*. Then:*

(1) $B + \gamma I$ is accretive in E;

(2) $R(I + \lambda B) \supset \widetilde{E}$ for every $\lambda > 0$ with $\lambda\gamma < 1$.

Proof. (1) Let $\lambda > 0$ with $\lambda\gamma < 1$, and $\phi_i \in D(B)$, $i = 1, 2$. Then

$$\begin{aligned}(\phi_1 - \phi_2) - \lambda\frac{d(\phi_1 - \phi_2)}{dt} &= \Big(\phi_1 - \lambda\frac{d\phi_1}{dt}\Big) - \Big(\phi_2 - \lambda\frac{d\phi_2}{dt}\Big),\\ (\phi_1 - \phi_2)(0) &= \phi_1(0) - \phi_2(0).\end{aligned}$$

From (E2),

$$\begin{aligned}&\|\phi_1 - \phi_2\|_E \\ &\le \max\Big\{\|\phi_1(0) - \phi_2(0)\|, \Big\|\Big(\phi_1 - \lambda\frac{d\phi_1}{dt}\Big) - \Big(\phi_2 - \lambda\frac{d\phi_2}{dt}\Big)\Big\|_E\Big\}.\end{aligned} \tag{2.5.3}$$

In the case that the second term of the right side in (2.5.3) is the maximum

$$\begin{aligned}\|\phi_1 - \phi_2\|_E &\le \Big\|\Big(\phi_1 - \lambda\frac{d\phi_1}{dt}\Big) - \Big(\phi_2 - \lambda\frac{d\phi_2}{dt}\Big)\Big\|_E \\ &\le \frac{1}{1 - \lambda\gamma}\|(\phi_1 - \phi_2) + \lambda(B\phi_1 - B\phi_2)\|_E.\end{aligned}$$

In the case that the first term of the right side in (2.5.3) is the maximum

$$\|\phi_1 - \phi_2\|_E \le \|\phi_1(0) - \phi_2(0)\|.$$

By (1) of (E1)

$$\|\phi_1 - \phi_2\|_E = \|\phi_1(0) - \phi_2(0)\|.$$

Since $\phi_i \in D(B)$ using the definition of B, $\phi_i(0) \in D(A)$ and

$$\frac{d\phi_i}{dt}(0) \in G(\phi_i) - \alpha\phi_i(0) - A\phi_i(0),$$

and then

$$\phi_i(0) = \Big(I + \frac{\lambda}{1 + \lambda\alpha}A\Big)^{-1}\Big(\frac{1}{1 + \lambda\alpha}\Big(\phi_i(0) - \lambda\frac{d\phi_i}{dt} + \lambda G(\phi_i)\Big)\Big) \tag{2.5.4}$$

for $i = 1, 2$. Thus from (1) of (E1), (3) and (4) of (A1), and (2.5.4)

$$\begin{aligned}\|\phi_1 - \phi_2\|_E &= \|\phi_1(0) - \phi_2(0)\| \\ &\le \frac{1}{1 + \lambda\alpha}\Big(\Big\|\Big(\phi_1 - \lambda\frac{d\phi_1}{dt}\Big) - \Big(\phi_2 - \lambda\frac{d\phi_2}{dt}\Big)\Big\|_E \\ &\qquad + \lambda\beta\|\phi_1 - \phi_2\|_E\Big),\end{aligned}$$

and thus

$$(1+\lambda(\alpha-\beta))\|\phi_1-\phi_2\|_E \le \Big\|\Big(\phi_1-\lambda\frac{d\phi_1}{dt}\Big)-\Big(\phi_2-\lambda\frac{d\phi_2}{dt}\Big)\Big\|_E.$$

Hence

$$\|\phi_1-\phi_2\|_E \le \frac{1}{1-\lambda\gamma}\|(\phi_1-\phi_2)+\lambda(B\phi_1-B\phi_2)\|_E.$$

Therefore $B+\gamma I$ is accretive in E.

(2) Let $\lambda>0$ with $\lambda\gamma<1$. Let $\phi\in\widetilde{E}$ and let $x\in\widetilde{X}$. We define an operator $Q:\widetilde{X}\to X$ by

$$Qx=\Big(I+\frac{\lambda}{1+\lambda\alpha}A\Big)^{-1}\Big(\frac{1}{1+\lambda\alpha}(\phi(0)+\lambda G(\psi))\Big), \tag{2.5.5}$$

where $\psi\equiv\psi_x$ is the solution in (E2) for $x\in\widetilde{X}$. Then from (3) of (A1), (A3) and (E3) Q is a well defined mapping from $\widetilde{X}$ into $\widetilde{X}$, and

$$\begin{aligned}\|Qx_1-Qx_2\| &\le \frac{\lambda\beta}{1+\lambda\alpha}\|\psi_{x_1}-\psi_{x_2}\|_E\\ &\le \frac{\lambda\beta}{1+\lambda\alpha}\|x_1-x_2\|\end{aligned}$$

for every $x_i\in\widetilde{X}$, $i=1,2$. Since $0<\dfrac{\lambda\beta}{1+\lambda\alpha}<1$ Q is a strict contraction. From Banach's contraction principle there exists $x_0\in\widetilde{X}$ uniquely such that $Qx_0=x_0$. By (2.5.5)

$$x_0=\Big(I+\frac{\lambda}{1+\lambda\alpha}A\Big)^{-1}\Big(\frac{1}{1+\lambda\alpha}(\phi(0)+\lambda G(\psi_{x_0}))\Big).$$

In particular,

$$\psi_{x_0}(0)=x_0\in D(A)\cap\widetilde{X},$$
$$\psi_{x_0}-\lambda\frac{d\psi_{x_0}}{dt}=\phi$$

and

$$\begin{aligned}\frac{d\psi_{x_0}}{dt}(0) &= \frac{1}{\lambda}(x_0-\phi(0))\\ &\in G(\psi_{x_0})-\alpha x_0-Ax_0\\ &= G(\psi_{x_0})-\alpha\psi_{x_0}(0)-A\psi_{x_0}(0).\end{aligned}$$

Thus $\psi_{x_0}\in D(B)$ and $(I+\lambda B)\psi_{x_0}=\phi$. Therefore $R(I+\lambda B)\supset\widetilde{E}$.

not uniformly in general on $(-\infty, 0]$ for every $\phi \in E_0$ except possibly at the point $\tau = -t$ where there may be a discontinuity.

Indeed, an inductive argument shows from (b) of (2) of Lemma 2.5.1 that iteration of $J_\lambda^{B_0}$ results in

$$\begin{aligned}\left(J_\lambda^{B_1}\right)^k \phi(\tau) &= \left(J_\lambda^{B_0}\right)^k (\phi - \phi(0))(\tau) \\ &\quad + \phi(0)\Big(1 - e_\lambda(\tau) \sum_{i=0}^{k-1} \frac{1}{i!}\Big(\frac{-\tau}{\lambda}\Big)^i\Big) \qquad (2.5.11)\end{aligned}$$

for every $\phi \in E$ and $k = 1, 2, \cdots$, and thus, putting $\lambda = \dfrac{t}{n}$ in (2.5.11) for every $t > 0$

$$\begin{aligned}\left(J_{\frac{t}{n}}^{B_1}\right)^n \phi(\tau) &= \left(J_{\frac{t}{n}}^{B_0}\right)^n (\phi - \phi(0))(\tau) \\ &\quad + \phi(0)\Big(1 - e_{\frac{t}{n}}(\tau) \sum_{i=0}^{n-1} \frac{1}{i!} n^i \Big(\frac{-\tau}{t}\Big)^i\Big) \qquad (2.5.12)\end{aligned}$$

for every $t > 0$, $\phi \in E$ and $n = 1, 2, \cdots$. Since $\phi - \phi(0) \in E_0$ by (2.5.8)

$$S_0(t)(\phi - \phi(0)) = \lim_{n \to \infty} \left(J_{\frac{t}{n}}^{B_0}\right)^n (\phi - \phi(0))$$

uniformly on $(-\infty, 0]$ for every $t \geq 0$.

Now we define a function $p_n : (0, \infty) \to \mathbf{R}$ by

$$p_n(c) = \sum_{i=0}^{n-1} \frac{1}{i!} (cn)^i e^{cn}$$

for every $c \in (0, \infty)$ and $n = 1, 2, \cdots$. Then using (2.5.12)

$$\left(J_{\frac{t}{n}}^{B_1}\right)^n \phi(\tau) = \left(J_{\frac{t}{n}}^{B_0}\right)^n (\phi - \phi(0))(\tau) + \phi(0)\Big(1 - p_n\Big(-\frac{\tau}{t}\Big)\Big) \qquad (2.5.13)$$

for every $\tau \in (-\infty, 0]$, $t > 0$, $\phi \in E$ and $n = 1, 2, \cdots$. We know that

$$\lim_{n \to \infty} p_n(c) = \begin{cases} 1, & 0 < c < 1 \\ 0, & c > 1. \end{cases} \qquad (2.5.14)$$

If $\tau < -t$ for every $t \geq 0$ (2.5.13) and (2.5.14) imply that

$$\lim_{n \to \infty} \left(J_{\frac{t}{n}}^{B_1}\right)^n \phi(\tau) = S_0(t)(\phi - \phi(0))(\tau)$$

for every $\phi \in E$. Hence

$$\lim_{n\to\infty}\left(J^{B_1}_{\frac{t}{n}}\right)^n \phi = S_0(t)\phi$$

(nonuniformly) on $(-\infty, 0]$, except possibly at $\tau = -t$ for every $t \geq 0$ and $\phi \in E_0$, which is the proof of (2.5.10).

Next we show (2.5.9). By the iterates of J^B_λ, letting $\lambda = \dfrac{t}{n}$ for every $t \geq 0$ and $n = 1, 2, \cdots$, we have from (c) of (2) of Lemma 2.5.1 that

$$\left(J^B_{\frac{t}{n}}\right)^n \phi = \left(J^{B_1}_{\frac{t}{n}}\right)^n \phi + \sum_{i=0}^{n-1}\left(\left(J^B_{\frac{t}{n}}\right)^{n-i}\phi\right)(0)\left(J^{B_1}_{\frac{t}{n}}\right)^i e_{\frac{t}{n}}$$

for every $\phi \in E$. By (b) of (2) of Lemma 2.5.1

$$\left(\left(J^{B_1}_{\frac{t}{n}}\right)^i e_{\frac{t}{n}}\right)(\tau) = \frac{n^i}{i!}\Big(-\frac{\tau}{t}\Big)^i e^{\frac{n\tau}{t}}. \tag{2.5.15}$$

To show that $\left\{\left(\left(J^B_{\frac{t}{n}}\phi\right)^i\right)(0)\right\}$ is uniformly bounded for $i = 1, 2, \cdots$ on every bounded interval of $[0, \infty)$ let $\phi \in D(B)$. Using (4) of Proposition 1.1.4 for $i \leq n-1$

$$\begin{aligned}
&\left\|\left(\left(J^B_{\frac{t}{n}}\right)^i\phi\right)(0) - \phi(0)\right\| \\
&\leq \left\|\left(J^B_{\frac{t}{n}}\right)^i\phi - \phi\right\|_E \leq \frac{it}{n}\Big(1 - \frac{\gamma t}{n}\Big)^{-i+1}\|B\phi\|_E \\
&\leq t\Big(1 - \frac{\gamma t}{n}\Big)^{-n}\|B\phi\|_E < te^{2\gamma t}\|B\phi\|_E.
\end{aligned}$$

Thus from (2.5.15)

$$\left\|\left(J^B_{\frac{t}{n}}\right)^n\phi(\tau) - \left(J^{B_1}_{\frac{t}{n}}\right)^n\phi(\tau)\right\| \leq (te^{2\gamma t}\|B\phi\|_E + \|\phi(0)\|)p_n\Big(-\frac{\tau}{t}\Big)$$

for every $\phi \in D(B)$. Thus (2.5.10) holds for every $\phi \in D(B)$ and $\tau < -t$ on every bounded interval of $[0, \infty)$. Hence from (2.5.6) and (2.5.10)

$$(S(t)\phi)(\tau) = \begin{cases} 0, & -t \leq \tau \leq 0 \\ \phi(t+\tau), & \tau < -t \end{cases} \tag{2.5.16}$$

on every bounded interval of $[0, \infty)$ for every $\phi \in D(B)$. From the continuity of $S(t)\phi$ and (2) of (E1), (2.5.16) holds for every $\phi \in \overline{D(B)}$.

By (2.5.7) and (2.5.16), for every $t \geq 0$ and $\tau \leq 0$ with $\tau < -t$,

$$\begin{aligned}
x_t(\tau) &= x(t+\tau) = S(t+\tau)\phi(0) \\
&= \phi(t+\tau) = S(t)\phi(\tau),
\end{aligned}$$

and for every $t \geq 0$ and $\tau \leq 0$ with $-t \leq \tau$,

$$x_t(\tau) = x(t+\tau) = S(t+\tau)\phi(0) = 0 = S(t)\phi(\tau),$$

for every $\phi \in \overline{D(B)}$. Therefore $S(t)\phi = x_t$ for every $t \geq 0$ and $\phi \in \overline{D(B)}$.

So we have shown that Proposition 2.5.2 holds in the case of $I = (-\infty, 0]$. Also we can show that Proposition 2.5.2 holds in the case of $I = [-r, 0]$, $r > 0$, with

$$S_0(t)\psi(\tau) = \begin{cases} 0, & -t \leq \tau \leq 0 \\ \psi(t+\tau), & -r \leq \tau \leq -t \end{cases}$$

for every $t \geq 0$ and $\tau \in I$, instead of (2.5.6).

PROPOSITION 2.5.3 *Assume that (E1)–(E3) and (A1)–(A3) are satisfied. Let B be the operator defined as in* (2.6.2). *We put*

$$D_A = \cup_{\mu>0}(\cap_{0<\lambda<\mu} R(I+\lambda A)).$$

Then we have

$$\begin{aligned} \{\phi \in \widetilde{E} \mid \phi(0) \in \overline{D(A) \cap D_A}\} &\subset \overline{D(B)} \qquad (2.5.17) \\ &\subset \{\phi \in \widetilde{E} \mid \phi(0) \in \overline{D(A)}\}. \end{aligned}$$

In particular, if there exists $\mu > 0$ such that $R(I+\lambda A) \supset D(A)$ for every $\lambda > 0$ with $\lambda < \mu$ then

$$\overline{D(B)} = \{\phi \in \widetilde{E} \mid \phi(0) \in \overline{D(A)}\}. \qquad (2.5.18)$$

Proof. We show the first inclusion of (2.5.17). Let $\phi \in \widetilde{E}$ with $\phi(0) \in \overline{D(A) \cap D_A}$. Then from Proposition 2.5.1 and its proof $\phi_\lambda = J_\lambda^B \phi \in D(B)$, and

$$\phi_\lambda(0) = \Big(I + \frac{\lambda}{1+\lambda\alpha}A\Big)^{-1}\Big(\frac{1}{1+\lambda\alpha}(\phi(0) + \lambda G(\phi_\lambda))\Big).$$

Thus

$$\begin{aligned} &\|\phi_\lambda(0) - \phi(0)\| \\ &\quad \leq \Big\| \Big(I + \frac{\lambda}{1+\lambda\alpha}A\Big)^{-1}\Big(\frac{1}{1+\lambda\alpha}(\phi(0) + \lambda G(\phi_\lambda))\Big) \\ &\qquad - \Big(I + \frac{\lambda}{1+\lambda\alpha}A\Big)^{-1}\phi(0)\Big\| + \end{aligned}$$

$$
\begin{aligned}
&+\left\|\left(I+\frac{\lambda}{1+\lambda\alpha}A\right)^{-1}\phi(0)-\phi(0)\right\| \\
\leq &\left\|\frac{-\lambda\alpha}{1+\lambda\alpha}\phi(0)+\frac{\lambda}{1+\lambda\alpha}G(\phi_\lambda)\right\| \\
&+\left\|\left(I+\frac{\lambda}{1+\lambda\alpha}A\right)^{-1}\phi(0)-\phi(0)\right\| \\
\leq &\frac{\lambda|\alpha|}{1+\lambda\alpha}\|\phi(0)\|+\frac{\lambda\beta}{1+\lambda\alpha}\|\phi_\lambda-\phi\|_E+\frac{\lambda}{1+\lambda\alpha}\|G(\phi)\| \\
&+\left\|\left(I+\frac{\lambda}{1+\lambda\alpha}A\right)^{-1}\phi(0)-\phi(0)\right\|, \qquad (2.5.19)
\end{aligned}
$$

and by (2) of Lemma 2.5.1

$$
\begin{aligned}
&\|\phi_\lambda-\phi\|_E \\
&= \left\|J_\lambda^{B_0}(\phi-\phi(0))+(1-e_\lambda)\phi(0)+e_\lambda\phi_\lambda-\phi\right\|_E \\
&\leq \left\|J_\lambda^{B_0}(\phi-\phi(0))-(\phi-\phi(0))\right\|_E+\|\phi_\lambda(0)-\phi(0)\|. \qquad (2.5.20)
\end{aligned}
$$

Combining (2.5.19) and (2.5.20)

$$
\begin{aligned}
&\frac{1+\lambda(\alpha-\beta)}{1+\lambda\alpha}\|\phi_\lambda-\phi\|_E \\
&\quad\leq \|J_\lambda^{B_0}(\phi-\phi(0))-(\phi-\phi(0))\|_E \\
&\quad+\left\|\left(I+\frac{\lambda}{1+\lambda\alpha}A\right)^{-1}\phi(0)-\phi(0)\right\|, \\
&\quad+\frac{\lambda}{1+\lambda\alpha}(|\alpha|\|\phi(0)\|+\|G(\phi)\|).
\end{aligned}
$$

Hence using (1) of Lemma 2.5.1 and (3) of Proposition 1.1.4

$$
\lim_{\lambda\to 0+}\|\phi_\lambda-\phi\|=0,
$$

whereby $\phi\in\overline{D(B)}$.

Next the second inclusion is obvious. In fact, let $\phi\in\overline{D(A)}$. Then there exists $\phi_n\in D(B)$ for $n=1,2,\cdots$ such that $\lim_{n\to\infty}\phi_n=\phi$. Since $\widetilde{E}$ is closed, from (2) of (E1), $\phi\subset\widetilde{E}$ and $\phi(0)\in\overline{D(A)}$.

From the assumption and (2.5.17) we obtain (2.5.18), and the proof is complete.

We now turn to the problem that, actually, the function $x(t)$ in (2.5.7) is the unique solution of (2.5.1), for every $\phi\in\overline{D(B)}$.

Definition 2.5.1 *1. A function $x(t): I\cup[0,\infty)\to X$ is called a strong solution of* (2.5.1) *on $I\cup[0,\infty)$ if:*

(1) $x(t) = \phi(t)$ *for every* $t \in I$*;*

(2) $x(t)$ *is continuous on* $[0, \infty)$*;*

(3) $x(t)$ *is absolutely continuous on every compact interval of* $[0, \infty)$*;*

(4) $x(t)$ *is differentiable a.e.* $t \in (0, \infty)$*;*

(5) $x_t \in \widetilde{E}$ *for every* $t \geq 0$*;*

(6) $x(t) \in D(A)$ *for every* $t \geq 0$*;*

(7) $x(t)$ *satisfies*

$$\frac{dx}{dt}(t) + \alpha x(t) + Ax(t) \ni G(x_t)$$

a.e. $t \in (0, \infty)$.

2. A function $x(t) : I \cup [0, \infty) \to X$ *is called an integral solution of type* α*, of* (2.5.1) *on* $I \cup [0, \infty)$ *if:*

(1) $x(t) = \phi(t)$ *for every* $t \in I$*;*

(2) $x(t)$ *is continuous on* $[0, \infty)$*;*

(3) $x(t)$ *satisfies*

$$\begin{aligned} &\|x(t) - x\| - \|x(s) - x\| \\ &\leq \alpha \int_s^t \|x(\sigma) - x\| d\sigma + \int_s^t (G(x_\sigma) - y, x(\sigma) - x)_+ d\sigma \end{aligned}$$

for every $[x, y] \in A$ *and* $0 \leq s \leq t$.

3. A function $x(t) : I \cup [0, \infty) \to X$ *is called a limit solution of* (2.5.1) *on* $I \cup [0, \infty)$ *if:*

(1) $x(t) = \phi(t)$ *for every* $t \in I$*;*

(2) $x(t)$ *is continuous on* $[0, \infty)$*;*

(3) $x(t) = \lim_{n \to \infty} x_n(t)$ *uniformly on every compact interval of* $[0, \infty)$.

Here

$$\begin{aligned} x_n(0) &= x_0^n, \\ x_n(t) &= x_i^n, \quad t_{i-1}^n < t < t_i^n \end{aligned}$$

and

$$\frac{x_i^n - x_{i-1}^n}{t_i^n - t_{i-1}^n} + \alpha x_i^n + Ax_i^n \ni G(x_{t_i^n})$$

for $i = 1, 2, \cdots, N(n)$; $n = 1, 2, \cdots$, *with*

$0 = t_0^n < t_1^n < \cdots < t_{N(n)}^n = T$ for every $T \in (0, \infty)$, $n = 1, 2, \cdots$,

$\lim_{n\to\infty} \max_{1\le i\le N(n)}(t_i^n - t_{i-1}^n) = 0$,

$\{x_i^n \in X \mid i = 1, 2, \cdots, N(n);\ n = 1, 2, \cdots\}$.

We now deal with existence for limit solutions and integral solutions of (2.5.1).

THEOREM 2.5.1 *Assume that (E1)–(E3) and (A1)–(A3) be satisfied. Let* $\{S(t) \mid t \ge 0\}$ *be the nonlinear semigroup on* $\overline{D(B)}$ *of type* γ, *generated by* $-B$ *with the operator* B *defined in* (2.5.2). *Then*

$$x(t) \equiv x(t;\phi) : I \cup [0,\infty) \to X$$

defined by

$$x(t) = \begin{cases} \phi(t), & t \in I \\ (S(t)\phi)(0), & t \ge 0 \end{cases}$$

for every $\phi \in \overline{D(B)}$, *is a limit solution of* (2.5.1) *on* $I \cup [0,\infty)$, *with* $x_t \in \widetilde{E}$ *and* $x(t) \in \widetilde{X} \cap \overline{D(A)}$ *for every* $t \ge 0$.

Let $x(t) \equiv x(t,\phi)$ *and* $y(t) \equiv y(t,\psi)$ *be two limit solutions of* (2.5.1) *on* $I \cup [0,\infty)$ *for* ϕ, $\psi \in \overline{D(B)}$. *Then*

$$\begin{aligned} &e^{-\alpha t}\|x(t) - y(t)\| - e^{-\alpha s}\|x(s) - y(s)\| \\ &\le \int_s^t e^{-\alpha\sigma}[G(S(\sigma)\phi) - G(S(\sigma)\psi), x(\sigma) - y(\sigma)]_+ d\sigma \end{aligned}$$

for every $0 \le s \le t$.

Proof. Referring to that of Theorem 2.2.3 the proof is completed.

We now consider existence and uniqueness of strong solutions.

PROPOSITION 2.5.4 *Let (E1)–(E3) and (A1)–(A3) are satisfied. Let* $\phi \in E$. *Then there exists at most one strong solution of* (2.5.1) *on* $I \cup [0,\infty)$.

Proof. Let $x_i(t)$ be a strong solution of (2.5.1) on $I \cup [0,\infty)$ satisfying $x_i(t) = \phi(t)$ for every $t \in I$, $i = 1, 2$. Fixing $T > 0$ from Definition 2.5.1 $x_1(t) - x_2(t)$ is absolutely continuous on $[0,T]$ and differentiable a.e. $t \in (0,T)$, and thus $\|x_1(t) - x_2(t)\|^2$ is also differentiable a.e. $t \in [0,T]$. Let $N \subset [0,T]$ be a set of Lebesgue measure 0 such that $x_1(t) - x_2(t)$

and $\|x_1(t) - x_2(t)\|^2$ are differentiable for every $t \in [0,T] - N$, and $x_1(t) - x_2(t)$ satisfies (2.5.1) for every $t \in [0,T] - N$. Since A is accretive from (4) of Theorem 1.1.1 there exists $x^*(t) \in F(x_1(t) - x_2(t))$ such that

$$\Big(\Big(\frac{dx_1}{dt}(t)+\alpha x_1(t)-G((x_1)_t)\Big)-\Big(\frac{dx_2}{dt}(t)+\alpha x_2(t)-G((x_2)_t)\Big), x^*(t)\Big) \leq 0$$

for every $t \in [0,T] - N$. Thus by (12) of Lemma 1.1.2

$$\begin{aligned}
&\frac{1}{2}\frac{d}{dt}\|x_1(t) - x_2(t)\|^2 \\
&\quad = \Big(\frac{dx_1}{dt}(t) - \frac{dx_2}{dt}(t), x^*(t)\Big) \\
&\quad \leq (G((x_1)_t) - G((x_2)_t), x^*(t)) - \alpha(x_1(t) - x_2(t), x^*(t)) \\
&\quad \leq \beta\|x_1(t) - x_2(t)\|\|(x_1)_t - (x_2)_t\| - \alpha\|x_1(t) - x_2(t)\|^2
\end{aligned}$$

for every $t \in [0,T] - N$. Thus

$$\begin{aligned}
&\frac{1}{2}\|x_1(t_1) - x_2(t_1)\|^2 - \frac{1}{2}\|x_1(t_0) - x_2(t_0)\|^2 \\
&\qquad \leq \beta \int_{t_0}^{t_1} \|x_1(s) - x_2(s)\|\|(x_1)_s - (x_2)_s\| ds \\
&\qquad\quad -\alpha \int_{t_0}^{t_1} \|x_1(s) - x_2(s)\|^2 ds
\end{aligned} \tag{2.5.21}$$

for every $0 \leq t_0 \leq t_1 \leq T$.

We suppose that $x_1 \neq x_2$. Noting that $x_1(0) = x_2(0)$, we put

$$t_0 = \sup\{t \in [0,\infty) \mid x_1(s) = x_2(s) \text{ for every } s \in [0,t]\}.$$

Then $t_0 < \infty$ and $x_1(t_0) = x_2(t_0)$. Furthermore, considering (2) of (E3) we put

$$K = \sup\{M_1(s) \mid 0 \leq s \leq t_0 + 1\}$$

and let

$$\delta = \min\{(3(\beta K + |\alpha|) + 1)^{-1}, 1\},$$

and since $\|x_1(t_1) - x_2(t_1)\| > 0$ we take $t_1 \in (t_0, t_0 + \delta)$ such that

$$\|x_1(t_1) - x_2(t_1)\| \geq \|x_1(s) - x_2(s)\|$$

for every $s \in [t_0, t_0 + \delta]$. But from (2.5.21) with $T = t_0 + \delta$ and (2) of (E3)

$$\begin{aligned}
&\frac{1}{2}\|x_1(t_1) - x_2(t_1)\|^2 \\
&\quad \le \beta K \int_{t_0}^{t_1} \|x_1(s) - x_2(s)\| \max_{0 \le u \le s} \|x_1(u) - x_2(u)\| ds \\
&\qquad + |\alpha| \int_{t_0}^{t_1} \|x_1(s) - x_2(s)\|^2 ds \\
&\quad \le (\beta K + |\alpha|)\delta \|x_1(t_1) - x_2(t_1)\|^2,
\end{aligned}$$

which contradicts our choice of K and δ. Hence the proof of uniqueness is completed.

We now state our main result on semigroup solutions of (2.5.1) on $I \cup [0, \infty)$.

THEOREM 2.5.2 *Assume that (E1)-(E3) and (A1)-(A3) be satisfied. Let $\{S(t) \mid t \ge 0\}$ be the nonlinear semigroup on $\overline{D(B)}$ of type γ, generated by $-B$ with the operator B as defined in* (2.5.2). *Then*

$$x(t) \equiv x(t; \phi) : I \cup [0, \infty) \to X$$

defined by

$$x(t) = \begin{cases} \phi(t), & t \in I \\ (S(t)\phi)(0), & t \ge 0 \end{cases} \tag{2.5.22}$$

for every $\phi \in \overline{D(B)}$, is the unique strong solution of (2.5.1) *if one of the followings holds:*

(1) X is reflexive, the norm $\|\cdot\|$ of X is Fréchet differentiable at every $x \in X - \{0\}$, $A : X \supset D(A) \to 2^X$ is maximal accretive, and $\phi \in \widehat{D}(B)$;

(2) X has the Radon-Nikodym property, $D(A)$ is closed, $A : D(A) \subset X \to X$ is norm-weakly continuous, and $\phi \in \widehat{D}(B)$;

(3) $D(A)$ is closed, $A : D(A) \subset X \to X$ is continuous, and $\phi \in \widehat{D}(B)$ or $\phi \in \overline{D(B)}$ and $A : D(A) \subset X \to X$ is bounded;

(4) X is reflexive, $A : D(A) \subset X \to X$ is demiclosed, and $\phi \in \widehat{D}(B)$.

Proof. (1) From (2) of (A1) and (2) of Proposition 2.5.1,

$$R(I + \lambda B) \supset \widetilde{E} \supset \overline{\text{co}} D(B)$$

for every $\lambda > 0$ with $\lambda\gamma < 1$. By Theorem 1.2.4, let $\{S_\lambda(t) \mid t \geq 0\}$ be the nonlinear semigroup on $\overline{\text{co}}D(B)$, generated by $-B_\lambda$ for every $\lambda > 0$, and then

$$\frac{dS_\lambda}{dt}(t)\phi + B_\lambda S_\lambda(t)\phi = 0, \quad t \geq 0 \tag{2.5.23}$$

and

$$\lim_{\lambda \to 0+} S_\lambda(t)\phi = S(t)\phi \tag{2.5.24}$$

uniformly on every bounded interval of $[0, \infty)$ for every $\phi \in \overline{D(B)}$. Moreover, using (2.5.24)

$$\lim_{\lambda \to 0+} J_\lambda^B S_\lambda(t)\phi = S(t)\phi \tag{2.5.25}$$

uniformly on every bounded interval of $[0, \infty)$ for every $\phi \in \overline{D(B)}$.

Let $T > 0$. Then (2.5.24) and (2.5.25) imply that there exists $(c_1)_T > 0$ such that

$$\|S_\lambda(t)\phi\|_E, \quad \|J_\lambda^B S_\lambda(t)\phi\|_E, \quad \|G(J_\lambda^B S_\lambda(t)\phi)\|_E \leq (c_1)_T \tag{2.5.26}$$

for every $t \in [0, T]$ and $\lambda > 0$ with $\lambda\gamma < 1$. Moreover, by (2.5.23), (2) of (E1), and the definition of $D(B)$

$$(S_\lambda(t)\phi)(0) = \phi(0) - \int_0^t (BJ_\lambda^B S_\lambda(s)\phi)(0)ds \tag{2.5.27}$$

for every $t \geq 0$ and $\phi \in \overline{D(B)}$.

Let $\phi \in \widehat{D}(B)$. Using (10) of Proposition 1.1.4 there exists $(c_2)_T > 0$ such that

$$\|B_\lambda S_\lambda(t)\phi\|_E \leq (c_2)_T |B\phi| < \infty \tag{2.5.28}$$

for every $t \geq 0$ and $n = 1, 2, \cdots$. From (2.5.25), (2.5.26), and (2.5.28)

$$\{(B_\lambda S_\lambda(t)\phi)(0) + G(J_\lambda^B S_\lambda(t)\phi)(0) - \alpha(B_\lambda S_\lambda(t)\phi)(0)\} \tag{2.5.29}$$

are uniformly bounded in λ for every $t \in [0, T]$. Since X is reflexive (2.5.29) has a subsequence $\{y_n(t)\}$ weakly convergent to $y(t)$ uniformly on $[0, T]$, where

$$y_n(t) = (B_{\lambda_n} S_{\lambda_n}(t)\phi)(0) + G(J_{\lambda_n}^B S_{\lambda_n}(t)\phi)(0) - \alpha(B_{\lambda_n} S_{\lambda_n}(t)\phi)(0).$$

We put

$$\begin{aligned} x(t) &= (S(t)\phi)(0), \\ x_n(t) &= (J_{\lambda_n}^B S_{\lambda_n}(t)\phi)(0), \\ y_n(t) &= (B_{\lambda_n} S_{\lambda_n}(t)\phi)(0) + G(J_{\lambda_n}^B S_{\lambda_n}(t)\phi)(0) \end{aligned}$$

for every $t \in [0,T]$. Then $[x_n(t), y_n(t)] \in A$ for $n = 1, 2, \cdots$, and $\lim_{n\to\infty} x_n(t) = x(t)$ uniformly on $[0,T]$.

From (2.5.27) and the definition of B we have

$$\begin{aligned}(S_{\lambda_n}(t)\phi)(0) &= \phi(0) - \int_0^t y_n(s)ds \\ &\quad + \int_0^t (G(J^B_{\lambda_n} S_{\lambda_n}(t)\phi) - \alpha x_n(s))ds \qquad (2.5.30)\end{aligned}$$

for every $t \in [0,T]$. For every $x^* \in X^*$ we have

$$\begin{aligned}(S_{\lambda_n}(t)\phi)(0), x^*) &= (\phi(0), x^*) - \int_0^t (y_n(s), x^*)(s)ds \\ &\quad + \int_0^t (G(J^B_{\lambda_n} S_{\lambda_n}(s)\phi) - \alpha x_n(s), x^*)ds,\end{aligned}$$

and letting $n \to \infty$, by (2.5.24)–(2.5.26) and (2.5.30)

$$\begin{aligned}((S(t)\phi)(0), x^*) &= (\phi(0), x^*) - \int_0^t (y(s), x^*)ds \\ &\quad + \int_0^t ((S(s)\phi) - \alpha(S(s)\phi)(0), x^*)ds\end{aligned}$$

for every $t \in [0,T]$. Thus

$$\begin{aligned}(S(t)\phi)(0) &= \phi(0) - \int_0^t y(s)ds \\ &\quad + \int_0^t (G(S(s)\phi) - \alpha(S(s)\phi)(0))ds \qquad (2.5.31)\end{aligned}$$

for every $t \in [0,T]$. Hence $x(t) = (S(t)\phi)(0)$ is locally absolutely continuous on $[0,\infty)$ and differentiable a.e. $t \in (0,\infty)$. But since A is maximal $[x(t), y(t)] \in A$ for every $t \in [0,T]$. Thus using (2.5.31)

$$\frac{dx}{dt}(t) + \alpha x(t) - G(x_t) = -y(t) \in -Ax(t)$$

a.e. $t \in (0,\infty)$. Since $x(t) = \phi(t)$ for every $t \in I$, $x(t) \equiv x(t;\phi)$ is a strong solution of (2.5.1) on $I \cup [0,\infty)$ for every $\phi \in \widehat{D}(B)$.

(2) In the case A is single valued from (2.5.27) we have

$$\begin{aligned}&(S_\lambda(t)\phi)(0) \\ &\quad = \phi(0) + \int_0^t \Big(G(J^B_\lambda S_\lambda(s)\phi) - \alpha(J^B_\lambda S_\lambda(s)\phi)(0) \\ &\qquad\qquad -A(J^B_\lambda S_\lambda(s)\phi)(0)\Big)ds \qquad (2.5.32)\end{aligned}$$

for every $t \in [0, \infty)$ and $\phi \in \overline{D(B)}$. By (2.5.32) with (2.5.24)–(2.5.26) and (2.5.28)

$$(x(t), x^*) = (\phi(0), x^*) + \int_0^t (G(x_s) - \alpha x(s) - Ax(s), x^*) ds \qquad (2.5.33)$$

for every $t \in [0, T]$, $\phi \in \widehat{D}(B)$ and $x^* \in X^*$.

Using the continuity assumptions, there exists a separable closed linear subspace X_0 of X such that

$$\{x(t) \mid t \in [0, T]\},\ \{G(x_t) - \alpha x(t) - Ax(t) \mid t \in [0, T]\} \subset X_0.$$

Since from (10) of Proposition 1.1.4 $x(t) = (S(t)\phi)(0)$ is Lipschitz continuous on $[0, T]$, and thus, since X has the Radon–Nikodym property $x(t)$ is differentiable a.e. $t \in (0, T)$.

Consider $\{x_n^*\} \subset B_X^*$ such that $\{x_n^* \,|_{X_0}\}$ is w^*–dense in $B_{X_0}^*$, and let $N \subset [0, T]$ be a set of Lebesgue measure 0 such that $x(t)$ is differentiable at $t \in [0, T] - N$ and

$$\frac{d}{dt} \int_0^t (G(x_s) - \alpha x(s) - Ax(s), x_n^*) ds = (G(x_t) - \alpha x(t) - Ax(t), x_n^*)$$

for every $t \in [0, T] - N$ and $n = 1, 2, \cdots$. From (2.5.33) we have

$$\left(\frac{dx}{dt}(t), x_n^* \right) = (G(x_t) - \alpha x(t) - Ax(t), x_n^*)$$

for every $t \in [0, T] - N$ and $n = 1, 2, \cdots$. Thus from our choice of X_0 and $\{x_n^*\}$ we have

$$\frac{dx}{dt}(t) + \alpha x(t) + Ax(t) = G(x_t)$$

for a.e. $t \in [0, T]$. Since $x(t) = \phi(t)$ for every $t \in I$ $x(t) \equiv x(t; \phi)$ is a strong solution of (2.5.1) on $I \cup [0, \infty)$.

(3) In view of (2.5.28) assuming either $\phi \in \widehat{D}(B)$ or $\phi \in \overline{D(B)}$ and A is bounded, (2.5.32) implies that

$$x(t) = \phi(0) + \int_0^t (G(x_s) - \alpha x(s) - Ax(s)) ds$$

for every $t \in [0, T]$. Thus

$$\frac{dx}{dt}(t) + \alpha x(t) + Ax(t) = G(x_t)$$

for a.e. $t \in [0,T]$. Since $x(t) = \phi(t)$ for every $t \in I$ $x(t) \equiv x(t;\phi)$ is a strong solution of (2.5.1) on $I \cup [0,\infty)$.

(4) From the definition of B

$$A(J^B_{\lambda_n} S_{\lambda_n}(t)\phi)(0)) = (B_{\lambda_n} S_{\lambda_n}(t)\phi)(0) \\ +G(J^B_{\lambda_n} S_{\lambda_n}(t)\phi)(0) - \alpha(J^B_{\lambda_n} S_{\lambda_n}(t)\phi)(0)$$

for every $t \in [0,T]$ and $n = 1, 2, \cdots$, and using (2.5.28) by the assumptions there exist a weakly compact subset K and a closed separable linear subspace X_0 of X such that

$$\{AJ^B_{\lambda_n} S_{\lambda_n}(t)\phi)(0) \mid t \in [0,T],\ n = 1, 2, \cdots\} \subset K \subset X_0.$$

Hence, given $t \in [0,T]$ every subsequence of $\{A(J^B_{\lambda_n} S_{\lambda_n}(t)\phi)(0))\}$ has a further subsequence that is weakly convergent. From (2.6.25) and the fact that A is demiclosed, $\{A(J^B_{\lambda_n} S_{\lambda_n}(t)\phi)(0))\}$ converges weakly to $Ax(t)$ for every $t \in [0,T]$. Thus starting from (2.5.32) we can reach (2.5.33), and by repeating the proof of (3) the proof is completed.

2.6. Stability for Solutions

In this section we consider the stability and the asymptotic stability for solutions of (2.2.1) on $[-r,\infty)$.

First of all we give definitions of the stability and the asymptotic stability

DEFINITION 2.6.1 *Let $x(t) \equiv x(t;\phi)$ be the (classical) solution of (2.2.1) on $[-r,\infty)$ for $\phi \in E \subset C([-r,0];X)$.*

(1) $x(t)$ is said to be stable if

$$\|x(t;\phi_1) - x(t;\phi_2)\| \le \|\phi_1 - \phi_2\|_C$$

for every $t \ge 0$ and $\phi_i \in E$, $i = 1, 2$;

(2) $x(t)$ is said to be asymptotically stable if there exists $\overline{x} \in X$ uniquely such that

$$\lim_{t\to\infty} x(t;\phi) = \overline{x}$$

for every $\phi \in E$.

We define a weighted norm $\|\cdot\|_\tau$ on $C([-r,0];X)$ for the parameter $\tau \in \mathbf{R}$ by

$$\|\phi\|_\tau = \sup_{\sigma\in[-r,0]} e^{-\tau\sigma}\|\phi(\sigma)\| \tag{2.6.1}$$

for every $\phi \in C([-r,0];X)$.

We note that the weighted noms in (2.6.1) are equivalent norms on $C([-r,0];X)$.

Indeed, if $\sigma < \tau$ then from (2.6.1) we have

$$\|\phi\|_\sigma \le \|\phi\|_\tau \le e^{(\tau-\sigma)r}\|\phi\|_\sigma \tag{2.6.2}$$

for every $\phi \in C([-r,0];X)$.

LEMMA 2.6.1 *Assume that $G : C([-r,0];X) \to X$ satisfies*

$$\|G(\psi_1) - G(\psi_2)\| \le \beta\|\psi_1 - \psi_2\|_C$$

for every $\psi_i \in C([-r,0];X)$, $i = 1,2$, and for some $\beta > 0$.

Let β_τ be the Lipschitz constant of G in $\|\cdot\|_\tau$, that is, β_τ is the smallest real number satisfying

$$\|G(\psi_1) - G(\psi_2)\| \le \beta_\tau\|\psi_1 - \psi_2\|_\tau \tag{2.6.3}$$

for every $\psi_i \in C([-r,0];X)$, $i = 1,2$, and $\tau \in \boldsymbol{R}$.

(1) If $\sigma < \tau$ then

$$\beta_\tau \le \beta_\sigma \le e^{(\tau-\sigma)r}\beta_\tau; \tag{2.6.4}$$

(2) $\beta_\sigma \le \beta\max\{1, e^{-r\sigma}\}$ for every $\sigma \in \boldsymbol{R}$;

(3) $\beta_\sigma : \boldsymbol{R} \to \boldsymbol{R}$ is continuous in $\sigma \in \boldsymbol{R}$.

Proof. (1) Let $\sigma < \tau$. Since from (2.6.2) and (2.6.3)

$$\|G(\psi_1) - G(\psi_2)\| \le \beta_\sigma\|\psi_1 - \psi_2\|_\sigma \le \beta_\sigma\|\psi_1 - \psi_2\|_\tau$$

for every $\psi_i \in C([-r,0];X)$, $i = 1,2$, and $\beta_\tau \le \beta_\sigma$. Since also by (2.6.2) and (2.6.3),

$$\|G(\psi_1) - G(\psi_2)\| \le \beta_\tau\|\psi_1 - \psi_2\|_\tau \le \beta_\tau e^{(\tau-\sigma)r}\|\psi_1 - \psi_2\|_\sigma$$

for every $\psi_i \in C([-r,0];X)$, $i = 1,2$, and $\beta_\sigma \le e^{(\tau-\sigma)r}\beta_\tau$.

(2) Let $\sigma \in \mathbf{R}$. If $\sigma \ge 0$ then from the first inequality of (2.6.4) $\beta_\sigma \le \beta$. If $\sigma < 0$ then from the second inequality of (2.6.4) $\beta_\sigma \le \beta e^{-\sigma r}$. Thus (2) holds.

(3) From (1) the limit $\lim_{\tau\to\sigma}\beta_\tau$ exists. Thus we put $\lim_{\tau\to\sigma}\beta_\tau = \overline{\beta}$. If $\sigma < \tau$ then letting $\tau \to \sigma+$ in (2.6.4) $\overline{\beta} \le \beta_\sigma \le \overline{\beta}$. Thus $\beta_\sigma = \overline{\beta}$. If $\sigma > \tau$, then

$$\beta_\tau \le e^{(\sigma-\tau)r}\beta_\sigma \le e^{(\sigma-\tau)r}\beta_\tau.$$

Letting $\tau \to \sigma-$ then $\beta_\sigma = \overline{\beta}$.

We now give a proposition to get the stability theorem.

PROPOSITION 2.6.1 *Let $A : X \supset D(A) \to 2^X$ be an operator and $\alpha \in \mathbf{R}$ such that $A + \alpha I$ is accretive and $R(I + \lambda A) = X$ for every $\lambda > 0$ with $\lambda\alpha < 1$. Let $G : C([-r, 0]; X) \to X$ satisfy*

$$\|G(\psi_1) - g(\psi_2)\| \le \beta\|\psi_1 - \psi_2\|_C$$

for every $\psi_i \in C([-r, 0]; X)$, $i = 1, 2$ and for some $\beta > 0$. Let B be the operator defined as in (2.2.2). Let $\{S(t) \mid t \ge 0\}$ be the nonlinear semigroup on $\overline{D(B)}$ of the type γ, generated by $-B$, where $\gamma = max\{0, \alpha+\beta\}$. Then

$$\|S(t)\phi_1 - S(t)\phi_2\|_\tau \le e^{\gamma_\tau t}\|\phi_1 - \phi_2\|_\tau$$

for every $\phi_i \in \overline{D(B)}$, $i = 1, 2$, and $t \ge 0$, where $\gamma_\tau = max\{\tau, \alpha + \beta_\tau\}$.

Proof. We note that if $(I + \lambda B)\psi \ni \phi$ for every $\phi \in C([-r, 0]; X)$ then from the proof of Proposition 2.2.1

$$\psi(t) = e^{\frac{t}{\lambda}}\psi(0) + \frac{1}{\lambda}\int_t^0 e^{-\frac{t-\sigma}{\lambda}}\phi(\sigma)d\sigma, \tag{2.6.5}$$

$$\psi(0) = J_\lambda^A(\phi(0) + \lambda G(\psi)) \tag{2.6.6}$$

for every $t \in [-r, 0]$ and $\lambda > 0$ with $\lambda\alpha < 1$.

Let $\phi_i \in C([-r, 0]; X)$ and $\psi_i = J_\lambda^A\phi_i$, $i = 1, 2$. From (2.6.5)

$$\begin{aligned}
&\|\psi_1(t) - \psi_2(t)\| \\
&\quad \le e^{\frac{t}{\lambda}}\|\psi_1(0) - \psi_2(0)\| \\
&\qquad + \frac{1}{\lambda}\int_t^0 e^{-\frac{t-\sigma}{\lambda}}\|\phi_1(\sigma) - \phi_2(\sigma)\|d\sigma \\
&\quad < e^{\frac{t}{\lambda}}\|\psi_1(0) - \psi_2(0)\| + \frac{1}{\lambda}\int_t^0 e^{(\tau\sigma - \frac{t-\sigma}{\lambda})}d\sigma\|\phi_1 - \phi_2\|_\tau \\
&\quad \le e^{\frac{t}{\lambda}}\|\psi_1(0) - \psi_2(0)\| + \frac{1}{1-\lambda\tau}(e^{t\tau} - e^{\frac{t}{\lambda}})\|\phi_1 - \phi_2\|_\tau,
\end{aligned}$$

and thus

$$\begin{aligned}
&\|e^{-\sigma\tau}(\psi_1(t) - \psi_2(t))\| \\
&\quad \le \frac{1}{1-\lambda\tau}\|\phi_1 - \phi_2\|_\tau \\
&\qquad + e^{\frac{(1-\lambda\tau)t}{\lambda}}\left(\|\psi_1(0) - \psi_2(0)\| - \frac{1}{1-\lambda\tau}\|\phi_1 - \phi_2\|_\tau\right)
\end{aligned} \tag{2.6.7}$$

for every $t \in [-r, 0]$ and $\lambda > 0$ with $\lambda\sigma < 1$. Hence by (2.6.7)

$$\|\psi_1 - \psi_2\|_\tau \leq \max\left\{\|\psi_1(0) - \psi_2(0)\|, \frac{1}{1-\lambda\tau}\|\phi_1 - \phi_2\|_\tau\right\} \tag{2.6.8}$$

for every $\lambda > 0$ with $\lambda\tau < 1$. On the other hand, by (2.6.6)

$$\begin{aligned} &\|\psi_1(0) - \psi_2(0)\| \\ &\quad = \|J_\lambda^A(\phi_1(0) + \lambda G(\psi_1)) - J_\lambda^A(\phi_2(0) + \lambda G(\psi_2))\| \\ &\quad \leq \frac{1}{1-\lambda\tau}(\|\phi_1 - \phi_2\|_\tau + \lambda\beta_\tau\|\psi_1 - \psi_2\|_\tau) \end{aligned} \tag{2.6.9}$$

for every $\lambda > 0$ with $\lambda\tau < 1$.

If

$$\|\psi_1(0) - \psi_2(0)\| \geq \frac{1}{1-\lambda\tau}\|\phi_1 - \phi_2\|_\tau$$

then (2.6.8) and (2.6.9) imply

$$\begin{aligned} \|\psi_1 - \psi_2\|_\tau &\leq \|\psi_1(0) - \psi_2(0)\| \\ &\leq \frac{1}{1-\lambda\tau}(\|\phi_1 - \phi_2\|_\tau + \lambda\beta_\tau\|\psi_1 - \psi_2\|_\tau) \end{aligned}$$

for every $\lambda > 0$ with $\lambda\tau < 1$, and thus

$$\|\psi_1 - \psi_2\|_\tau \leq \frac{1}{1-\lambda(\alpha + \beta_\tau)}\|\phi_1 - \phi_2\|_\tau. \tag{2.6.10}$$

for every $\lambda > 0$ with $\lambda(\alpha + \beta_\tau) < 1$.

If

$$\|\psi_1(0) - \psi_2(0)\| \leq \frac{1}{1-\lambda\tau}\|\phi_1 - \phi_2\|_\tau$$

then from (2.6.8)

$$\|\psi_1 - \psi_2\|_\tau \leq \frac{1}{1-\lambda\tau}\|\phi_1 - \phi_2\|_\tau \tag{2.6.11}$$

for every $\lambda > 0$ with $\lambda\tau < 1$.

We put $\gamma_\tau = \max\{\tau, \alpha + \beta_\tau\}$. Then by (2.6.10) and (2.6.11)

$$\|\psi_1 - \psi_2\|_\tau \leq \frac{1}{1-\lambda\gamma_\tau}\|\phi_1 - \phi_2\|_\tau$$

and thus

$$\|J_\lambda^B\phi_1 - J_\lambda^B\phi_2\|_\tau \leq \frac{1}{1-\lambda\gamma_\tau}\|\phi_1 - \phi_2\|_\tau$$

for every $\lambda > 0$ with $\lambda\gamma_\tau < 1$. Hence

$$\left\| \left(J^B_{\frac{t}{n}}\right)^n \phi_1 - \left(J^B_{\frac{t}{n}}\right)^n \phi_2 \right\|_\tau \leq \left(1 - \frac{\gamma_\tau t}{n}\right)^{-n} \|\phi_1 - \phi_2\|_\tau \tag{2.6.12}$$

for every $t \geq 0$ and for sufficiently large n. Letting $n \to \infty$ in (2.6.12) from Theorem 1.2.1

$$\|S(t)\phi_1 - S(t)\phi_2\|_\tau \leq e^{\gamma_\tau t}\|\phi_1 - \phi_2\|_\tau$$

for every $t \geq 0$ and $\phi_i \in \overline{D(B)}$, $i = 1, 2$.

COROLLARY 2.6.1 *Under the assumptions of Proposition 2.6.1,*

$$\|S(t)\phi_1 - S(t)\phi_2\|_\tau \leq e^{\tau t}\|\phi_1 - \phi_2\|_\tau, \tag{2.6.13}$$

for every $t \geq 0$ and $\phi_i \in \overline{D(B)}$, $i = 1, 2$, where $\tau \in [-r, 0]$ is the unique solution of

$$\tau = \alpha + \beta_\tau. \tag{2.6.14}$$

Proof. From (2.6.14) $\gamma_\tau = \tau$. Thus from Proposition 2.6.1 (2.6.13) holds.

COROLLARY 2.6.2 *Suppose that the assumptions of Proposition 2.6.1 are satisfied.*

(1) If $\alpha + \beta < 0$ then (2.6.13) *holds with $\tau \in \boldsymbol{R}$ being the unique solution of*

$$\tau = \alpha + \beta e^{-\tau r}; \tag{2.6.15}$$

(2) If $\alpha + \beta \geq 0$ then

$$\|S(t)\phi_1 - S(t)\phi_2\|_\tau \leq e^{(\alpha+\beta)t}\|\phi_1 - \phi_2\|_\tau \tag{2.6.16}$$

for every $t \geq 0$ and $\phi_i \in \overline{D(B)}$, $i = 1, 2$.

Proof. (1) From (2.6.15) and (2) of Lemma 2.6.1

$$\begin{aligned} \gamma_\tau &= \max\{\tau, \alpha + \beta_\tau\} \\ &\leq \max\{\tau, \alpha + \beta \max\{1, e^{-r\tau}\}\}. \end{aligned} \tag{2.6.17}$$

If $1 \leq e^{-r\tau}$ then

$$\gamma_\tau \leq \max\{\tau, \alpha + \beta e^{-r\tau}\} = \max\{\tau, \tau\} = \tau.$$

If $1 \geq e^{-r\tau}$ then $\tau \geq 0$, and thus since $\alpha + \beta < 0$

$$\gamma_\tau \leq \max\{\tau, \alpha + \beta\} = \tau.$$

Using Proposition 2.6.1 (2.6.13) holds for every $\tau \in \mathbf{R}$.

(2) Let $\alpha + \beta \geq 0$. If $1 \leq e^{-r\tau}$ then $\tau \leq 0$, and thus from (2.6.17)

$$\gamma_\tau \leq \max\{\tau, \alpha + \beta e^{-r\tau}\} = \tau \leq \alpha + \beta.$$

If $1 \geq e^{-r\tau}$ then by (2.6.15)

$$\tau = \alpha + \beta e^{-r\tau} \leq \alpha + \beta$$

and thus,

$$\gamma_\tau \leq \max\{\tau, \alpha + \beta\} = \alpha + \beta.$$

Hence using (2.6.15), (2.6.16) holds.

THEOREM 2.6.1 *Let $A : X \supset D(A) \to 2^X$ be an operator and $\alpha \in \mathbf{R}$ such that $A + \alpha I$ is accretive and $R(I + \lambda A) = X$ for every $\lambda > 0$ with $\lambda\alpha < 1$. Let $G : C([-r, 0]; X) \to X$ satisfy*

$$\|G(\psi_1) - G(\psi_2)\| \leq \beta\|\phi - \psi\|_C$$

for every $\psi_i \in C([-r, 0]; X)$, $i = 1, 2$, and for some $\beta > 0$. Let B be the operator defined as in (2.2.2). Let $\{S(t) \mid t \geq 0\}$ be the nonlinear semigroup on $\overline{D(B)}$ of type γ, generated by $-B$, where $\gamma = max\{0, \alpha + \beta\}$. Let $x(t) \equiv x(t; \phi)$ be the unique solution of (2.2.1) on $[-r, \infty)$ for every $\phi \in \overline{D(B)}$ with the type

$$x(t) = \begin{cases} \phi(t), & -r \leq t \leq 0 \\ (S(t)\phi)(0), & t \geq 0. \end{cases}$$

(1) If $\alpha + \beta = 0$ then $x(t)$ is stable;

(2) If $\alpha + \beta < 0$ then $x(t)$ is asymptotically stable;

(3) If $\alpha + \beta > 0$ then

$$\|x(t; \phi_1) - x(t; \phi_2)\| \leq e^{(\alpha+\beta)t}\|\phi_1 - \phi_2\|_C$$

for every $t \geq 0$ and $\phi_i \in \overline{D(B)}, i = 1, 2$.

Proof. (1) Let $\alpha + \beta = 0$. Then $\tau = 0$ is the solution of (2.6.14). Thus from (2.6.13)

$$\|S(t)\phi_1 - S(t)\phi_2\| \leq \|\phi_1 - \phi_2\|_\tau$$

for every $t \geq 0$ and $\phi_i \in \overline{D(B)}$, $i = 1, 2$. By (2.2.14)

$$\begin{aligned}\|x(t;\phi_1) - x(t;\phi_2)\| &= \|S(t)\phi_1(0) - S(t)\phi_2(0)\| \\ &\leq \|S(t)\phi_1 - S(t)\phi_2\|_C \\ &\leq \|\phi_1 - \phi_2\|_C\end{aligned}$$

for every $t \geq 0$ and $\phi_i \in \overline{D(B)}$, $i = 1, 2$. Hence $x(t)$ is stable.

(2) Let $\alpha + \beta < 0$. There exists a unique solution τ of (2.6.15), and $\tau < 0$. Indeed, let $\tau \geq 0$. Then $0 \leq \tau \leq \alpha + \beta e^{-\tau r} \leq \alpha + \beta < 0$. We define a mapping $Q : X \to X$ by

$$Qu = (I + A)^{-1}(u + G(u \cdot \mathbf{1})) \tag{2.6.18}$$

for every $u \in X$, where $\mathbf{1}(\tau) = 1$ for every $\tau \in [-r, 0]$. Then

$$\|Qu_1 - Qu_2\| \leq \frac{1+\beta}{1-\alpha}\|u_1 - u_2\|$$

for every $u_i \in X$, $i = 1, 2$. Since $0 < \dfrac{1+\beta}{1-\alpha} < 1$ Q is a strict contraction. From Banach's contraction principle there exists $\tilde{u} \in X$ such that $Q\tilde{u} = \tilde{u}$. From (2.6.18)

$$\tilde{u} = (I + A)^{-1}(\tilde{u} + G(\tilde{u} \cdot \mathbf{1})).$$

Thus $\tilde{u} \in D(A)$ and $A\tilde{u} \ni G(\tilde{u} \cdot \mathbf{1})$. We put $\tilde{\phi} = \tilde{u} \cdot \mathbf{1}$. Then $\tilde{\phi} \in D(B)$ and $B\tilde{\phi} = 0$ and thus $J_\lambda^B \tilde{\phi} = \tilde{\phi}$ for every $\lambda > 0$. By Theorem 1.2.1 $S(t)\tilde{\phi} = \tilde{\phi}$ for every $t \geq 0$. (1) of Corollary 2.6.2 implies

$$\begin{aligned}\|x(t;\phi) - \tilde{\phi}(0)\| &= \|S(t)\phi(0) - \tilde{\phi}(0)\| \\ &\leq \|S(t)\phi - \tilde{\phi}\|_C \\ &= \|S(t)\phi - S(t)\tilde{\phi}\|_C \\ &\leq e^{(\alpha+\beta)t}\|\phi - \tilde{\phi}\|_C\end{aligned}$$

for every $\phi \in \overline{D(B)}$. Letting $t \to \infty$ in this inequality $\lim_{t\to\infty} x(t;\phi) = \tilde{\phi}(0)$. Hence $x(t)$ is asymptotically stable.

(3) Let $\alpha + \beta > 0$. From (2) of Corollary 2.6.2, (3) holds.

2.7. Examples and Applications

In this section we give some examples for (2.2.1) as an illustration of how the results of previous sections may be applied.

EXAMPLE 2.7.1 (Burch–Goldstein [18]) *We consider the following nonlinear functional heat equation with the delay:*

$$\begin{cases} \dfrac{\partial x}{\partial t}(t,u) - k(\dfrac{\partial x}{\partial u}(t,u)) = f(x(t-r,u)), \quad t \geq 0,\ u \in (0,1) \\ x(t,u) = \phi(t,u), \quad t \in [-r,0],\ u \in (0,1) \\ x(t,0) = a\dfrac{\partial x}{\partial t}(t,0), \quad t \geq 0 \\ x(t,1) = -b\dfrac{\partial x}{\partial t}(t,1), \quad t \geq 0. \end{cases} \tag{2.7.1}$$

Here $x(t,u) : (0,1) \times [-r,\infty) \to \mathbf{R}$, $k : \mathbf{R} \to \mathbf{R}$, $f : \mathbf{R} \to \mathbf{R}$, $r > 0$, $a, b \geq 0$ and $\phi : (0,1) \times [-r,0] \to \mathbf{R}$.

(2.7.1) represents a problem in heat conduction, where the thermal conductivity depends on the temperature gradient, i.e., the function k depends on the thermal conductivity.

THEOREM 2.7.1 *Let $k : \boldsymbol{R} \to \boldsymbol{R}$ be continuous and there exists $c > 0$ such that $k(v) \geq c$ for every $v \in \boldsymbol{R}$. Let $f : \boldsymbol{R} \to \boldsymbol{R}$ be Lipshitzian, i.e.,*

$$|f(v_1) - f(v_2)| \leq \beta |v_1 - v_2|$$

for every $v_i \in \boldsymbol{R}$, $i = 1, 2$ and some $\beta > 0$. Let $\phi \in C([-r,0]; L^p(0,1))$ and let $1 < p < \infty$. Then (2.7.1) has at most one solution $x(t,u) \equiv x(t,u;\phi)$ on $[-r,\infty)$ a.e. $u \in (0,1)$.

Proof. Let $1 < p < \infty$. Now we define the operator $\widehat{A} : D(\widehat{A}) \subset L^p(0,1) \to L^p(0,1)$ by

$$\begin{aligned} D(\widehat{A}) &= \Big\{ x \in L^p(0,1) \ \Big|\ x \in C^2([0,1]),\ x(0) = a\frac{dx}{du}(0), \\ &\qquad\qquad x(1) = -\beta\frac{dx}{du}(1) \Big\}, \\ (\widehat{A}x)(u) &= -k\Big(\frac{dx}{du}\Big)\frac{d^2x}{du^2}, \quad x \in D(\widehat{A}), \end{aligned}$$

and let A be the closure of $\widehat{A}$. We define $G : C([-r,0]; L^p(0,1)) \to L^p(0,1)$ by

$$G(\psi) = f(\psi(-r))$$

for every $\psi \in C([-r,0]; X)$. Then

$$G(x_t) = f(x_t(-r)) = f(x(t-r))$$

for every $t \geq 0$. Thus we can rewrite (2.7.1) in the form (2.2.1). On the other hand, from Burch–Goldstein [18, Theorem 2.2] A is m-accretive in X. We have

$$\begin{aligned}
&\|G(\psi_1) - G(\psi_2)\|_p \\
&= \left(\int_0^1 | f(\psi_1(u,-r)) - f(\psi_2(u,-r)) |^p \, du\right)^{\frac{1}{p}} \\
&\leq \beta \left(\int_0^1 | \psi_1(u,-r) - \psi_2(u,-r) |^p \, du\right)^{\frac{1}{p}} \\
&= \beta \|\psi_1 - \psi_2\|_{C([-r,0];X)}
\end{aligned}$$

for every $\psi_i \in C([0,1]; L^p(0,1))$, $i = 1, 2$. Thus by Theorem 2.2.2 there exists a strong solution $x(t) \equiv x(t;\phi)$ of (2.2.1) on $[-r,\infty)$ for every $\phi \in \widehat{D}(B)$. Hence $x(t,u) = (x(t))(u)$ satisfies (2.7.1) a.e. $t \in [0,\infty)$ and a.e. $u \in (0,1)$ for every $\phi \in \widehat{D}(B)$.

EXAMPLE 2.7.2 *Let Ω be a bounded domain in $\boldsymbol{R}^n$ with smooth boundary $\partial\Omega$ for $n = 1, 2, \cdots$. We consider the following nonlinear partial differential equation with the finite delay:*

$$\begin{cases}
\dfrac{\partial x}{\partial t}(t,u) - \Delta\gamma(x(t,u)) = f(t, x(t-r,u)), & t \in [0,\infty),\ u \in \Omega \\
x(t,u) = 0, & t \in [0,\infty) \\
x(u,t) = \phi(u,t), & t \in [-r,0],\ u \in \Omega
\end{cases} \tag{2.7.2}$$

Here $r > 0$, $T > 0$, $x(t,u) : [-r,\infty) \times \Omega \to \mathbf{R}$, Δ is the Laplacian differential operator, $\gamma : \mathbf{R} \to \mathbf{R}$, $f : [0,\infty) \times \mathbf{R} \to \mathbf{R}$, and $\phi : [-r,0] \times \Omega \to \mathbf{R}$.

Now we have the following existence theorem of an integral solution of (2.7.2):

THEOREM 2.7.2 *Let $\gamma : \boldsymbol{R} \to \boldsymbol{R}$ be a nondecreasing function with $\gamma(0) = 0$. Assume that $\gamma \in C^1(\boldsymbol{R} - \{0\}; \boldsymbol{R})$, and there exist $c > 0$ and $p > \dfrac{n-2}{n}$ if $n > 2$, $p > 0$ if $n \leq 2$, such that*

$$\frac{d\gamma}{du}(u) \geq c|u|^{p-1}$$

for every $u \in \boldsymbol{R} - \{0\}$. Assume, in addition, that there exists a constant $\beta > 0$ such that

$$|f(t_1,u_1) - f(t_2,u_2)| \leq \frac{\beta}{m(\Omega)}(|t_1 - t_2| + |u_1 - u_2|)$$

for every $t_i \in [0, T]$ *and* $u_i \in \boldsymbol{R}$, $i = 1, 2$. *Then* (2.7.2) *has at least one local integral solution on* $[-r, \infty)$ *for every* $\phi \in C([-r, 0]; L^1(\Omega))$.

Proof. We define an operator $A : D(A) \subset L^1(\Omega) \to L^1(\Omega)$ by

$$\begin{aligned} D(A) &= \{x \in L^1(\Omega) \mid \gamma(x) \in W_0^{1,1}(\Omega),\ \Delta\gamma(x) \in L^1(\Omega)\}, \\ Ax &= -\Delta\gamma(x),\ x \in D(A). \end{aligned}$$

Also we define $G : [0, \infty) \times C([-r, 0]; L^1(\Omega)) \to L^1(\Omega)$ by

$$G(t, \psi) = f(t, \psi(-r))$$

for every $t \in [0, \infty)$ and $\psi \in C([-r, 0]; L^1(\Omega))$. Then

$$G(t, x_t) = f(t, x_t(-r)) = f(t, x(t - r))$$

for every $t \in [0, \infty)$. Thus we can also rewrite (2.7.2) in the form (2.3.1). From Bénilan [6] A is m-accretive in $L^1(\Omega)$ and from Badii–Diaz–Tesei [3] $-A$ generates a nonlinear compact semigroup on $\overline{D(A)}$. And also we have

$$\begin{aligned} &\|G(t_1, \psi_1) - G(t_2, \psi_2)\|_1 \\ &= \int_\Omega |f(t_1, \psi_1(-r))(u) - f(t_2, \psi_2(-r))(u)| du \\ &\leq \frac{\beta}{m(\Omega)} \int_\Omega (|t_1 - t_2| + |\psi_1(-r) - \psi_2(-r)|) du \\ &\leq \beta(|t_1 - t_2| + \|\psi_1 - \psi_2\|_{C([-r,0];L^1(\Omega))}) \end{aligned}$$

for every $t_i \in [0, \infty)$ and $\psi_i \in C([-r, 0]; L^1(\Omega))$, $i = 1, 2$. Thus since G is continuous on $[0, \infty) \times C([-r, 0]; L^1(\Omega))$, from Theorem 2.3.1 (2.3.1) has a local integral solution $x(t) \equiv x(t; \phi)$ on $[-r, \infty)$. We put $x(t, u) = (x(t))(u)$ for $(t, u) \in [-r, T_0] \times \Omega$. Then $x(t, u)$ is a local solution of (2.7.2) on $[0, \infty)$, a.e. $u \in \Omega$ and $\phi \in C([-r, 0]; L^1(\Omega))$.

EXAMPLE 2.7.3 *We consider a single species population model based on the delay logistic equation*

$$\begin{cases} \dfrac{dx}{dt}(t) = x(t)\left(a - bx(t) - \displaystyle\int_{-\infty}^{0} k(-\tau)x_t(\tau)d\tau\right), & t \geq 0 \\ x(t) = \phi(t),\ t \leq 0. \end{cases} \tag{2.7.3}$$

Here $x(t) : (-\infty, \infty) \to \mathbf{R}$, $a, b > 0$, $k \in L^1(0, \infty) \cap C([0, \infty))$ and $\phi \in BUC((-\infty, 0])$ under the usual supremum norm $\|\cdot\|_\infty$. But the integral term in (2.7.3) is can be replaced by a more general operator $K : BUC((-\infty, 0]) \to \mathbf{R}$ under the following conditions:

(C1) Given $c > 0$ there exists a constant $L(c) \geq 0$ such that

$$|K(\psi_1) - K(\psi_2)| \leq L(c)\|\psi_1 - \psi_2\|_\infty$$

for every $\psi_i \in BUC((-\infty, 0])$ with $\|\psi_i\|_\infty \leq c$, $i = 1, 2$;

(C2) There exists a continuous non-decreasing function $l : [0, \infty) \to [0, \infty)$ with $l(0) = 0$ such that

$$|K(\psi)| \leq l(\|\psi\|_\infty)$$

for every $\psi \in BUC((-\infty, 0])$.

Thus given $\phi \in BUC((-\infty, 0])$ the problem (2.7.3) takes the form

$$\begin{cases} \dfrac{dx}{dt}(t) = x(t)(a - bx(t) - K(x_t)), & t \geq 0 \\ x(t) = \phi(t), & t \leq 0. \end{cases} \tag{2.7.4}$$

THEOREM 2.7.3 *Assume that $K : BUC((-\infty, 0]) \to \mathbf{R}$ satisfies (C1) and (C2). Let a, $b > 0$. Furthermore, assume that there exists $c \geq \dfrac{a}{b}$ such that (i) $l(c) \leq a$, and (ii) $a + l(c) \leq bc$. We put*

$$\widehat{E} = \{\phi \in BUC((-\infty, 0]) \mid 0 \leq \phi \leq c\}$$

and

$$\widehat{E}_\alpha = \left\{\phi \in \widehat{E} \;\middle|\; \frac{\alpha}{2b} \leq \phi(0) \leq c\right\}.$$

Then:

(1) For every $\phi \in \widehat{E}_\alpha$ there exists a strong solution $x(t) \equiv x(t; \phi)$ of (2.7.4) uniquely on $(-\infty, \infty)$ into $\mathbf{R}$ with $x(t) \in [0, c]$ for every $t \geq 0$, and $x(t) > 0$ for every $t \geq 0$ in case $\phi(0) > 0$;

(2) For every $\alpha \in [0, 2(a - l(c))]$ there exists a continuous semigroup $\{S_\alpha(t) \mid t \geq 0\}$ on $\widehat{E}_\alpha$ such that $x(t) = (S_\alpha(t)\phi)(0) \in [\dfrac{\alpha}{2b}, c]$ for every $\phi \in \widehat{E}_\alpha$ and $t \geq 0$;

(3) Putting $\beta(c) = a + l(c) + cL(c)$, if $\beta(c) \leq \alpha$ for some $\alpha \in [0, 2(a - l(c))]$ then

$$\|S_\alpha(t)\phi_1 - S_\alpha(t)\phi_2\|_\infty \leq \|\phi_1 - \phi_2\|_\infty$$

for every $t \geq 0$ and $\phi_i \in \widehat{E}_\alpha$, $i = 1, 2$.

Before we give the proof of Theorem 2.7.3 we show that $BUC((-\infty,0])$ satisfies (E1)–(E3). For this we consider a particular class of spaces of initial functions satisfying (E1)–(E3).

PROPOSITION 2.7.1 *Let X be a real Banach space with norm $\|\cdot\|$. Assume that $v:(-\infty,0]\to(0,1]$ is a function satisfying the following properties:*

(1) v is continuous, non-decreasing, and $v(0)=1$;

(2) $\lim_{\tau\to 0-}\dfrac{v(\sigma+\tau)}{v(\sigma)}=1$ uniformly over $\sigma\in(-\infty,0]$.

We then put

$$E_v=\{\phi\in C((-\infty,0];X)\mid v\phi \text{ is bounded and uniformly continuous}\}$$

with the weighted sup-norm $\|\cdot\|_v$ defined on E_v by

$$\|\phi\|_v=\sup_{\tau\le 0}v(\tau)\|\phi(\tau)\|$$

for every $\phi\in E_v$. Then E_v satisfies (E1)–(E3).

Proof. E_v is a real Banach space with respect to the norm $\|\cdot\|_v$.

(E1) (1) (a) From the definition of the norm

$$\|\phi(0)\|=v(0)\|\phi(0)\|\le\sup_{\tau\le 0}v(\tau)\|\phi(\tau)\|\le\|\phi\|_v$$

for every $\phi\in E_v$.

(b) We define a function $\tilde\phi:(-\infty,0]\to\mathbf{R}$ by $\tilde\phi(\tau)=\phi(0)$ for every $\tau\in(-\infty,0]$. Then $\tilde\phi\in E_v$.

(2) Let ϕ, $\phi_n\in E_v$ for $n=1,2,\cdots$ with $\|\phi_n-\phi\|_v\to 0$ as $n\to\infty$. Then for every $\tau\in(-\infty,0]$

$$v(\tau)\|\phi_n(\tau)-\phi(\tau)\|\le\|\phi_n-\phi\|_v\to 0$$

as $n\to\infty$. Thus for every $\tau\in(-\infty,0]$, $\|\phi_n(\tau)-\phi(\tau)\|\to 0$ as $n\to\infty$.

(E2) Let $\lambda>0$, $x\in X$, $\phi\in E_v$ and $\psi\in C^1((-\infty,0];X)$ with

$$\psi-\lambda\frac{d\psi}{d\tau}=\phi,\quad \psi(0)=x.$$

Then

$$\psi(\tau)=e^{\frac{\tau}{\lambda}}x+\frac{1}{\lambda}e^{\frac{\tau}{\lambda}}\int_\tau^0 e^{-\frac{\sigma}{\lambda}}\phi(\sigma)d\sigma.$$

Thus $\psi \in C((-\infty, 0]; X)$. Since $v\psi$ is bounded and uniformly continuous, $\psi \in E_v$, and

$$\begin{aligned} v(\tau)\|\psi(\tau)\| &\leq e^{\frac{\tau}{\lambda}} v(\tau)\|x\| + \frac{1}{\lambda} e^{\frac{\tau}{\lambda}} \int_\tau^0 e^{-\frac{\sigma}{\lambda}} v(\tau)\|\phi(\sigma)\| d\sigma \\ &\leq e^{\frac{\tau}{\lambda}}\|x\| + (1 - e^{\frac{\tau}{\lambda}})\|\phi\|_v \\ &\leq \max\{\|x\|, \|\phi\|_v\} \end{aligned}$$

for every $\tau \in I$. Hence $\|\psi\|_v \leq \max\{\|x\|, \|\phi\|_v\}$.

(E3) (1) (a) Let $x(t) : (-\infty, \infty) \to X$ be continuous and $x_{|(-\infty,0]} \in BUC((-\infty, 0]; X)$. Then $x_t \in E_v$ for every $t \geq 0$.

(b) $x(t)$ and $v(t)$ are uniformly continuous on every compact interval of $(-\infty, \infty)$ and $(-\infty, 0]$, respectively. Let $\epsilon > 0$ and $t \in [0, \infty)$. Then there exists $\delta \equiv \delta(\epsilon) > 0$ such that for every t, $s \in [0, t_0]$ and for some $t_0 > 0$ with $\|t - s| < \delta$ we have $|(t + \tau) - (s + \tau)| < \delta$ with $t + \tau$, $s + \tau \in (-\infty, t_0]$ for every $\tau \in (-\infty, 0]$ and thus

$$\|x(t + \tau) - x(s + \tau)\| < \epsilon \leq \frac{\epsilon}{v(\tau)}$$

for every $\tau \in (-\infty, 0]$. Hence $v(\tau)\|x(t + \tau) - x(s + \tau)\| < \epsilon$ and $v(\tau)\|x_t(\tau) - x_s(\tau)\| < \epsilon$. Therefore $\|x_t - x_s\| \leq \epsilon$ and then $t \to x_t$ is continuous.

(2) Let $x(t) : (-\infty, \infty) \to X$ be as in (1). Then

$$\begin{aligned} \|x_t\|_v &= \sup_{\sigma \leq 0} v(\sigma)\|x(t + \sigma)\| \\ &\leq \sup_{\sigma \leq -t} v(\sigma)\|x(t + \sigma)\| + \sup_{-t \leq \sigma \leq 0} v(\sigma)\|x(t + \sigma)\| \\ &\leq \sup_{\sigma \leq 0} v(\sigma)\|x(\sigma)\| + \sup_{-t \leq \sigma \leq 0} \frac{v(\sigma)}{v(t + \sigma)} v(t + \sigma)\|x(t + \sigma)\| \\ &\leq \|x_0\|_v + \sup_{-t \leq \sigma \leq 0} \frac{v(\sigma)}{v(t + \sigma)} \sup_{-t \leq \sigma \leq 0} v(t + \sigma)\|x(\sigma)\| \\ &\leq M_0\|x_0\|_v + M_1(t) \sup_{0 \leq \sigma \leq t} \|x(\sigma)\| \end{aligned}$$

for every $t \geq 0$, where $M_0 \equiv 1$ and

$$M_1(t) \equiv \sup_{-t \leq \sigma \leq 0} \frac{v(\sigma)}{v(t + \sigma)} \to 1$$

as $t \to 0-$.

Proof of Theorem 2.7.3. Let $\alpha \in [0, 2(a - l(c))]$. We define $A_\alpha : \mathbf{R} \to \mathbf{R}$ by

$$A_\alpha x = bx^2 - \alpha x$$

for every $x \in \mathbf{R}$, and also we define $G : E \to \mathbf{R}$ by

$$G(\phi) = \phi(0)(a - K(\phi))$$

for every $\phi \in E$. Then (2.7.4) can be formulated as

$$\begin{cases} \dfrac{dx}{dt}(t) + \alpha x(t) + A_\alpha x(t) = G(x_t), & t \geq 0 \\ x(t) = \phi(t), \quad t \leq 0 \end{cases} \tag{2.7.5}$$

for every $\phi \in BUC((-\infty, 0])$. Since $BUC((-\infty, 0]) = E_v$ with $v(\tau) \equiv 1$ for every $\tau \in (-\infty, 0]$, from Proposition 2.7.1, E satisfies (E1)–(E3). We put $\widehat{X}_\alpha = \left[\dfrac{\alpha}{2b}, c\right]$, and we note that A_α is accretive in $\left[\dfrac{\alpha}{2b}, \infty\right)$. In fact,

$$\begin{aligned} &(A_\alpha x_1 - A_\alpha x_2, x_1 - x_2)(x_1 - x_2) \\ &\quad = ((bx_1^2 - \alpha x_1) - (bx_2^2 - \alpha x_2))(x_1 - x_2) \\ &\quad = (b(x_1 + x_2) - \alpha)|x_1 - x_2|^2 \\ &\quad \geq 0 \end{aligned}$$

for every $x_i \in \left[\dfrac{\alpha}{2b}, \infty\right)$, $i = 1, 2$. From (C1) and (C2) we have

$$\begin{aligned} &|G(\psi_1) - G(\psi_2)| \\ &\quad = |\psi_1(0)(a - K(\psi_1)) - \psi_2(0)(a - K(\psi_2))| \\ &\quad \leq |\psi_1(0)(a - K(\psi_1)) - \psi_1(0)(a - K(\psi_2))| \\ &\qquad + |\psi_1(0)(a - K(\psi_2)) - \psi_2(0)(a - K(\psi_2))| \\ &\quad \leq |\psi_1(0)||K(\psi_1) - K\psi_2)| + |\psi_1(0) - \psi_2(0)||a - K(\psi_2)| \\ &\quad \leq (a + l(c) + cL(c))\|\psi_1 - \psi_2\|_\infty \\ &\quad = \beta\|\psi_1 - \psi_2\|_\infty \end{aligned}$$

for every $\psi_i \in \widehat{E}_\alpha$, $i = 1, 2$, where

$$\beta = \beta(c) = a + l(c) + cL(c).$$

Thus (A1) is satisfied. For (A2) let $x \in \widehat{X}_\alpha$, $\phi \in \widehat{E}_\alpha$ and $\lambda > 0$ with $\lambda\gamma < 1$ $\gamma = \max\{0, \beta - \alpha\}$. Let ψ be the solution of

$$\begin{cases} \psi - \lambda \dfrac{d\psi}{dt} = \phi \\ \psi(0) = x. \end{cases}$$

Then from the proof of (E2) $\psi \in \widehat{E}_\alpha$. Thus the proof of (A2) is completed. For the proof of (A3) let $x \in \widehat{X}_\alpha$, $\phi \in \widehat{E}_\alpha$, and $\lambda > 0$ with $\lambda\gamma < 1$. Since $\alpha \in [0, 2(a - l(c))]$ from (1), (2), and (C2) we have

$$\phi(0) + \lambda G(\psi) \leq c + \lambda c(a + l(c)) \leq c + \lambda bc^2 \tag{2.7.6}$$

and

$$\begin{aligned} \phi(0) + \lambda G(\psi) &\geq \frac{\alpha}{2b} + \lambda(a - l(c))\frac{\alpha}{2b} \\ &\geq \frac{\alpha}{2b} + \lambda\frac{\alpha^2}{4b}. \end{aligned} \tag{2.7.7}$$

Since the function

$$(1 + \lambda\alpha)I + \lambda A_\alpha : X_\alpha \to \left[\frac{\alpha}{2b} + \lambda\frac{\alpha_2}{4b}, c + \lambda c^2\right]$$

is continuous and non-decreasing, and since from (2.7.6) and (2.7.7)

$$\phi(0) + \lambda G(\psi) \in \left[\frac{\alpha}{2b} + \lambda\frac{\alpha_2}{4b}, c + \lambda c^2\right]$$

there exists $x \in \widehat{X}_\alpha$ such that

$$\phi(0) + \lambda G(\psi) = (1 + \lambda\alpha)x + \lambda A_\alpha x,$$

and thus

$$\frac{1}{1 + \lambda\alpha}(\phi(0) + \lambda G(\psi)) \in \left(I + \frac{\lambda}{1 + \lambda\alpha}A_\alpha\right)\widehat{X}_\alpha.$$

Thus (A3) holds.

The strong solution $x(t)$ on $(-\infty, \infty)$ of (2.7.5) satisfies

$$x(t) = \phi(0)\exp\left(\int_0^t (a - bx(s) - K(x_s))ds\right) \tag{2.7.8}$$

for every $t \geq 0$. Indeed, from (2.7.5)

$$\frac{d}{dt}(x(t)\exp\left(-\int_0^t (a - bx(s) - K(x_s))ds)\right) = 0,$$

and thus

$$x(t)\exp\left(-\int_0^t (a - bx(s) - K(x_s))ds\right) = C$$

for every $t \geq 0$. Since $C = x(0) = \phi(0)$ (2.7.8) holds. Thus (1)–(3) can now be read from (3) of Theorem 2.6.1, Proposition 2.6.1, Proposition 2.6.3 and (2.7.8).

2.8. Notes for References

Theorem 2.1.1 is the existence and the uniqueness theorem for solutions of (2.1.1) given by using Picard's iterative method for ordinary differential equations appeared in Coddington–Levinson [20] and Hale [64]. Theorem 2.1.1, Theorem 2.1.2 and Theorem 2.1.3 are adapted from Brewer [11] with $v \equiv 0$ and $\phi \in C([-r,0];X)$. Chapter 3 of Ha [59] is referred. In Brewer [11] the nonlinear functional evolutions of the type

$$\begin{cases} \dfrac{dx}{dt}(t) = G(x_t), & t \geq 0 \\ x(t) = \phi(t), & t \leq 0 \end{cases} \tag{2.8.1}$$

were considered. Here $\phi \in E$, where

$$E = \{\phi : (-\infty, 0] \to X \mid \phi \text{ is continuous on } [-r,0] \text{ and } v\phi \text{ is integrable on } (-\infty,-r)\}$$

with norm $\|\cdot\|_E$ defined by

$$\|\phi\|_E = \sup_{t\in[-r,0]} \|\phi(t)\| + \int_{-\infty}^{-r} v(t)\|\phi(t)\|dt$$

with $v : (-\infty,-r] \to [0,\infty)$ is a non-decreasing function. As a form similar to Theorem 2.1.1 we get the following result:

THEOREM 2.8.1 *Let $G : E \to X$. Assume that there exists a constant $\beta > 0$ such that*

$$\|G(\psi_1) - G(\psi_2)\| \leq \beta\|\psi_1 - \psi_2\|_E$$

for every $\psi_i \in E$, $i = 1,2$. Let $\phi \in E$. Then (2.8.1) *has a unique (classical) solution $x(t) \equiv x(t;\phi)$ on $(-\infty,\infty)$ in the sense that*

$$x(t) = \begin{cases} \phi(0) + \displaystyle\int_0^t G(x_s)ds, & t \geq 0 \\ \phi(t), & t \leq 0. \end{cases}$$

Lemma 2.1.1 and Proposition 2.1.3 are due to Plant [118]. Dyson–Villella–Bressan [43] is referred.

Many mathematicians considered (2.1.1) with infinite delay or finite delay in finite-dimensional space, i.e., Flaska–Leitman [52] in **R**, Hale [65], [66], [67], Kappel-Schappacher [73], [74], [75] and Atkinson–Haddock [2], Corduneanu–Lakshmikantham [21], Haddock–Hornor [63],

Mackey [103], Ntouyas–Sficas–Tsamatos [121], Sawano [127] and Webb [136] in $\mathbf{R}^n$ for $n = 1, 2, \cdots$, etc.

Theorem 2.2.1 and Theorem 2.2.2 are from Proposition 3 and the remark in the final part of Webb [137]. Theorem 2.2.3 is owed to Ruess [124]. In Brewer [12], [13] nonlinear functional evolutions of the type

$$\begin{cases} \dfrac{dx}{dt}(t) + \alpha x(t) + Ax(t) = G(x_t), \quad t \geq 0 \\ x_0 = \phi \end{cases}$$

are considered.

One can refer Bothe [10] in which the initial value problem of the type

$$\begin{cases} \dfrac{dx}{dt}(t) + Ax \in F(t, x), \quad 0 \leq t \leq T \\ x(0) = x_0 \end{cases}$$

was considered. For nonlinear evolutions one can refer Vrabie [134]. For operator equations, Calvert–Kartsatos [19], Crandall–Evans [25], Morales [107] and Schöneberg [129] are referred. For linear evolutions Travis–Webb [132], [133] are referred.

Propositions 2.3.1–2.3.3 are from Vrabie [135] while Theorem 2.3.1 and Theorem 2.3.2 are taken from Mitidieri–Vrabie [105]. In Dyson–Villella–Bressan [44] the existence of integral solutions of (2.3.1) was studied with using Theorem 1 of Crandall–Nohel [28]. one may refer Dyson–Villella–Bressan [42]. Lemma 2.3.1 is due to Theorem 1 of Arino–Gautier–Penot [1]. Badii–Diaz–Tesei [3] is referred. For nonlinear evolutions, Baras [4] and Gutman [55], [56] and Shioji [128] are referred. For operator equations, one can refer Kartsatos [80], [82], [83].

As in Lemma 2.4.1 results in Proposition 2.1 of Webb [138] hold also in real Banach spaces. Hence results in Theorem 2.4.1 are the versions in real Banach spaces of those in Webb [136]. Theorem 2.4.2 is from Proposition 5.8 for $p > 1$ of Webb [138]. One can refer Dyson–Villella–Bressan [44].

The results in Section 2.5 containing Theorem 2.5.1 and Theorem 2.5.2 are due to Ruess–Summers [125]. They studied nonlinear functional evolutions of the type

$$\begin{cases} \dfrac{dx}{dt}(t) + \alpha x(t) + Ax(t) \ni G(x_t), \quad t \geq 0 \\ x_0 = \phi. \end{cases}$$

are considered. One may refer Ruess [123] and Ruess–Summers [126].

Theorem 2.6.1 is owed to Plant [120]. Webb [137] also treats the stability for solutions of (2.2.1). But the estimate (3.16) in Webb [137] is complicated, and does not tell us if we can take $\gamma < 0$ in

$$\|S(t)\psi_1 - S(t)\psi_2\|_C \leq e^{\gamma t}\|\psi_1 - \psi_2\|_C$$

for every $t \geq 0$ and $\psi_i \in \overline{D(A)}$ for $i = 1, 2$. Ruess [124] and Ruess-Summers [126] are refered.

Theorem 2.7.1 and Theorem 2.7.2 are taken from Burch-Goldstein [18] and Mitidieri–Vrabie [105], respectively, while, Theorem 2.7.3 is owed to Ruess–Summers [125].

Chapter 3

NON–AUTONOMOUS NONLINEAR FUNCTIONAL EVOLUTIONS

The object of this chapter is to consider non-autonomous nonlinear functional evolutions associated with accretive operators in real Banach spaces. In Section 3.1 we study the existence of solutions by Kato's methods applied to that of nonlinear evolutions. Section 3.2 deals with evolution operator methods in which the evolution operators are generated by the accretive operators. Section 3.3 investigates monotonicity methods, and Section 3.4 contains methods of lines. Section 3.5 and Section 3.6 are owed to compactness methods and L^p-space methods, respectively, whilst Section 3.7 is devoted to examples and applications. Finally, in Section 3.8 comments and notes for references are presented.

We treat non-autonomous nonlinear functional evolutions of the type

$$\begin{cases} \dfrac{dx}{dt}(t) + A(t)x(t) \ni G(t, x_t), & s \leq t \leq T \\ x_s(t) = \phi(t), \quad t \in I \end{cases}$$

associated with a non-autonomous nonlinear operator $A(t) : X \supset D(A(t)) \to 2^X$ for every $t \in [0,T]$ in a real Banach space X, where $T > 0$, $s \in [0,T]$, $x(t) : I \cup [0,T] \to X$ is an unknown function, $G : [0,T] \times C(I;X) \to X$, $\phi \in C(I;X)$ or $\phi \in L^p(I;X)$ for $1 \leq p < \infty$, and $x_t : I \to X$ defined by $x_t(\tau) = x(t+\tau)$ for every $\tau \in I$ and $t \in [0,T]$. Here, $I = [-r, 0]$ for $r > 0$ or $I = (-\infty, 0]$.

This chapter is devoted to considering only the case of $I = [-r, 0]$. In the case of $I = (-\infty, 0]$, we may choose $\phi \in E_v$, where E_v is the space in the beginning of Chapter 2.

3.1. Kato's Methods

In this section we study non-autonomous nonlinear functional evolutions of the type

$$\begin{cases} \dfrac{dx}{dt}(t) + A(t)x(t) = G(t, x_t), & 0 \le t \le T \\ x(t) = \phi(t), & -r \le t \le 0 \end{cases} \tag{3.1.1}$$

associated with an operator $A(t) : D(A(t)) \subset X \to X$ for every $t \in [0, T]$, in a real Banach space X, where $T > 0$, $r > 0$, $x(t) : [-r, T] \to X$ is an unknown function, $G : [0, T] \times C([-r, 0]; X) \to X$, and $\phi \in C([-r, 0]; X)$.

We are now in a position to give definitions for solutions of (3.1.1).

DEFINITION 3.1.1 *1. A function $x(t) : [-r, T] \to X$ is called a (classical) solution of* (3.1.1) *on $[-r, T]$ if:*

(1) $x(t) = \phi(t)$ for every $t \in [-r, 0]$;

(2) $x(t)$ is continuously differentiable on $[0, T]$;

(3) $x(t)$ satisfies

$$\frac{dx}{dt}(t) + A(t)x(t) = G(t, x_t)$$

for every $t \in [0, T]$.

2. A function $x(t) : [-r, T] \to X$ is called a strong solution of (3.1.1) *on $[-r, T]$ if:*

(1) $x(t) = \phi(t)$ for every $t \in [-r, 0]$;

(2) $x(t)$ is continuous on $[0, T]$;

(3) $x(t)$ is absolutely continuous on every compact interval of $[0, T]$;

(4) $x(t)$ is differentiable a.e. $t \in (0, T)$;

(5) $x(t)$ satisfies

$$\frac{dx}{dt}(t) + A(t)x(t) \ni G(t, x_t)$$

a.e. $t \in (0, T)$.

To show existence for solutions, we impose the following conditions.

$(A1)_1$ $A(t) : D \equiv D(A(t)) \subset X \to X$ is m-accretive in X for every $t \in [0, T]$.

$(A2)_1$ There exists a non-decreasing function $L : [0, \infty) \to [0, \infty)$ such that

$$\|A(t_1)u - A(t_2)u\| \le |t_1 - t_2| L(\|u\|)(1 + \|A(t_2)u\|)$$

for every $u \in D$ and $t_i \in [0,T]$, $i = 1,2$.

$(G1)_1$ There exists a constant $\beta > 0$ such that

$$\|G(t,\psi_1) - G(t,\psi_2)\| \leq \beta\|\psi_1 - \psi_2\|_C$$

for every $t \in [0,T]$ and $\psi_i \in C([-r,0];X)$, $i = 1,2$.

$(G2)_1$ There exists an increasing function $l : [0,\infty) \to [0,\infty)$ such that

$$\|G(t_1,\psi) - G(t_2,\psi)\| \leq |t_1 - t_2| l(\|\psi\|_C)$$

for every $t_i \in [0,T]$, $i = 1,2$ and $\psi \in C([-r,0];X)$.

Following Kato's approach in Kato [95], we show that the unique strong solution of (3.1.1) exists as a uniform limit of a sequence of the unique (classical) solutions of the approximating non-autonomous nonlinear functional evolutions.

We put

$$J_n(t) = \left(I + \frac{1}{n}A(t)\right)^{-1}, \quad A_n(t) = n(I - J_n(t))$$

for every $t \in [0,T]$ and $n = 1,2,\cdots$.

Now we start to consider non-autonomous nonlinear functional approximating evolutions of the type

$$\begin{cases} \dfrac{dx_n}{dt} + A_n(t)x_n(t) = G(t,(x_n)_t), \ 0 \leq t \leq T \\ x_n(t) = \phi(t), \ -r \leq t \leq 0 \end{cases} \tag{3.1.2}$$

associated with a single-valued operator $A_n(t) : X \to X$ for every $t \in [0,T]$ and $n = 1,2,\cdots$, in a real Banach space X.

We need the next lemma in the proof of Proposition 3.1.1.

LEMMA 3.1.1 *We have the following (1) and (2):*

(1) Let $x \in C^1([0,T];X)$. *Then*

$$\lim_{h\to 0+} \sup_{\tau\in[-t,0]} \frac{\|x(t+\tau+h) - x(t+\tau)\|}{h}$$

exists and is equal to

$$\sup_{\tau\in[-t,0]} \left\|\frac{dx}{dt}(t+\tau)\right\|$$

PROPOSITION 3.1.1 *Let X^* be uniformly convex. Assume that $(A1)_1$, $(A2)_1$, $(G1)_1$, and $(G2)_1$ are satisfied. Let $\phi \in C^1([-r,0];X)$ with $\phi(0) \in D$. Then:*

(1) There exists a unique (classical) solution $x_n(t)$ of (3.1.2) on $[-r,T]$;

(2) There exists a constant $K > 0$ such that

$$\|x_n(t)\| \leq K$$

for every $t \in [0,T]$ and $n = 1,2,\cdots$;

(3) There exists a constant $N > 0$ such that

$$\left\|\frac{dx_n}{dt}(t)\right\| \leq N$$

for every $t \in [0,T]$ and $n = 1,2\cdots$.

Proof. (1) From Proposition 1.4.4 there exists a nondecreasing function $L_1 : [0,\infty) \to [0,\infty)$ such that

$$\|A_n(t_1)u - A_n(t_2)u\| \leq |t_1 - t_2|L_1(\|u\|)(1 + \|A_n(t_2)u\|) \qquad (3.1.3)$$

for every $u \in D$ and $t_i \in [0,T]$, $i = 1,2$. This shows that $A_n(t)u$ is Lipschitz continuous in $t \in [0,T]$ for every $u \in D$ and $n = 1,2\cdots$, and thus, so is

$$J_n(t)u = u - \frac{1}{n}A_n(t)u.$$

We put

$$C_T([-r,T];X) = \{x \in C([-r,T];X) \mid x(t) = \phi(t), \ -r \leq t \leq 0\}.$$

Then $C_T([-r,T];X)$ is a real Banach space with norm $\|\cdot\|_T$ defined by

$$\|x\|_T = \sup_{t\in[-r,T]} \|x(t)\|$$

for every $x \in C_T([-r,T];X)$. We define an operator $Q_n : C_T([-r,T];X) \to C([-r,0];X)$ by

$$(Q_n x)(t) = \begin{cases} e^{-nt}\phi(0) + n\displaystyle\int_0^t e^{n(\sigma-t)}(J_n(\sigma)x(\sigma) \\ \qquad\qquad + \dfrac{1}{n}G(\sigma,x_\sigma))d\sigma, \quad 0 \leq t \leq T \\ \phi(t), \quad -r \leq t \leq 0 \end{cases}$$

for every $x \in C_T([-r,T];X)$ and $n = 1, 2, \cdots$. Then $Q_n : C_T([-r,T];X) \to C_T([-r,T];X)$ for $n = 1, 2, \cdots$. Let x, $y \in C_T([-r,T];X)$. Thus by the definition of Q_n

$$\begin{aligned} \|Q_n x(t) - Q_n y(t)\| &\leq (n+\beta) \int_{-r}^{t} \|x(\sigma) - y(\sigma)\| d\sigma \\ &\leq (n+\beta)(t+r)\|x-y\|_T, \end{aligned}$$

for every $t \in [-r,T]$ and

$$\begin{aligned} \|(Q_n)^2 x(t) - (Q_n)^2 y(t)\| &\leq (n+\beta) \int_{-r}^{t} \|Q_n x(\sigma) - Q_n y(\sigma)\| d\sigma \\ &\leq \frac{(n+\beta)}{2!}(t+r)^2 \|x-y\|_T. \end{aligned}$$

for every $t \in [-r,T]$. Iterating for $i = 3, 4, \cdots$ we have

$$\|(Q_n)^i x(t) - (Q_n)^i y(t)\| \leq \frac{(n+\beta)^i}{i!}(t+r)^i \|x-y\|_T$$

for every $t \in [-r,T]$, and then

$$\begin{aligned} \|(Q_n)^i x - (Q_n)^i y\|_T &\leq \frac{(n+\beta)^i}{i!}(t+r)^i \|x-y\|_T \\ &\leq \frac{(n+\beta)^i}{i!}(T+r)^i \|x-y\|_T \end{aligned}$$

for every $t \in [-r,T]$, x, $y \in C_T([-r,T];X)$ and $n = 1, 2, \cdots$. Since there exists a sufficiently large number i such that

$$\frac{(n+\beta)^i}{i!}(T+r)^i < 1$$

Banach's contraction principle implies that there exists $x \in C_T([-r,T];X)$ such that $(Q_n)^i x = x$, and thus $Q_n x = x$ for $n = 1, 2, \cdots$. If we put $Q_n x_n = x_n$ then it follows from the definition of Q_n that $x_n \in C_T([-r,T];X)$ is a (classical) solution of (3.1.2) on $[-r,T]$.

(2) First of all, we obtain

$$\begin{aligned} &\|G(t,\psi)\| \\ &\quad \leq \|G(t,\psi) - G(t,\bar{0})\| + \|G(t,\bar{0}) - G(0,\bar{0})\| + \|G(0,\bar{0}\| \\ &\quad \leq \beta\|\psi\|_C + Tl(0) + \|G(0,\bar{0}\| \\ &\quad = K_1 \end{aligned}$$

for every $t \in [0,T]$ and $\psi \in C([-r,0];X)$.

Let F be the duality mapping on X. By the definition of (classical) solutions, $(A1)_1$ and $(G1)_1$

$$\begin{aligned}
&\Big(\frac{dx_n}{dt}, F(x_n(t)-\phi(0))\Big)\\
&\quad = -(A_n(t)x_n(t) - G(t,(x_n)_t), F(x_n(t)-\phi(0)))\\
&\quad = -(A_n(t)x_n(t) - A_n(t)\phi(0), F(x_n(t)-\phi(0)))\\
&\qquad -(A_n(t)\phi(0), F(x_n(t)-\phi(0))) + (G(t,(x_n)_t), F(x_n(t)-\phi(0)))\\
&\quad \le -(A_n(t)\phi(0), F(x_n(t)-\phi(0))) + (G(t,(x_n)_t), F(x_n(t)-\phi(0)))\\
&\quad = -(A_n(t)\phi(0), F(x_n(t)-\phi(0)))\\
&\qquad +(G(t,(x_n)_t) - G(t,\phi), F(x_n(t)-\phi(0)))\\
&\qquad +(G(t,\phi), F(x_n(t)-\phi(0)))\\
&\quad \le \|A_n(t)\phi(0)\|\|x_n(t)-\phi(0)\| + \beta\|(x_n)_t-\phi\|_C\|x_n(t)-\phi(0)\|\\
&\qquad +K_1\|x_n(t)-\phi(0)\|.
\end{aligned}$$

From (3.1.3) and (10) of Proposition 1.1.4 we obtain

$$\begin{aligned}
&\|A_n(t)\phi(0) - A_n(0)\phi(0)\|\\
&\quad \le tL_1(\|\phi(0)\|)(1+\|A_n(0)\phi(0)\|)\\
&\quad \le TL_1(\|\phi(0)\|)(1+\|A(0)\phi(0)\|),
\end{aligned}$$

which yields

$$\begin{aligned}
&\|A_n(t)\phi(0)\|\\
&\quad \le TL_1(\|\phi(0)\|)(1+\|A(0)\phi(0)\|) + \|A(0)\phi(0)\|\\
&\quad \equiv K_2
\end{aligned}$$

for every $t \in [0,T]$. Thus

$$\begin{aligned}
&\Big(\frac{dx_n}{dt}, F(x_n(t)-\phi(0))\Big)\\
&\quad \le K_2\|x_n(t)-\phi(0)\| + \beta\|(x_n)_t-\phi\|_C\|x_n(t)-\phi(0)\|\\
&\qquad +K_1\|x_n(t)-\phi(0)\|
\end{aligned}$$

for every $t \in [0,T]$. Since $x_n(t)$ is absolutely continuous so is $\|x_n(t) - \phi(0)\|$ on $[0,T]$. Thus $\|x_n(t)-\phi(0)\|$ is differentiable a.e. $t \in (0,T)$. By (1) of Lemma 1.1.2, (12) and the above inequality

$$\begin{aligned}
\frac{d}{dt}\|x_n(t)-\phi(0)\| &\le K_1 + \beta\|(x_n)_t-\phi\|_C + K_2\\
&= K_3 + \beta\|(x_n)_t-\phi\|_C \qquad (3.1.4)
\end{aligned}$$

for a.e. $t \in (0,T)$, where $K_3 = K_1 + K_2$. Now we integrate (3.1.4) to obtain

$$\|x_n(t) - \phi(0)\| \leq K_3T + \beta \int_0^t \|(x_n)_s - \phi(0)\|_C ds \tag{3.1.5}$$

for every $t \in [0,T]$.

Suppose $t \geq r$. Then since $t + \tau \geq 0$ for every $\tau \in [-r, 0]$ (3.1.5) implies that

$$\begin{aligned} \|x_n(t+\tau) - \phi(0)\| &\leq K_3T + \beta \int_0^{t+\tau} \|(x_n)_s - \phi(0)\|_C ds \\ &\leq K_3T + \beta \int_0^t \|(x_n)_s - \phi(0)\|_C ds. \end{aligned}$$

Thus we have

$$\begin{aligned} &\|x_n(t+\tau) - \phi(\tau)\| \\ &\quad \leq \|x_n(t+\tau) - \phi(0)\| + \|\phi(0) - \phi(\tau)\| \\ &\quad \leq \|\phi(0)\| + \|\phi\|_C + K_3T + \beta \int_0^t \|(x_n)_s - \phi\|_C ds, \end{aligned}$$

and then

$$\begin{aligned} \|(x_n)_t - \phi\|_C &= \sup_{-r\leq\tau\leq 0} \|x_n(t+\tau) - \phi(\tau)\| \\ &\leq \|\phi(0)\| + \|\phi\|_C + K_3T \\ &\quad + \beta \int_0^t \|(x_n)_s - \phi\|_C ds. \end{aligned} \tag{3.1.6}$$

Suppose $0 \leq t < r$. Then since $t + \tau < 0$ for every $\tau \in [-r, -t)$

$$\begin{aligned} \|x_n(t+\tau) - \phi(0)\| &= \|\phi(t+\tau) - \phi(0)\| \\ &\leq \|\phi\|_C + \|\phi(0)\|, \end{aligned} \tag{3.1.7}$$

and since $t + \tau \leq 0$ for every $\tau \in [-t, 0]$ from (3.1.5)

$$\|x_n(t+\tau) - \phi(0)\| \leq K_3T + \beta \int_0^t \|(x_n)_s - \phi\|_C ds \tag{3.1.8}$$

for every $t \in [0,T]$. By (3.1.7) and (3.1.8) we conclude that

$$\begin{aligned} \|x_n(t+\tau) - \phi(0)\| &\leq \|\phi\|_C + \|\phi(0)\| + K_3T \\ &\quad + \beta \int_0^t \|(x_n)_s - \phi\|_C ds \end{aligned}$$

for every $\tau \in [-r, 0]$ with $t \in [0, r)$ from which it follows that

$$\begin{aligned}\|(x_n)_t - \phi\|_C &\leq 2(\|\phi\|_C + \|\phi(0)\|) + K_3 T \\ &\quad + \beta \int_0^t \|(x_n)_s - \phi\|_C ds \end{aligned} \tag{3.1.9}$$

for every $t \in [0, T]$. As a consequence of (3.1.6) and (3.1.9), we arrive at

$$\begin{aligned}\|(x_n)_t - \phi\|_C &\leq 2(\|\phi\|_C + \|\phi(0)\|) + K_3 T \\ &\quad + \beta \int_0^t \|(x_n)_s - \phi\|_C ds\end{aligned}$$

for every $t \in [0, T]$. By Gronwall's inequality

$$\|(x_n)_t - \phi\|_C \leq (2(\|\phi\|_C + \|\phi(0)\|) + K_3 T) e^{\beta T} < \infty.$$

Since

$$\|x_n(t) - \phi(0)\| \leq \|(x_n)_t - \phi\|_C$$

for every $t \in [0, T]$ we have

$$\|x_n(t)\| \leq K$$

for every $t \in [0, T]$ and $n = 1, 2, \cdots$, where

$$K = \|\phi(0)\| + (2(\|\phi\|_C + \|\phi(0)\|) + K_3 T) e^{\beta T}.$$

(3) Let $0 < h < \min\{r, T\}$. It follows from (12) of Lemma 1.1.2, $(A1)_1$, $(A2)_1$, $(G1)_1$, and $(G2)_1$ that

$$\begin{aligned}
&\|x_n(t+h) - x_n(t)\| \frac{d}{dt} \|x_n(t+h) - x_n(t)\| \\
&= \Big(\frac{dx_n}{dt}(t+h) - \frac{dx_n}{dt}(t), F(x_n(t+h) - x_n(t))\Big) \\
&= -(A_n(t+h)x_n(t+h) - A_n(t)x_n(t), F(x_n(t+h) - x_n(t))) \\
&\quad + (G(t+h, (x_n)_{t+h}) - G(t, (x_n)_t), F(x_n(t+h) - x_n(t))) \\
&= -(A_n(t+h)x_n(t+h) - A_n(t+h)x_n(t), F(x_n(t+h) - x_n(t))) \\
&\quad + (A_n(t)x_n(t) - A_n(t+h)x_n(t), F(x_n(t+h) - x_n(t))) \\
&\quad + (G(t+h, (x_n)_{t+h}) - G(t+h, (x_n)_t), F(x_n(t+h) - x_n(t))) \\
&\quad + (G(t+h, (x_n)_t) - G(t, (x_n)_t), F(x_n(t+h) - x_n(t))) \\
&\leq h L_1(\|x_n(t)\|)(1 + \|A_n(t)x_n(t)\|) \|x_n(t+h) - x_n(t)\| \\
&\quad + \beta \|(x_n)_{t+h} - (x_n)_t\|_C \|x_n(t+h) - x_n(t)\| \\
&\quad + h l(\|(x_n)_t\|_C) \|x_n(t+h) - x_n(t)\|
\end{aligned}$$

a.e. $t \in (0,T)$. By (2) and its proof, there exists a constant $K' > 0$ such that $\|(x_n)_t - \phi\|_C \le K'$, $\|(x_n)_t\|_C \le K'$ and $\|x_n(t)\| \le K'$. Since

$$\begin{aligned} \|A_n(t)x_n(t)\| &\le \left\|\frac{dx_n}{dt}(t)\right\| + \|G(t,(x_n)_t)\| \\ &\le \left\|\frac{dx_n}{dt}(t)\right\| + \beta\|(x_n)_t - \phi\|_C + \|G(t,\phi)\| \\ &\le \left\|\frac{dx_n}{dt}(t)\right\| + \beta K' + K_1 \end{aligned}$$

we have

$$\begin{aligned} &\frac{d}{dt}\|x_n(t+h) - x_n(t)\| \\ &\quad \le hC_1 + hC_2\left\|\frac{dx_n}{dt}(t)\right\| + \beta\|(x_n)_{t+h} - (x_n)_t\|_C, \end{aligned}$$

where

$$C_1 = L_1(K')(1 + \beta K" + K_1) + l(K')$$

and $C_2 = L_1(K')$. Integrating this inequality on $[0,t]$ we have

$$\begin{aligned} &\frac{\|x_n(t+h) - x_n(t)\|}{h} \\ &\quad \le \frac{\|x_n(h) - x_n(0)\|}{h} + C_1T + C_2\int_0^t \left\|\frac{dx_n}{ds}(s)\right\|ds \\ &\qquad + \beta\int_0^t \frac{\|(x_n)_{s+h} - (x_n)_s\|_C}{h}ds \end{aligned} \tag{3.1.10}$$

for every $t \in [0,T]$.

Suppose $t < r$. Let $t + h < r$ for sufficiently small $h \ne 0$. Then we have

$$\begin{aligned} &\int_0^t \frac{\|(x_n)_{s+h} \quad (x_n)_s\|_C}{h}ds \\ &\quad = \int_0^t \sup_{\tau\in[-r,0]} \frac{\|x_n(s+h+\tau) - x_n(s+\tau)\|}{h}ds \\ &\quad \le \int_0^t \sup_{\tau\in[-r,-(s+h))} \frac{\|x_n(s+h+\tau) - x_n(s+\tau)\|}{h}ds \\ &\qquad + \int_0^t \sup_{\tau\in[-(s+h),-s]} \frac{\|x_n(s+h+\tau) - x_n(s+\tau)\|}{h}ds \\ &\qquad + \int_0^t \sup_{\tau\in[-s,0]} \frac{\|x_n(s+h+\tau) - x_n(s+\tau)\|}{h}ds. \end{aligned} \tag{3.1.11}$$

If $\tau \in [-r, -(s+h))$ then $s+h+\tau < 0$ and $s+h < 0$. Thus since $\frac{d\phi}{ds}(s) \in C([-r,0];X)$

$$\begin{aligned} &\frac{\|x_n(s+h+\tau) - x_n(s+\tau)\|}{h} \\ &\quad = \frac{\|\phi(s+h+\tau) - \phi(s+\tau)\|}{h} \\ &\quad \leq \frac{N_1 h}{h} = N_1 \end{aligned} \tag{3.1.12}$$

for some $N_1 > 0$.

If $\tau \in [-(s+h), -s]$ then $s+h+\tau \geq 0$ but $s+\tau \leq 0$. Since $x_n \in C^1([0,T];X)$ and $\phi \in C^1([-r,0];X)$, (2) of Lemma 3.1.1 implies

$$\begin{aligned} &\overline{\lim}_{h\to 0+} \sup_{\tau \in [-(s+h),-s]} \frac{\|x_n(s+h+\tau) - x_n(s+\tau)\|}{h} \\ &\qquad \leq \left\|\frac{dx_n}{ds}(0)\right\| \leq \left\|\frac{d\phi}{ds}(0)\right\|. \end{aligned} \tag{3.1.13}$$

If $\tau \in [-s, 0]$ then $s+\tau \geq 0$. By (1) of Lemma 3.1.1

$$\begin{aligned} &\lim_{h\to 0+} \sup_{\tau\in[-s,0]} \frac{\|x_n(s+h+\tau) - x_n(s+\tau)\|}{h} \\ &\quad = \sup_{\tau\in[-s,0]} \lim_{h\to 0+} \frac{\|x_n(s+h+\tau) - x_n(s+\tau)\|}{h} \\ &\quad = \sup_{\tau\in[-s,0]} \left\|\frac{dx_n}{ds}(s+\tau)\right\|. \end{aligned} \tag{3.1.14}$$

Thus from (3.1.10)–(3.1.14), and Fatou's lemma

$$\begin{aligned} &\overline{\lim}_{h\to 0+} \frac{\|x_n(t+h) - x_n(t)\|}{h} \\ &\quad = \lim_{h\to 0+} \frac{\|x_n(t+h) - x_n(t)\|}{h} \\ &\quad \leq \left\|\frac{dx_n}{ds}(0)\right\| + C_1 T + C_2 \int_0^t \left\|\frac{dx_n}{ds}(s)\right\| ds + \beta \int_0^t N_1 ds \\ &\quad +\beta \int_0^t \left(\left\|\frac{dx_n}{ds}(0)\right\| + \left\|\frac{d\phi}{ds}(0)\right\|\right) ds \\ &\quad +\beta \sup_{\tau\in[-s,0]} \left\|\frac{dx_n}{ds}(s+\tau)\right\| ds. \end{aligned}$$

Since

$$\begin{aligned}\left\|\frac{dx_n}{ds}(s)\right\| &= \lim_{h\to 0+}\frac{\|x_n(s+h)-x_n(s)\|}{h}\\ &\le \lim_{h\to 0+}\sup_{\tau\in[-s,0]}\frac{\|x_n(t+h+\tau)-x_n(t+\tau)\|}{h}\end{aligned}$$

we have

$$\begin{aligned}&\lim_{h\to 0+}\frac{\|x_n(s+h)-x_n(s)\|}{h}\\ &\quad=\left\|\frac{dx_n}{ds}(s)\right\|\\ &\quad\le(1+\beta T)\left\|\frac{dx_n}{dt}(0)\right\|+\Big(C_1+\beta\Big(N_1+\left\|\frac{d\phi}{dt}(0)\right\|\Big)\Big)T\\ &\quad\quad+(C_2+\beta)\int_0^t\sup_{\tau\in[-s,0]}\left\|\frac{dx_n}{dt}(s+\tau)\right\|ds. \qquad (3.1.15)\end{aligned}$$

Suppose $t\ge r$. Then for $r\le t+h\le T$ we have

$$\begin{aligned}&\int_0^t\frac{\|(x_n)_{s+h}-(x_n)_s\|_C}{h}ds\\ &\quad=\int_0^t\sup_{\tau\in[-r,0]}\frac{\|x_n(s+h+\tau)-x_n(s+\tau)\|}{h}ds\\ &\quad=\int_0^{r-h}\sup_{\tau\in[-r,0]}\frac{\|x_n(s+h+\tau)-x_n(s+\tau)\|}{h}ds\\ &\quad\quad+\int_{r-h}^t\sup_{\tau\in[-r,0]}\frac{\|x_n(s+h+\tau)-x_n(s+\tau)\|}{h}ds. \qquad (3.1.16)\end{aligned}$$

Thus we obtain

$$\begin{aligned}&\int_0^t\frac{\|(x_n)_{s+h}-(x_n)_s\|_C}{h}ds\\ &\quad\le\int_0^{r-h}\sup_{\tau\in[-r,-(s+h))}\frac{\|x_n(s+h+\tau)-x_n(s+\tau)\|}{h}ds\\ &\quad\quad+\int_0^{r-h}\sup_{\tau\in[-(s+h),-s]}\frac{\|x_n(s+h+\tau)-x_n(s+\tau)\|}{h}ds\\ &\quad\quad+\int_0^{r-h}\sup_{\tau\in[-s,0]}\frac{\|x_n(s+h+\tau)-x_n(s+\tau)\|}{h}ds\\ &\quad\quad+\int_{r-h}^t\sup_{\tau\in[-r,0]}\frac{\|x_n(s+h+\tau)-x_n(s+\tau)\|}{h}ds. \qquad (3.1.17)\end{aligned}$$

If $\tau \in [-r, -(s+h))$ with $s+h \leq r$ then $s+h+\tau < 0$, and $s+\tau < 0$. As in the case $t < r$

$$\frac{\|x_n(s+h+\tau) - x_n(s+\tau)\|}{h} \leq N_1$$

so that

$$\int_0^{r-h} \sup_{\tau \in [-r,-(s+h))} \frac{\|x_n(s+h+\tau) - x_n(s+\tau)\|}{h} ds \leq N_1 T. \quad (3.1.18)$$

If $\tau \in [-(s+h), -s]$ then $s+h+\tau \geq 0$ but $s+\tau \leq 0$. Also as in the case $t < r$ we find

$$\overline{\lim}_{h\to 0+} \sup_{\tau \in [-(s+h),-s]} \frac{\|x_n(s+h+\tau) - x_n(s+\tau)\|}{h} \leq \left\|\frac{dx_n}{ds}(0)\right\| \leq \left\|\frac{d\phi}{ds}(0)\right\|.$$

Since

$$\begin{aligned}\int_0^{r-h} \sup_{\tau \in [-s,0]} & \frac{\|x_n(s+h+\tau) - x_n(s+\tau)\|}{h} ds \\ &\leq \int_0^t \sup_{\tau \in [-s,0]} \frac{\|x_n(s+h+\tau) - x_n(s+\tau)\|}{h} ds \\ &\quad - \int_r^t \sup_{\tau \in [-s,0]} \frac{\|x_n(s+h+\tau) - x_n(s+\tau)\|}{h} ds\end{aligned}$$

and

$$\begin{aligned}\int_{r-h}^t \sup_{\tau \in [-r,0]} & \frac{\|x_n(s+h+\tau) - x_n(s+\tau)\|}{h} ds \\ &= \int_r^t \sup_{\tau \in [-r,0]} \frac{\|x_n(s+h+\tau) - x_n(s+\tau)\|}{h} ds \\ &\quad + \int_{r-h}^r \sup_{\tau \in [-s,0]} \frac{\|x_n(s+h+\tau) - x_n(s+\tau)\|}{h} ds \\ &\leq \int_r^t \sup_{\tau \in [-s,0]} \frac{\|x_n(s+h+\tau) - x_n(s+\tau)\|}{h} ds \\ &\quad + \int_{r-h}^r \sup_{\tau \in [-s,0]} \frac{\|x_n(s+h+\tau) - x_n(s+\tau)\|}{h} ds\end{aligned}$$

we arrive at

$$\begin{aligned}
&\int_0^{r-h} \sup_{\tau\in[-s,0]} \frac{\|x_n(s+h+\tau)-x_n(s+\tau)\|}{h} ds \\
&\quad + \int_{r-h}^{t} \sup_{\tau\in[-r,0]} \frac{\|x_n(s+h+\tau)-x_n(s+\tau)\|}{h} ds \\
&\leq \int_0^{t} \sup_{\tau\in[-s,0]} \frac{\|x_n(s+h+\tau)-x_n(s+\tau)\|}{h} ds \\
&\quad + \int_{r-h}^{r} \sup_{\tau\in[-r,0]} \frac{\|x_n(s+h+\tau)-x_n(s+\tau)\|}{h} ds.
\end{aligned}$$

We now show that

$$\lim_{h\to 0+} \int_{r-h}^{r} \sup_{\tau\in[-r,0]} \frac{\|x_n(s+h+\tau)-x_n(s+\tau)\|}{h} ds = 0.$$

In fact, since $-s-h \leq -r$

$$\begin{aligned}
&\sup_{\tau\in[-r,0]} \frac{\|x_n(s+h+\tau)-x_n(s+\tau)\|}{h} \\
&\quad \leq \sup_{\tau\in[-s-h,0]} \frac{\|x_n(s+h+\tau)-x_n(s+\tau)\|}{h} \\
&\quad \leq \sup_{\tau\in[-s-h,-s]} \frac{\|x_n(s+h+\tau)-x_n(s+\tau)\|}{h} \\
&\qquad + \sup_{\tau\in[-s,0]} \frac{\|x_n(s+h+\tau)-x_n(s+\tau)\|}{h}.
\end{aligned} \tag{3.1.19}$$

From the proof of (2) of Lemma 3.1.1

$$\begin{aligned}
&\sup_{\tau\in[-(s+h),-s]} \frac{\|x_n(s+h+\tau)-x_n(s+\tau)\|}{h} \\
&\qquad \leq \left\|\frac{dx_n}{dt}(\mu h)\right\| + \left\|\frac{d\phi}{dt}(-\gamma h)\right\|,
\end{aligned}$$

where $\mu,\ \gamma \in (0,1)$. Thus

$$\begin{aligned}
&\lim_{h\to 0+} \int_{r-h}^{r} \sup_{\tau\in[-(s+h),-s]} \frac{\|x_n(s+h+\tau)-x_n(s+\tau)\|}{h} ds \\
&\quad \leq \lim_{h\to 0+} \int_{r-h}^{r} \left(\left\|\frac{dx_n}{dt}(\mu h)\right\| + \left\|\frac{d\phi}{dt}(-\gamma h)\right\|\right) ds \\
&\quad = \lim_{h\to 0+} \left(\left\|\frac{dx_n}{dt}(\mu h)\right\| + \left\|\frac{d\phi}{dt}(-\gamma h)\right\|\right) h \\
&\quad = 0.
\end{aligned} \tag{3.1.20}$$

There exist $\tau_1 \in [-s, 0]$ and $\lambda \in (0, 1)$ such that

$$\begin{aligned}\sup_{\tau\in[-s,0]} \frac{\|x_n(s+h+\tau) - x_n(s+\tau)\|}{h} &= \frac{\|x_n(s+h+\tau_1) - x_n(s+\tau_1)\|}{h} \\ &\le \left\|\frac{dx_n}{dt}(s+\tau_1+\lambda h)\right\| \\ &\le \sup_{s\in[0,T]} \left\|\frac{dx_n}{dt}(s)\right\| \\ &= N_2(n).\end{aligned}$$

Thus

$$\begin{aligned}\lim_{h\to 0+} \int_{r-h}^{r} \sup_{\tau\in[-r,0]} \frac{\|x_n(s+h+\tau) - x_n(s+\tau)\|}{h} ds \\ \le \lim_{h\to 0+} N_2(n)h = 0. \qquad (3.1.21)\end{aligned}$$

Hence by (3.1.19)–(3.1.21) we obtain

$$\lim_{h\to 0+} \int_{r-h}^{r} \sup_{\tau\in[-s,0]} \frac{\|x_n(s+h+\tau) - x_n(s+\tau)\|}{h} ds = 0. \qquad (3.1.22)$$

We now use (3.1.15)–(3.1.18), (3.1.22), and (1) of Lemma 3.1.1 to obtain

$$\begin{aligned}\overline{\lim_{h\to 0+}} \int_0^t \frac{\|(x_n)_{s+h} - (x_n)_s\|_C}{h} ds \\ \le N_1 T + \left(\left\|\frac{dx_n}{dt}(0)\right\| + \left\|\frac{d\phi}{dt}(0)\right\|\right)T \\ + \int_0^t \sup_{\tau\in[-s,0]} \left\|\frac{dx_n}{ds}(s+\tau)\right\| ds. \qquad (3.1.23)\end{aligned}$$

It follows from (3.1.10) and (3.1.23) that

$$\begin{aligned}&\overline{\lim_{h\to 0+}} \frac{\|x_n(t+h) - x_n(t)\|}{h} \\ &= \lim_{h\to 0+} \frac{\|x_n(t+h) - x_n(t)\|}{h} = \left\|\frac{dx_n}{dt}(t)\right\| \\ &\le (1+\beta T)\left\|\frac{dx_n}{dt}(0)\right\| + \left(C_1 + \beta\left(N_1 + \left\|\frac{d\phi}{dt}(0)\right\|\right)\right)T \\ &\quad + (C_2+\beta) \int_0^t \sup_{\tau\in[-s,0]} \left\|\frac{dx_n}{ds}(s+\tau)\right\| ds \qquad (3.1.24)\end{aligned}$$

for every $t \geq r$. Taking into consideration (3.1.15) we see that (3.1.24) holds for every $t \in [0, T]$. By (3.1.24) and

$$\left\|\frac{dx_n}{dt}(0)\right\| \leq \|A_n(0)x_n(0)\| + \|A(0)\phi(0)\| + K_1$$

we conclude that

$$\lim_{h\to 0+} \frac{\|x_n(t+h) - x_n(t)\|}{h} \leq C_3 + C_4 \int_0^t \sup_{\tau\in[-s,0]} \left\|\frac{dx_n}{ds}(s+\tau)\right\| ds$$

for every $\tau \leq 0$ with $t + \tau \in [0, T]$, where

$$C_3 = (1 + \beta T)(\|A(0)\phi(0)\| + K_1) + \Big(C_1 + \beta\Big(N_1 + \left\|\frac{d\phi}{dt}(0)\right\|\Big)\Big)T$$

and $C_4 = C_2 + \beta$. Hence from Gronwall's inequality

$$\sup_{\tau\in[-t,0]} \lim_{h\to 0+} \frac{\|x_n(t+h) - x_n(t)\|}{h} = \sup_{\tau\in[-t,0]} \left\|\frac{dx_n}{dt}(t+\tau)\right\| \leq C_3 e^{C_4 T} = N.$$

Therefore,

$$\left\|\frac{dx_n}{dt}(t)\right\| \leq \sup_{\tau\in[-t,0]} \left\|\frac{dx_n}{dt}(t+\tau)\right\| \leq N.$$

Now we state our main result.

THEOREM 3.1.1 *Let X^* be uniformly convex. Assume that $(A1)_1$, $(A2)_1$, $(G1)_1$, and $(G2)_1$ are satisfied. Let $\phi \in C^1([-r, 0], X)$ with $\phi(0) \in D$. Then there exists a unique strong solution $x(t)$ of (3.1.1) on $[-r, T]$ with the following properties:*

(1) the limit $x(t) = \lim_{n\to\infty} x_n(t)$ exists uniformly on $[0, T]$, where $x_n(t)$ is a unique (classical) solution of (3.1.2) on $[-r, T]$;

(2) $x(t) \in D$ for every $t \in [0, T]$ and $A(t)x(t)$ is bounded and weakly continuous on $[0, T]$;

(3) $G(t, x_t) - A(t)x(t)$ is Bochner integrable on $[0, T]$ and

$$x(t) = \phi(0) + \int_0^t (G(s, x_s) - A(s)x(s))ds$$

for every $t \in [0, T]$;

(4) *Let* $x(t)$, $y(t)$ *be a strong solution of* (3.1.1) *on* $[-r, T]$ *and let* $x(t) = \phi(t)$, $y(t) = \psi(t)$ *for every* $t \in [-r, 0]$ *and* ϕ, $\psi \in C^1([-r, 0]; X)$ *with* $\phi(0)$, $\psi(0) \in D$. *Then*

$$\|x(t) - y(t)\| \leq e^{\beta T}\|\phi - \psi\|_C$$

for every $[0, T]$.

Proof. (1) Since from (12) of Lemma 1.1.2, $A_n(t) = A(t)J_n(t)x_n(t)$ and the accretivity of $A(t)$ for every $t \in [0, T]$

$$\begin{aligned}
&\frac{1}{2}\frac{d}{dt}\|x_n(t) - x_m(t)\|^2 \\
&\quad = \Big(\frac{dx_n}{dt} - \frac{dx_m}{dt}, F(x_n(t) - x_m(t))\Big) \\
&\quad = -(A_n(t)x_n(t) - A_m(t)x_m(t), F(x_n(t) - x_m(t))) \\
&\qquad +(G(t, (x_n)_t) - G(t, (x_m)_t), F(x_n(t) - x_m(t))) \\
&\quad \leq (A_n(t)x_n(t) - A_m(t)x_m(t), \\
&\qquad\quad F(J_n(t)x_n(t) - J_n(t)x_n(t)) - F(x_n(t) - x_m(t))) \\
&\qquad +(G(t, (x_n)_t) - G(t, (x_m)_t), F(x_n(t) - x_m(t))) \\
&\quad \leq \|A_n(t)x_n(t) - A_m(t)x_m(t)\| \\
&\qquad\quad \times\|F(J_n(t)x_n(t) - J_n(t)x_n(t)) - F(x_n(t) - x_m(t))\| \\
&\qquad +\beta\|(x_n)_t - (x_m)_t\|_C\|x_n(t) - x_m(t)\|
\end{aligned}$$

a.e. $t \in [0, T)$, and from Lemma 3.1.1 and (3.1.2)

$$\begin{aligned}
\|A_n(t)x_n(t)\| &\leq \left\|\frac{dx_n}{dt}(t)\right\| + \|G(t, (x_n)_t)\| \\
&\leq \left\|\frac{dx_n}{dt}(t)\right\| + \beta\|(x_n)_t - \phi\|_C + \|G(t, \phi)\| \\
&\leq N + \beta K' + K_1 = K_0 \qquad (3.1.25)
\end{aligned}$$

$$\begin{aligned}
&\frac{d}{dt}\|x_n(t) - x_m(t)\|^2 \\
&\qquad \leq 2K_0\|F(J_n(t)x_n(t) - J_n(t)x_n(t)) - F(x_n(t) - x_m(t))\| \\
&\qquad\quad +\beta\|(x_n)_t - (x_m)_t\|_C^2.
\end{aligned}$$

Integrating this inequality over $[0, T]$ we have from the absolute continuity of $\|x_n(t) - x_m(t)\|^2$ and $x_n(0) - x_m(0) = 0$

$$
\begin{aligned}
&\|x_n(t) - x_m(t)\|^2 \\
&\quad \leq 4K_0 \int_0^t \|F(J_n(s)x_n(s) - J_n(s)x_n(s)) - F(x_n(s) - x_m(s))\| ds \\
&\quad +2\beta \int_0^t \|(x_n)_s - (x_m)_s\|_C^2 ds
\end{aligned}
$$

and thus

$$
\begin{aligned}
&\|(x_n)_t - (x_m)_t\|_C^2 \\
&\quad \leq 4K_0 \int_0^T \|F(J_n(s)x_n(s) - J_n(s)x_n(s)) - F(x_n(s) - x_m(s))\| ds \\
&\quad +2\beta \int_0^t \|(x_n)_s - (x_m)_s\|_C^2 ds,
\end{aligned}
$$

from which we conclude by Gronwall's inequality that

$$
\begin{aligned}
&\|(x_n)_t - (x_m)_t\|_C^2 \\
&\quad \leq 4K_0 e^{2\beta T} \int_0^T \|F(J_n(s)x_n(s) - J_n(s)x_n(s)) \\
&\qquad\qquad -F(x_n(s) - x_m(s))\| ds. \qquad (3.1.26)
\end{aligned}
$$

On the other hand, by (3.1.25)

$$
\begin{aligned}
&\|(J_n(s)x_n(s) - J_n(s)x_n(s)) - (x_n(s) - x_m(s))\| \\
&\quad \leq \|J_n(s)x_n(s) - x_n(s)\| + \|J_m(s)x_m(s) - x_m(s)\| \\
&\quad \leq \frac{1}{n}\|A_n(t)x_n(t)\| + \frac{1}{m}\|A_m(t)x_m(t)\| \\
&\quad \leq \left(\frac{1}{n} + \frac{1}{m}\right) K_0.
\end{aligned}
$$

Thus

$$\lim_{n,m\to\infty} \|(J_n(s)x_n(s) - J_n(s)x_n(s)) - (x_n(s) - x_m(s))\| = 0.$$

From the uniform continuity of F on bounded subsets of X

$$\lim_{n,m\to\infty} \|F(J_n(s)x_n(s) - J_n(s)x_n(s)) - F(x_n(s) - x_m(s))\| = 0$$

uniformly on $[0, T]$. Thus from (3.1.26)

$$\lim_{n,m\to\infty} \|(x_n)_t - (x_m)_t\|_C^2 = 0$$

a.e. $t \in (0, T)$.

From $\lim_{n\to\infty}(x_n)_t = x_t$ for every $t \in [0, T]$, we also conclude that $x(t) = \phi(t)$ for every $t \in [-r, 0]$. Hence, $x(t)$ is a strong solution of (3.1.1) on $[-r, T]$.

(4) Let $x(t)$, $y(t)$ be two strong solution of (3.1.1) on $[-r, T]$. Then $\|x(t) - y(t)\|$ is Lipschitz continuous on $[0, T]$, and thus $\|x(t) - y(t)\|$ is differentiable a.e. $t \in (0, T)$. From $(A1)_1$, $(A2)_1$, $(G2)_1$, and (12) of Lemma 1.1.2

$$\begin{aligned}
&\frac{1}{2}\frac{d}{dt}\|x(t) - y(t)\|^2 \\
&\quad = \Big(\frac{dx}{dt} - \frac{dy}{dt}, F(x(t) - y(t))\Big) \\
&\quad = -(A(t)x(t) - A(t)y(t), F(x(t) - y(t))) \\
&\qquad +(G(t, x_t) - G(t, y_t), F(x(t) - y(t))) \\
&\quad \le \|G(t, x_t) - G(t, y_t)\|\|F(x(t) - y(t))\| \\
&\quad \le \beta\|x_t - y_t\|_C^2.
\end{aligned}$$

Integrating this inequality on $[0, t]$ we obtain

$$\|x(t) - y(t)\|^2 \le \|\phi(0) - \psi(0)\|^2 + 2\beta \int_0^t \|x_s - y_s\|_C^2 ds$$

for every $t \in [0, T]$. Thus we have

$$\|x_t - y_t\|_C^2 \le \|\phi - \psi\|_C^2 + 2\beta \int_0^t \|x_s - y_s\|_C^2 ds$$

for every $t \in [0, T]$. By Gronwall's inequality

$$\|x_t - y_t\|_C^2 \le e^{2\beta T}\|\phi - \psi\|_C^2$$

for every $t \in [0, T]$. Hence

$$\|x(t) - y(t)\| \le \|x_t - y_t\|_C \le e^{\beta T}\|\phi - \psi\|_C$$

for every $t \in [0, T]$.

3.2. Evolution Operator Methods

In this section we consider non-autonomous nonlinear functional evolutions of the type

$$\begin{cases} \dfrac{dx}{dt}(t) + A(t)x(t) = G(t, x_t), & s \le t \le T \\ x_s(t) = \phi(t), & -r \le t \le 0 \end{cases} \tag{3.2.1}$$

associated with an operator $A(t) : D(A(t)) \subset X \to X$ for every $t \in [0, T]$ in a real Banach space X, where $T > 0$, $r > 0$, $s \in [0, T]$, $x(t) : [-r, T] \to X$ is an unknown function, $G : [0, T] \times C([-r, 0]; X) \to X$ and $\phi \in C([-r, 0]; X)$.

DEFINITION 3.2.1 *A function $x(t) : [-r, T] \to X$ is called a (classical) solution of* (3.2.1) *on* $[-r, T]$ *if:*

(1) $x_s(t) = \phi(t)$ *for every* $t \in [-r, 0]$;

(2) $x(t)$ *is continuously differentable on* $[s, T]$;

(3) $x(t)$ *satisfies*

$$\frac{dx}{dt}(t) + A(t)x(t) = G(t, x_t)$$

for every $t \in [s, T]$.

DEFINITION 3.2.2 *A function $x(t) : [-r, T] \to X$ is called a strong solution of* (3.2.1) *on* $[-r, T]$ *if:*

(1) $x_s(t) = \phi(t)$ *for every* $t \in [-r, 0]$;

(2) $x(t)$ *is continuous on* $[s, T]$

(3) $x(t)$ *is absolutely continuous on every compact interval of* $[s, T]$;

(4) $x(t)$ *is differentiable a.e. on* (s, T);

(5) $x(t)$ *satisfies*

$$\frac{dx}{dt}(t) + A(t)x(t) = G(t, x_t)$$

a.e. on (s, T).

We give the following conditions for existence of solutions of (3.2.1).

$(A1)_2$ For every $t \in [0, T]$ $A(t) : D_A \equiv D(A(t)) \subset X \to X$ is an operator such that $A(t) + \alpha I$ is accretive for some $\alpha \in \mathbf{R}$, and $R(I + \lambda A(t)) = X$ for every $\lambda > 0$ with $\lambda\alpha < 1$;

$(A2)_2$ There exists a non-decreasing function $L : [0, \infty) \to [0, \infty)$ such that

$$\|A(t_1)u - A(t_2)u\| \leq |t_1 - t_2| L(\|u\|)(1 + \|a(t_2)u\|)$$

for every $t_i \in [0, \infty)$, $i = 1, 2$, and $u \in D_A$;

$(G1)_2$ There exists a constant $\beta > 0$ such that

$$\|G(t, \psi_1) - (t, \psi_2)\| \leq \beta \|\psi_1 - \psi_2\|_C$$

for every $t \in [0, T]$ and $\psi_i \in C([-r, 0]; X)$, $i = 1, 2$;

$(G2)_2$ There exists a non-decreasing function $l : [0, \infty) \to [0, \infty)$ such that

$$\|G(t_1, \psi) - G(t_2, \psi)\| \leq |t_1 - t_2| l(\|\psi\|_C)$$

for every $t_i \in [0, T]$, $i = 1, 2$, and $\psi \in C([-r, 0]; X)$.

Now we define an operator

$$B(t) : D(B(t)) \subset C([-r, 0]; X) \to C([-r, 0]; X)$$

by

$$\begin{aligned} D(B(t)) &= \Big\{\psi \in C([-r,0];X) \ \Big| \ \frac{d\psi}{dt} \in C([-r,0];X), \ \psi(0) \in D(A(t)), \\ &\qquad \frac{d\psi}{dt}(0) + A(t)\psi(0) = G(t,\psi)\Big\}, \\ B(t)\psi &= -\frac{d\psi}{dt}, \quad \psi \in D(B(t)) \end{aligned} \tag{3.2.2}$$

for every $t \in [0, T]$.

Since $D_A \equiv D(A(t))$ for every $t \in [0, T]$ we put $D_B \equiv D(B(t))$ for every $t \in [0, T]$.

We put

$$J_\lambda^A(t) = (I + \lambda A(t))^{-1}, \quad J_\lambda^B(t) = (I + \lambda B(t))^{-1}$$

and

$$A_\lambda(t) = \frac{1}{\lambda}(I - J_\lambda^A(t)), \quad B_\lambda(t) = \frac{1}{\lambda}(I - J_\lambda^B(t))$$

for every $t \in [0, T]$ and $\lambda > 0$.

PROPOSITION 3.2.1 *Assume that* $(A1)_2$, $(A2)_2$, $(G1)_2$, *and* $(G2)_2$ *are satisfied. Let* $B(t)$ *be the operator defined as in* (3.2.2) *for every* $t \in [0, T]$. *Let* $\gamma = \max\{0, \alpha + \beta\}$. *Then:*

(1) $B(t) + \gamma I$ *is accretive in* $C([-r, 0]; X)$ *for every* $t \in [0, T]$;

(2) $R(I + \lambda B(t)) = C([-r, 0]; X)$ *for every* $t \in [0, T]$ *and* $\lambda > 0$ *with* $\lambda\gamma < 1$;

(3) There exists a non-decreasing function $L_0 : [0, \infty) \to [0, \infty)$ *such that*

$$\|B_\lambda(t_1)\psi - B_\lambda(t_2)\psi\| \leq |t_1 - t_2| L_0(\|\psi\|_C)(1 + \|B_\lambda(t_2)\psi\|)$$

for every $\psi \in C([-r, 0]; X)$ *and* $t_i \in [0, T]$, $i = 1, 2$.

Proof. We prove (1) and (2) as in the proof of (1) and (2) of Proposition 2.2.1 under $(A1)_2$ and $(G1)_2$.

(2) Let $\phi \in C([-r,0];X)$. We define an operator $Q : X \to X$ by

$$Qx = J_\lambda^A(t)\Big(\phi(0) + \lambda G\Big(e^{\frac{\tau}{\lambda}}x + \frac{1}{\lambda}\int_\tau^0 e^{-\frac{\tau-\sigma}{\lambda}}\phi(\sigma)d\sigma\Big)\Big) \tag{3.2.3}$$

for every $x \in X$, $t \in [0,T]$ and $\tau \in [-r,0]$. Then

$$\begin{aligned}
&\|Qx_1 - Qx_2\| \\
&\le \frac{\lambda}{1-\lambda\alpha}\Big\|G\Big(e^{\frac{t}{\lambda}}x_1 + \frac{1}{\lambda}\int_\tau^0 e^{-\frac{\tau-\sigma}{\lambda}}\phi(\sigma)d\sigma\Big) \\
&\qquad - G\Big(e^{\frac{t}{\lambda}}x_2 + \frac{1}{\lambda}\int_\tau^0 e^{-\frac{\tau-\sigma}{\lambda}}\phi(\sigma)d\sigma\Big)\Big\| \\
&\le \frac{\lambda\beta}{1-\lambda\alpha}\|e^{\frac{\tau}{\lambda}}x_1 - e^{\frac{\tau}{\lambda}}x_2\|_C \\
&= \frac{\lambda\beta}{1-\lambda\alpha}\sup_{\tau\in[-r,0]} e^{\frac{\tau}{\lambda}}\|x_1 - x_2\| \\
&\le \frac{\lambda\beta}{1-\lambda\alpha}\|x_1 - x_2\|
\end{aligned}$$

for every $x_i \in X$, $i = 1,2$ and $\tau \in [-r,0]$. Since $0 < \dfrac{\lambda\beta}{1-\lambda\alpha} < 1$ Q is a strict contraction. From Banach's contraction principle there exists $x_0 \in X$ such that $Qx_0 = x_0$. From (3.2.3)

$$x_0 = J_\lambda^A(t)\Big(\phi(0) + \lambda G\Big(e^{\frac{\tau}{\lambda}}x_0 + \frac{1}{\lambda}\int_\tau^0 e^{-\frac{\tau-\sigma}{\lambda}}\phi(\sigma)d\sigma\Big)\Big) \tag{3.2.4}$$

for every $\tau \in [-r,0]$ and $t \in [0,T]$. Thus $x_0 \in D_A$. We put

$$\psi(\tau) = e^{\frac{\tau}{\lambda}}x_0 + \frac{1}{\lambda}\int_\tau^0 e^{-\frac{\tau-\sigma}{\lambda}}\phi(\sigma)d\sigma \tag{3.2.5}$$

for every $\tau \in [-r,0]$. Then $\psi \in C([-r,0];X)$. By (3.2.5) $\psi(0) = x_0$ and $\dfrac{d\psi}{d\tau} \in C([-r,0];X)$. Using (3.2.4)

$$\begin{aligned}
\psi(0) &= J_\lambda^A(t)(\phi(0) + \lambda G(t,\psi)), &(3.2.6)\\
\frac{1}{\lambda}(\psi(0) - \phi(0)) &= G(t,\psi) - A(t)\psi(0) &(3.2.7)
\end{aligned}$$

for every $t \in [0,T]$. But then since from (3.2.5) for every $\tau \in [-r,0]$

$$\begin{aligned}
&\frac{d\psi}{d\tau}(\tau) = \frac{1}{\lambda}(\psi(\tau) - \phi(\tau)),\\
&\frac{d\psi}{d\tau}(0) = \frac{1}{\lambda}(\psi(0) - \phi(0))
\end{aligned} \tag{3.2.8}$$

it follows from (3.2.7) and (3.2.8) that we have

$$\frac{d\psi}{d\tau}(0) = G(t,\psi) - A(t)\psi(0)$$

for every $t \in [0,T]$. Thus $\psi \in D_B$. From

$$\psi(\tau) - \lambda\frac{d\psi}{d\tau}(\tau) = \phi(\tau)$$

for every $\tau \in [-r,0]$

$$(I + \lambda B(t))\psi = \psi - \lambda\frac{d\psi}{d\tau} = \phi$$

and then

$$R(I + \lambda B(t)) = C([-r,0];X)$$

for every $t \in [0,T]$ and $\lambda > 0$ with $\lambda\gamma < 1$.

(1) Let $\phi_i \in C([-r,0];X)$, $i = 1,2$. From (2) there exists $\psi_i \in D_B$ such that

$$R(I + \lambda B(t))\psi_i(t,\cdot) = \phi_i$$

for $i = 1,2$ and for every $t \in [0,T]$. Thus $(G1)_2$, (3.2.5) and (3.2.6) imply

$$\begin{aligned}
&\|\psi_1(t,\tau) - \psi_2(t,\tau)\| \\
&\quad\le e^{\frac{\tau}{\lambda}}\|\psi_1(t,0) - \psi_2(t,0)\| + \frac{1}{\lambda}\int_\tau^0 e^{\frac{\tau-\sigma}{\lambda}}\|\phi_1(\sigma) - \phi_2(\sigma)\|d\sigma \\
&\quad\le \frac{e^{\frac{\tau}{\lambda}}}{1-\lambda\alpha}(\|\phi_1(0) - \phi_2(0)\| + \lambda\beta\|\psi_1(t,\cdot) - \psi_2(t,\cdot)\|_C) \\
&\qquad+(1 - e^{\frac{\tau}{\lambda}})\|\phi_1 - \phi_2\|_C \\
&\quad\le \frac{1-\lambda\alpha+\lambda\alpha e^{\frac{\tau}{\lambda}}}{1-\lambda\alpha}\|\phi_1 - \phi_2\|_C + \frac{\lambda\beta e^{-\frac{\tau}{\lambda}}}{1-\lambda\alpha}\|\psi_1(t,\cdot) - \psi_2(t,\cdot)\|_C.
\end{aligned}$$

for every $t \in [0,T]$ and $\tau \in [-r,0]$. Thus

$$\|\psi_1(t,\cdot) - \psi_2(t,\cdot)\|_C \le \frac{1-\lambda\alpha+\lambda\alpha e^{\frac{\tau}{\lambda}}}{1-\lambda\alpha-\lambda\beta e^{\frac{\tau}{\lambda}}}\|\phi_1 - \phi_2\|_C \tag{3.2.9}$$

for every $t \in [0,T]$ and $\tau \in [-r,0]$.

If $\alpha + \beta \ge 0$

$$\frac{1-\lambda\alpha+\lambda\alpha e^{\frac{\tau}{\lambda}}}{1-\lambda\alpha-\lambda\beta e^{\frac{\tau}{\lambda}}} \le \frac{1}{1-\lambda(\alpha+\beta)} = \frac{1}{1-\lambda\gamma}$$

and if $\alpha + \beta \geq 0$

$$\frac{1 - \lambda\alpha + \lambda\alpha e^{\frac{\tau}{\lambda}}}{1 - \lambda\alpha - \lambda\beta e^{\frac{\tau}{\lambda}}} \leq 1 < \frac{1}{1 - \lambda\gamma}.$$

From (3.2.9)

$$\|\psi_1(t,\cdot) - \psi_2(t,\cdot)\|_C \leq \frac{1}{1 - \lambda\gamma}\|\phi_1 - \phi_2\|_C. \tag{3.2.10}$$

Thus $(I + \lambda B(t))^{-1}$ exists for every $t \in [0,T]$ and $\lambda > 0$ with $\lambda\gamma < 1$, and by (3.2.10)

$$\|(I + \lambda B(t))^{-1}\phi_1 - (I + \lambda B(t))^{-1}\phi_2)\|_C \leq \frac{1}{1 - \lambda\gamma}\|\phi_1 - \phi_2\|_C$$

for every $t \in [0,T]$ and $\lambda > 0$ with $\lambda\gamma < 1$. Hence $B(t) + \gamma I$ is accretive in $C([-r,0];X)$ for every $t \in [0,T]$ and $\lambda > 0$ with $\lambda\gamma < 1$.

(3) Let $t_i \in [0,T]$, $i = 1,2$, $\phi \in C([-r,0];X)$ and $\lambda > 0$.

We put $\psi(t,\cdot) = J_\lambda^B(t)\phi$. Then $\psi(t,\cdot) \in D_B$ and

$$\psi(t,\cdot) + \lambda B(t)\psi(t,\cdot) = \phi.$$

From (3.2.2)

$$\psi(t,\tau) - \lambda\frac{d\psi}{d\tau}(t,\tau) = \phi(\tau)$$

for every $\tau \in [-r,0]$. Solving this linear evolution we have

$$\psi(t,\tau) = e^{\frac{\tau}{\lambda}}\psi(t,0) + \frac{1}{\lambda}\int_\tau^0 e^{-\frac{\tau-\sigma}{\lambda}}\phi(\sigma)d\sigma \tag{3.2.11}$$

for every $\tau \in [-r,0]$. Using $\psi(t,\cdot) \in D_B$

$$\begin{aligned}\psi(t,0) &= \phi(0) + \lambda\frac{d\psi}{d\tau}(t,0)\\ &= \phi(0) + \lambda G(t,\psi(t,\cdot)) - \lambda A(t)\psi(t,0),\end{aligned}$$

and thus

$$\psi(t,0) = J_\lambda^A(t)(\phi(0) + \lambda G(t,\psi(t,\cdot))). \tag{3.2.12}$$

By (3.2.11), (3.2.12), $(A2)_2$ and Proposition 1.4.3 for every $\lambda \in (0,1)$ with $\lambda\alpha < \frac{1}{2}$

$$\begin{aligned}&\|\psi(t_1,\cdot) - \psi(t_2,\cdot)\|_C\\ &\qquad = \|\psi(t_1,0) - \psi(t_2,0)\| =\end{aligned}$$

$$\begin{aligned}
&= \|J_\lambda^A(t_1)(\phi(0) + \lambda G(t_1, \psi(t_2, \cdot))) \\
&\quad - J_\lambda^A(t_2)(\phi(0) + \lambda G(t_2, \psi(t_2, \cdot)))\| \\
&\le \|J_\lambda^A(t_1)(\phi(0) + \lambda G(t_1, \psi(t_2, \cdot))) \\
&\quad - J_\lambda^A(t_1)(\phi(0) + \lambda G(t_2, \psi(t_2, \cdot)))\| \\
&\quad + \|J_\lambda^A(t_1)(\phi(0) + \lambda G(t_2, \psi(t_2, \cdot))) \\
&\quad - J_\lambda^A(t_2)(\phi(0) + \lambda G(t_2, \psi(t_2, \cdot)))\| \\
&\le \|J_\lambda^A(t_1)(\phi(0) + \lambda G(t_1, \psi(t_2, \cdot))) \\
&\quad - J_\lambda^A(t_1)(\phi(0) + \lambda G(t_2, \psi(t_2, \cdot)))\| \\
&\quad + \lambda\|A_\lambda(t_1)(\phi(0) + \lambda G(t_2, \psi(t_2, \cdot))) \\
&\quad - A_\lambda(t_2)(\phi(0) + \lambda G(t_2, \psi(t_2, \cdot)))\| \\
&\le \frac{\lambda}{1-\lambda\alpha}\|G(t_1, \psi(t_1, \cdot)) - G(t_2, \psi(t_2, \cdot))\| \\
&\quad + \lambda|t_1 - t_2|L_1(\|\phi(0) + \lambda G(t_2, \psi(t_2, \cdot))\|) \\
&\quad \times(1 + \|A_\lambda(t_2)(\phi(0) + \lambda G(t_2, \psi(t_2, \cdot)))\|),
\end{aligned}$$

where $L_1 : [0, \infty) \to [0, \infty)$ is a non-decreasing function. Since from $(G1)_2$ and $(G2)_2$

$$\begin{aligned}
&\|G(t_1, \psi(t_1, \cdot)) - G(t_2, \psi(t_2, \cdot))\| \\
&\quad \le \|G(t_1, \psi(t_1, \cdot)) - G(t_1, \psi(t_2, \cdot))\| \\
&\qquad + \|G(t_1, \psi(t_2, \cdot)) - G(t_2, \psi(t_2, \cdot))\| \\
&\quad \le \beta\|\psi(t_2, \cdot) - \psi(t_2, \cdot)\|_C + |t_1 - t_2|l(\|\psi(t_2, \cdot)\|_C)
\end{aligned}$$

and

$$\begin{aligned}
&\|A_\lambda(t_2)(\phi(0) + \lambda G(t_2, \psi(t_2, \cdot)))\| \\
&\quad = \frac{1}{\lambda}\|\phi(0) + \lambda G(t_2, \psi(t_2, \cdot)) - J_\lambda^A(t_2)(\phi(0) + \lambda G(t_2, \psi(t_2, \cdot)))\| \\
&\quad \le \frac{1}{\lambda}\|\phi(0) - \psi(t_2, \cdot)\| + \|G(t_2, \psi(t_2, \cdot))\| \\
&\quad = \|B_\lambda(t_2)\phi\|_C + \|G(t_2, \psi(t_2, \cdot))\|
\end{aligned}$$

we have using (3.2.13)

$$\begin{aligned}
&\|\psi(t_1, \cdot) - \psi(t_2, \cdot)\|_C \\
&\qquad \le \frac{\lambda}{1-\lambda}(\beta\|\psi(t_1, \cdot) - \psi(t_2, \cdot)\|_C + |t_1 - t_2|l(\|\psi(t_2, \cdot)\|_C)) \\
&\qquad\quad + \lambda|t_1 - t_2|L_1(\|\phi(0)\| + \lambda\|G(t_2, \psi(t_2, \cdot))\|) \\
&\qquad\quad \times(1 + \|B_\lambda(t_2)\phi\|_C + \|G(t_2, \psi(t_2, \cdot))\|),
\end{aligned}$$

from which

$$\begin{aligned}
&\|\psi(t_1,\cdot)-\psi(t_2,\cdot)\|_C \\
&\quad\le \frac{\lambda}{1-\lambda(\alpha+\beta)}|t_1-t_2|l(\|\psi(t_2,\cdot)\|_C) \\
&\quad\quad+\frac{\lambda(1-\lambda\beta)}{1-\lambda(\alpha+\beta)}|t_1-t_2|L_1(\|\phi(0)\|+\lambda\|G(t_2,\psi(t_2,\cdot))\|) \\
&\quad\quad\quad\times(1+\|B_\lambda(t_2)\phi\|_C+\|G(t_2,\psi(t_2,\cdot))\|) \\
&\quad\le \lambda\kappa|t_1-t_2|(l(\|\psi(t_2,\cdot)\|_C)+\|G(t_2,\psi(t_2,\cdot))\|) \\
&\quad\quad\quad\times(1+\|B_\lambda(t_2)\phi\|_C+\|G(t_2,\psi(t_2,\cdot))\|)) \\
&\quad\le \lambda\kappa|t_1-t_2|(l(\|\psi(t_2,\cdot)\|_C)+(1+\|G(t_2,\psi(t_2,\cdot))\|) \\
&\quad\quad\quad\times L_1(\|\phi(0)\|+\lambda\|G(t_2,\psi(t_2,\cdot))\|)) \\
&\quad\quad\quad\quad\times(1+\|B_\lambda(t_2)\phi\|_C),
\end{aligned} \tag{3.2.13}$$

where

$$\kappa=\max\left\{\frac{\lambda}{1-\lambda(\alpha+\beta)},\frac{\lambda(1-\lambda\beta)}{1-\lambda(\alpha+\beta)}\right\}.$$

Let $\phi_0\in C([-r,0];X)$. Since, by $(G1)_2$ and $(G2)_2$,

$$\begin{aligned}
&\|G(t_2,\psi(t_2,\cdot))\| \\
&\quad\le \|G(t_2,\psi(t_2,\cdot))-G(t_2,\phi_0)\| \\
&\quad\quad+\|G(t_2,\phi_0)-G(0,\phi_0)\|+\|G(0,\phi_0)\| \\
&\quad\le \beta\|\psi(t_2,\cdot)-\phi_0\|_C+t_2l(\|\phi_0\|_C)+\|G(0,\phi_0)\| \\
&\quad\le \beta(\|\psi(t_2,\cdot)\|_C-\|\phi_0\|_C)+Tl(\|\phi_0\|_C)+\|G(0,\phi_0)\| \\
&\quad\equiv L_2(\|\psi(t_2,\cdot)\|_C)
\end{aligned}$$

for some non-decreasing function $L_2: l0,\infty,)\to[0,\infty)$, (3.2.14) implies

$$\begin{aligned}
&\|\psi(t_1,\cdot)-\psi(t_2,\cdot)\|_C \\
&\quad\le \lambda\kappa|t_1-t_2|(l(\|\psi(t_2,\cdot)\|_C)+(1+L_2(\|\psi(t_2,\cdot)\|_C) \\
&\quad\quad\times L_1(\|\phi(0)\|+\lambda L_2(\|\psi(t_2,\cdot)\|_C)) \\
&\quad\quad\quad\times(1+\|B_\lambda(t_2)\phi\|_C),
\end{aligned} \tag{3.2.14}$$

from which we have

$$\begin{aligned}
&\|\psi(t_2,\cdot)\|_C \\
&\quad\le \|\psi(0,\cdot)\|_C+\lambda\kappa t_2(l(\|\psi(0,\cdot)\|_C)+(1+L_2(\|\psi(0,\cdot)\|_C) \\
&\quad\quad\times L_1(\|\phi(0)\|+\lambda L_2(\|\psi(0,\cdot)\|_C))(1+\|B_\lambda(0)\phi\|_C) \\
&\quad\le \|\psi(0,\cdot)\|_C+\kappa T(l(\|\psi(0,\cdot)\|_C)+(1+L_2(\|\psi(0,\cdot)\|_C) \\
&\quad\quad\times L_1(\|\phi(0)\|+L_2(\|\psi(0,\cdot)\|_C))(1+\lambda\|B_\lambda(0)\phi\|_C).
\end{aligned}$$

as $n \to \infty$ for every $\tau \in [-r, -(t-s))$ and $\phi \in \widehat{D}_B$. Indeed, we note that

$$\left(J^{B_1}_{\frac{t-s}{n}}\phi\right)(\tau) = \frac{n}{t-s}e^{\frac{n\tau}{t-s}}\int_\tau^0 e^{\frac{ns}{t-s}}\phi(s)ds \tag{3.2.22}$$

and

$$\left(J^{B}_{\frac{t-s}{n}}(t)\phi\right)(\tau) = e^{\frac{n\tau}{t-s}}\left(J^{B}_{\frac{t-s}{n}}(t)\phi\right)(0) + \left(J^{B_1}_{\frac{t-s}{n}}\phi\right)(\tau) \tag{3.2.23}$$

for every $\phi \in C([-r,0];X)$ and $\tau \in [-r,0]$. Using (3.2.24) we can prove, by induction on m, that

$$\begin{aligned}
&\Big(\prod_{i=1}^{m} J^{B}_{\frac{t-s}{n}}\Big(s+i\frac{t-s}{n}\Big)\phi\Big)(\tau) \\
&\quad = \sum_{j=0}^{m-1}\Big(\Big(J^{B_1}_{\frac{n\tau}{t-s}}\Big)^j\Big(e^{\frac{ns}{t-s}}\prod_{i=1}^{m-j} J^{B}_{\frac{t-s}{n}}\Big(s+i\frac{t-s}{n}\Big)\phi\Big)\Big)(0) \\
&\qquad +\Big(\Big(J^{B_1}_{\frac{t-s}{n}}\Big)^m\phi\Big)(\tau)
\end{aligned}$$

for every $\tau \in [-r,0]$.

Let $\phi \in \widehat{D}_B$. It follows from (4) of Proposition 1.1.4

$$\begin{aligned}
&\Big\|\Big(\prod_{i=1}^{m} J^{B}_{\frac{t-s}{n}}\Big(s+i\frac{t-s}{n}\Big)\phi\Big)(0) - \phi(0)\Big\| \\
&\quad \le \Big\|\prod_{i=1}^{m} J^{B}_{\frac{t-s}{n}}\Big(s+i\frac{t-s}{n}\Big)\phi - \phi\Big\|_C \\
&\quad \le c_1(t)\frac{(m+1)n}{t-s}\sup_{1\le i\le m+1}\Big\|B_{\frac{t-s}{n}}\Big(s+i\frac{t-s}{n}\Big)\phi\Big\|_C \\
&\quad \le c_2(t) < \infty
\end{aligned}$$

for every $m \le n-1$. Hence

$$\Big\|\Big(\prod_{i=1}^{m} J^{B}_{\frac{t-s}{n}}\Big(s+i\frac{t-s}{n}\Big)\phi\Big)(0)\Big\|$$

is bounded for every $(t,s) \in \Delta$, where $m \le n-1$. It follows from (3.2.23) that

$$\left(J^{B_1}_{\frac{t-s}{n}}e^{\frac{ns}{t-s}}\right)(\tau) = \frac{n^k}{k!}\Big(\frac{-\tau}{t-s}\Big)^k e^{\frac{n\tau}{t-s}}$$

and thus

$$\Big\|\Big(\prod_{i=1}^{n} J^{B}_{\frac{t-s}{n}}\Big(s+i\frac{t-s}{n}\Big)\phi\Big)(\tau)-\Big(J^{B_1}_{\frac{t-s}{n}}\phi\Big)(\tau)\Big\|$$
$$\leq (c_2(t)+\|\phi(0)\|)p_n\Big(\frac{-\tau}{t-s}\Big),$$

where $p_n(c) = \sum_{j=0}^{n-1}\frac{1}{j!}(cn)^j e^{cn}$ for every $c \in \mathbf{R}$. Since $\frac{-\tau}{t-s} > 1$, $p_n\Big(\frac{-\tau}{t-s}\Big) \to 0$ as $n \to \infty$. Hence (3.2.22) holds.

From (3.2.22) we have

$$(U(t,s)\phi)(\tau) = \phi(t-s+\tau) \tag{3.2.24}$$

for every $\tau \in [-r, -(t-s)]$ and $\phi \in \widehat{D}_B$, where (3.2.25) also holds at $\tau = -(t-s)$, by continuity.

Now, we show that (3.2.21) holds. Let $t \geq s$ and let $\tau \in [-r, 0]$. Then from (3.2.25)

$$x_t(s;\phi)(\tau) = x(s:\phi)(t+\tau) = \phi(t+\tau-s) = (U(t,s)\phi)(\tau)$$

for $\tau \leq -(t-s)$, and from (3.2.20)

$$x_t(s;\phi)(\tau) = x(s:\phi)(t+\tau) = (U(t+\tau,s)\phi)(0) = (U(t,s)\phi)(\tau)$$

for $\tau \geq -(t-s)$, and thus

$$x_t(s;\phi) = U(t,s)\phi$$

for every $t \geq s$.

We now two existence theorems of solutions for (3.2.1).

THEOREM 3.2.1 *Assume that* $(A1)_2$, $(A2)_2$, $(G1)_2$, *and* $(G2)_2$ *are satisfied. Let* $B(t)$ *be the operator as in* (3.2.2). *Suppose that* $A(t): X \to X$ *is continuous for every* $t \geq 0$. *let* $x(t) \equiv x(t;\phi)$ *be defined as in* (3.2.20). *Then* $x(t)$ *is a unique strong solution of* (3.2.1) *on* $[-r, T]$ *for every* $\phi \in \widehat{D}_B$.

Proof. Let $B_n(t) = B_{\frac{1}{n}}(t)$ and $J^{B_n}_n(t) = \Big(I + \frac{1}{n}B_n(t)\Big)^{-1}$ for every $t \in [0, T]$ for $n = 1, 2, \cdots$. Since

$$\lim_{n\to\infty}(I+\lambda B_n(t))^{-1}\phi = (I+\lambda B(t))^{-1}\phi$$

for sufficiently small $\lambda > 0$ and for every $\phi \in \overline{D}_B$, from Theorem 1.5.3, there exists a unique (classical) solution $x_n(t) = U_n(t,s)\phi$ of

$$\begin{cases} \dfrac{dx_n}{dt}(t) + B_n(t)x_n(t) = 0, & s \le t \le T \\ (x_n)_s(t) = \phi(t), & -r \le t \le 0 \end{cases} \tag{3.2.25}$$

and

$$\lim_{n\to\infty} U_n(t,s)\phi = U(t,s)\phi \tag{3.2.26}$$

uniformly on $[s,T]$ for every $\phi \in \overline{D}_B$, where $\{U_n(t,s) \mid (t,s) \in \Delta\}$ is the evolution operator generatedy by $B_n(t)$. Thus from (3.2.26)

$$\begin{aligned} U_n(t,s)\phi &= \phi - \int_s^t B_n(\sigma)U_n(\sigma,s)\phi d\sigma \\ &= \phi - \int_s^t B(\sigma)J_n^{B_n}(\sigma)U_n(\sigma,s)\phi d\sigma. \end{aligned}$$

So, taking values at $\tau = 0$ and from the definition of $D(B(\sigma))$

$$\begin{aligned} &(U_n(t,s)\phi)(0) \\ &\quad = \phi(0) + \int_s^t (G(\sigma,(J_n^{B_n}(\sigma)U_n(\sigma,s)\phi) \\ &\qquad\qquad -A(\sigma)(J_n^{B_n}(\sigma)U_n(\sigma,s)\phi)(0))d\sigma. \end{aligned} \tag{3.2.27}$$

Let $\phi \in \widehat{D}_B$. Let $s \le \sigma \le T$. Then from

$$\begin{aligned} &\|J_n^{B_n}(\sigma)U_n(\sigma,s)\phi - U(\sigma,s)\phi\|_C \\ &\quad \le \frac{1}{1-\lambda\gamma}\|_C U_n(\sigma,s)\phi - U(\sigma,s)\phi\| \\ &\qquad +\|J_n^{B_n}(\sigma)U(\sigma,s)\phi - U(\sigma,s)\phi\|_C, \end{aligned}$$

(3.2.27), (3) of Proposition 1.1.4, and (2) of Theorem 1.5.2 we have

$$\lim_{n\to\infty} J_n^{B_n}(\sigma)U_n(\sigma,s)\phi = U(\sigma,s)\phi.$$

Since $A(\sigma)$ and $G(\sigma,\cdot)$ are continuous

$$\begin{aligned} &\lim\nolimits_{n\to\infty}(G(\sigma,J_n^{B_n}(\sigma)U_n(\sigma,s)\phi) - A(\sigma)J_n^B(\sigma)(U_n(\sigma,s)\phi)(0)) \\ &\quad = G(\sigma,U(\sigma,s)\phi) - A(\sigma)(U(\sigma,s)\phi)(0), \end{aligned}$$

and since

$$\begin{aligned}&\|G(\sigma, J_n^{B_n}(\sigma)U_n(\sigma,s)\phi) - A(\sigma)J_n^{B_n}(\sigma)(U_n(\sigma,s)\phi)(0)\|\\&\quad\leq \|B(\sigma)J_n^{B_n}(\sigma)U_n(\sigma,s)\phi\|_C = \|B_n(\sigma)U_n(\sigma,s)\phi\|_C\\&\quad\leq c|B(s)\phi| < \infty \qquad (3.2.28)\end{aligned}$$

for some $c > 0$

$$G(\sigma, J_n^{B_n}(\sigma)U_n(\sigma,s)\phi) - A(\sigma)J_n^{B_n}(\sigma)(U_n(\sigma,s)\phi)(0)$$

is bounded for n large enough. Letting $n \to \infty$ in (3.2.28) from Lebesgue's dominated convergence theorem

$$\begin{aligned}&(U(t,s)\phi)(0)\\&\quad= \phi(0) + \int_s^t (G(\sigma, U(\sigma,s)\phi) - A(\sigma)(U(\sigma,s)\phi)(0))d\sigma\end{aligned}$$

from which by differentating this inequality we have

$$\frac{d(U(t,s)\phi)(0)}{dt} + A(t)(U(t,s)\phi)(0) = G(t, U(t,s)\phi)$$

a.e. $t \in (s,T)$. From (3.2.20) and (3.2.21)

$$\frac{dx}{dt}(t) + A(t)x(t) = G(t, x_t)$$

a.e. $t \in (s,T)$ and $x_s(\tau) = \phi(\tau)$ for every $\tau \in [-r, 0]$.

The uniqueness holds as in the Proof of Theorem 3.1.1.

THEOREM 3.2.2 *Let X^* be uniformly convex. Assume that $(A1)_2$, $(A2)_2$, $(G1)_1$ and $(G2)_2$ are satisfied. Let $B(t)$ be the operator as in (3.2.2). Let $x(t) = x(t;\phi)$ be defined as in (3.2.20). Then $x(t)$ is a unique strong solution of (3.2.1) on $[-r,T]$ for every $\phi \in \widehat{D}_B$.*

Proof. From (3.2.29)

$$\begin{aligned}&\|A(\sigma)J_n^{B_n}(\sigma)(U_n(\sigma,s)\phi)(0)\|\\&\quad\leq \|G(\sigma, J_n^{B_n}(\sigma)U_n(\sigma,s)\phi)\| + c|B(s)\phi|\end{aligned}$$

for some $c > 0$, and thus $A(\sigma)J_n^{B_n}(\sigma)(U_n(\sigma,s)\phi)(0)$ is bounded as $n \to \infty$. Since X is reflexive $(U(\sigma,s)\phi)(0) \in D_A$ and

$$w - \lim_{n\to\infty} A(\sigma, J_n^{B_n}(\sigma)U_n(\sigma,s)\phi)(0) = A(\sigma)(U(\sigma,s)\phi)(0)).$$

From (3.2.28) we have

$$\begin{aligned}
&((U_n(t,s)\phi)(0),x^*)\\
&\quad = (\phi(0),x^*) + \int_s^t (G(\sigma,(J_n^{B_n}(\sigma)U_n(\sigma,s)\phi)\\
&\quad -A(\sigma)(J_n^{B_n}(\sigma)U_n(\sigma,s)\phi)(0),x^*)d\sigma
\end{aligned}$$

so that using Lebesgue's dominate convergence theorem we have

$$\begin{aligned}
&((U(t,s)\phi)(0),x^*)\\
&\quad = (\phi(0),x^*) + \int_s^t (G(\sigma,U(\sigma,s)\phi) - A(\sigma)(U(\sigma,s)\phi)(0),x^*)d\sigma,
\end{aligned}$$

and thus by (3.2.20) and (3.2.21)

$$(x(t),x^*) = (\phi(0),x^*) + \int_s^t (G(\sigma,x_t) - A(\sigma)x(t),x^*)d\sigma. \quad (3.2.29)$$

Hence $x(t)$ is weakly differentiable a.e. $t \in (s,T)$. But $U(t,s)\phi$ is Lipschitz continuous in $t \in [s,T]$, so $x(t)$ is strongly differentiable a.e. $t \in (s,T)$. Using (3.2.30)

$$x(t) = \phi(0) + \int_s^t (G(\sigma,x_t) - A(\sigma)x(t))d\sigma,$$

and thus

$$\frac{dx}{dt}(t) + A(t)x(t) = G(t,x_t)$$

a.e. $t \in (s,T)$ and $x_s(\tau) = \phi(\tau)$ for every $\tau \in [-r,0]$.

The uniqueness holds as in the Proof of Theorem 3.3.1. Therefore the proof of Theorem is completed.

3.3. Monotonicity Methods

Let V be a real separable reflexive Banach space with norm $\|\cdot\|$ and let the dual space V^* of V have norm $\|\cdot\|_*$. Let H be a real separable Hilbert space with inner product $(\cdot,\cdot)_H$ and norm $\|\cdot\|_H$.

In this section we consider non-autonomous nonlinear functional evolutions of the type

$$\begin{cases} \dfrac{dx}{dt}(t) + A(t)x(t) = G(t,x_t), & 0 \le t \le T \\ x(t) = \phi(t), & -r \le t \le 0 \end{cases} \quad (3.3.1)$$

associated with a single-valued operator $A(t) : V \to V^*$ for every $t \in [0,T]$, where $T > 0$, $r > 0$, $x(t) : [-r,T] \to V$ is an unknown function, $G : [0,T] \times PC([-r,0];H) \to H$, and $\phi \in PC([-r,0];H)$ is Lipschitz continuous with Lipschitz constant $L_0 > 0$.

In our cases $V \cap H$, assumed nonempty, is a real reflexive separable Banach space with norm $\|\cdot\|_{V\cap H} = \|\cdot\| + \|\cdot\|_H$. Also $(V \cap H)^*$ is a real reflexive separable Banach space with norm $\|\cdot\|_{(V\cap H)^*}$. We assume that $V \cap H$ is dense in V and H. Then, identifying H with its dual, we have $V \cap H \subset H \subset (V \cap H)^*$ with H which is dense in $(V \cap H)^*$.

We note that $(y,x)_H = (y,x)$ for every $y \in H$ and $x \in V \cap H$.

DEFINITION 3.3.1 *A function $x(t) \equiv x(t,\phi) : [-r,T] \to H$ is called a solution of* (3.3.1) *on* $[-r,T]$ *if:*

(1) $x(t) = \phi(t)$ *for every* $t \in [-r,0]$*;*

(2) $x(t)$ *is continuous on* $[0,T]$*;*

(3) $x(t) \in V \cap H$ *for every* $t \in [0,T]$*;*

(4) $x(t)$ *is weakly continuously differentiable on* $[0,T]$*;*

(5) $x(t)$ *satisfies*

$$\frac{dx}{dt}(t) + A(t)x(t) = G(t,x_t)$$

for every $t \in [0,T]$, *where* $\frac{dx}{dt}(t)$ *denotes the weak derivative of* $x(t)$.

First of all, we need some definitons:

DEFINITION 3.3.2 *Let* $A(t) : V \to V^*$ *be an operator for every* $t \in [0,T]$.

(1) $A(t)$ *is said to be monotone if*

$$(A(t)u_1 - A(t)u_2, u_1 - u_2) \geq 0$$

for every $t \in [0,T]$ *and* $u_i \in V$, $i = 1,2$.
A monotone operator $A(t)$ *is said to be maximal monotone if* $R(A(t) + \lambda F_V) = V^*$ *for every* $\lambda > 0$ *and* $t \in [0,T]$, *where* F_V *is the duality mapping on* V.

(2) $A(t)$ *is said to be demicontinuous if*

$$\lim_{n\to\infty} u_n = u \text{ implies } w - \lim_{n\to\infty} A(t)u_n = A(t)u$$

for every $t \in [0,T]$ *and* u, $u_n \in V$, $n = 1,2,\cdots$.

(3) $A(t)$ is said to be coercive if there exist $s_0 > 0$ and a function $L : [0, \infty) \to \boldsymbol{R}$ such that

$$(A(t)u, u) \geq \|u\| L(\|u\|)$$

for every $t \in [0, T]$ and $u \in V$, where $L : [0, \infty) \to [0, \infty)$ which is nondecreasing on $[s_0, \infty)$, is bounded on $[0, s_0]$ and $\lim_{s \to \infty} L(s) = \infty$.

REMARK 3.3.1 *For every $t \in [0, T]$ $A(t) : V \to V^*$ is maximal monotone if and only if $A(t) : V \to V^*$ is monotone and demicontinuous.*

Now we give the following conditions for the existence of solutions of (3.3.1).

$(A1)_3$ $A(t) : V \to V^*$ is monotone, demicontinuous, and coercive for every $t \in [0, T]$;

$(A2)_3$ there exist two constants $c_i > 0$, $i = 1, 2$ such that

$$\|A(t_1)u - A(t_2)u\|_* \leq |t_1 - t_2|(c_1 + c_2 L(\|u\|))$$

for every $t_i \in [0, T]$, $i = 1, 2$ and $u \in V$;

$(G1)_3$ There exists a constant $\beta > 0$ such that

$$\|G(t, \psi_1) - G(t, \psi_2)\|_H \leq \beta \|\psi_1 - \psi_2\|_{PC}$$

for every $t \in [0, T]$ and $\psi_i \in PC([-r, 0]; H)$, $i = 1, 2$;

$(G2)_3$ There exists a non-decreasing function $l : [0, \infty) \to [0, \infty)$ such that

$$\|G(t_1, \psi) - G(t_2, \psi)\|_H \leq |t_1 - t_2| l(\|\psi\|_{PC})$$

for every $t_i \in [0, T]$, $i = 1, 2$, and $\psi \in PC([-r, 0]; H)$.

REMARK 3.3.2 *The functions $L(s)$ in $(A2)_3$ and $sL(s)$ in $(A1)_3$ with the definition of coerciveness can be replaced by the functions $L_1(s)$ and $sL_1(s) - \mu$, respectively, where μ is a fixed positive constant and $L_1 : [0, \infty) \to [0, \infty)$ is a non-decreasing function such that $L_1(s) \geq 1$ for every $s \in [0, \infty)$ and $\lim_{s \to \infty} L_1(s) = \infty$.*

Proof. Since $L(s)$ is non-decreasing for $s \geq s_0 > 0$ and $\lim_{s \to \infty} L(s) = \infty$, there exists $s_1 \geq s_0$ such that $L(s) > 1$ for every $s \in [s_1, \infty)$ and $L(s)$ is non-decreasing on $[s_1, \infty)$. Since there exists a constant $C > 0$ such that $sL(s) \geq -C$ for every $s \in [0, s_1]$, letting a constant C_2 such that $C_2 \max_{t \in [0, s_1]} |L(t)| \leq L(s_1)$, we have

$$\begin{aligned} sL(s) &\geq C_2 s \max_{t \in [0,s]} |L(t)| - (C_2 s \max_{t \in [0,s]} |L(t)| + C) \\ &\geq C_2 s_1 \max_{t \in [0,s]} |L(t)| - C_3, \end{aligned}$$

where $C_3 = C_2 s_1 \max_{t\in[0,s_1]} |L(t)| + C$. Let

$$L_1(s) = \chi_{[0,s_1)}(C_2 \max_{t\in[0,s]} |L(t)| + 1) + \chi_{[s_1,\infty)} L(s),$$

where χ_I denotes the characteristic function of the interval I. Then we see that $L_1(s) \geq 1$ for every $s \in [0,\infty)$, $\lim_{s\to\infty} L_1(s) = \infty$ and $L_1(s) : [0,\infty) \to [0,\infty)$ is non-decreasing. We also obtain

$$sL(s) \geq C_2 \max_{t\in[0,s]} |L(t)| - C_3 \geq sL_1(s) - s - C_3 \geq sL_1(s) - \mu$$

for every $s \in [0, s_1)$, where $\mu = s_1 + C_3$, and

$$sL(s) = sL_1(s) \geq sL_1(s) - \mu$$

for every $s \in [s_1, \infty)$.

Hence $sL(s) \geq sL_1(s) - \mu$ for every $s \in [0,\infty)$. It can be also seen that $c_1 + c_2 L(s) \leq c_1 c_2' L_1(s)$ for every $s \in [0,\infty)$, where $c_2' = \max\{c_2, c_2/C_2\}$.

From Remark 3.3.2 we shall only consider the function $L_1(s)$ and the constant μ, but we shall maintain the notation c_1 and c_2 for the constant in $(A2)_3$. Our method of proving the existence of a unique solution of (3.3.1) is based on the 'methods of lines' for the approximation of the solution of (3.3.1). We construct some discrete approximation equations for (3.3.1).

PROPOSITION 3.3.1 *Assume that $(A1)_3$, $(A2)_3$, $(G1)_3$, and $(G2)_3$ are satisfied. Let $\phi \in PC([-r,0]; H)$ be Lipschitz continuous with Lipschitz constant $L_0 > 0$, such that $\phi(0) \in V \cap H$ and $A(0)\phi(0) \in H$. Then there exist a partition $\{t_i^n \mid i = 0,1,2,\cdots,n\}$ of $[0,T]$ and a sequence $\{x_i^n \in V \mid i = 0,1,2,\cdots,n\}$ such that*

$$\lim_{n\to\infty} \max_{1\leq i\leq n} (t_i^n - t_{i-1}^n) = 0$$

and

$$\frac{x_i^n - x_{i-1}^n}{t_i^n - t_{i-1}^n} + A(t_{i-1}^n) x_i^n = G(t_{i-1}^n, (\overline{x_i^n})_{t_{i-1}^n}),$$

where

$$\overline{x_i^n}(t) = \begin{cases} \phi(t), & t \in [-r, 0] \\ x_j^n, & t \in (t_{j-1}^n, t_j^n],\ j = 1,2,\cdots,i-1 \\ x_i^n, & t \in (t_{i-1}^n, T] \end{cases}$$

with properties

(1) there exists a constant $K > 0$ *and* $h_0 \in (0, 1/\beta)$ *such that*

$$\|x_i^n\|_T \le K$$

for every $i = 1, 2, \cdots, n$*;* $n = 1, 2, \cdots$ *and* $h \le h_0$ *;*

(2) there exists a constant $M > 0$ *and* $h_1 \in (0, h_0]$ *such that*

$$\left\| \frac{x_i^n - x_{i-1}^n}{t_i^n - t_{i-1}^n} \right\|_H \le M$$

for every $i = 1, 2, \cdots, n$*;* $n = 1, 2, \cdots$ *and* $h \le h_1$.

Proof. Let $\{t_i^n\}$ be a partition of $[0, T]$ with $t_0^n = 0$, $t_n^n = T$ and $t_i^n = ih = \frac{iT}{n}$ for $i = 0, 1, 2, \cdots, n$, where n is sufficiently large with $\beta h < 1$. We define the function $\overline{x_0^n}(t)$ by

$$\overline{x_0^n}(t) = \begin{cases} \phi(t), & t \in [-r, 0] \\ \phi(0), & t \in [0, T]. \end{cases}$$

We also define the operator $A_1(t) : D(A_1(t)) \subset V \to V^*$ by

$$\begin{aligned} D(A_1(t)) &= \{x \in V \cap H \mid A(t)x \in H\}, \\ A_1(t)x &= A(t)x, \quad x \in D(A_1(t)) \end{aligned}$$

for every $t \in [0, T]$. Then $A_1(t)$ is maximal monotone on $D(A_1(t))$. Thus there exists $x \in D(A_1(t))$ uniquely such that $A_1(t)x + \frac{1}{h}x = y$ for every $y \in H$ and $h > 0$. Moreover, the operator $\left(A_1(t) + \frac{1}{h}I\right)^{-1}$ is Lipschitz continuous with Lipschitz constant h. We define the space S_1 by

$$S_1 = \{x : [-r, T] \to H \mid x(t) = \phi(t),\ t \in [-r, 0], \\ x(t) \text{ is a constant on } (0, T]\}.$$

Then S_1 is a complete metric space with a metric induced by the norm $\|\cdot\|_T$ defined by

$$\|x\|_T = \sup_{t \in [-r, T]} \|x(t)\|_H \tag{3.3.2}$$

for every $x \in S_1$.

Let $x_0^n = \phi(0)$. Now, we define a mapping Q_1 on S_1 by

$$(Q_1 x)(t) = \begin{cases} \phi(t), & t \in [-r, 0] \\ \left(A_1(t_0^n) + \frac{1}{h}I\right)^{-1} \left(\frac{x_0^n}{h} + G(t_0^n, x_{t_0^n})\right), & t \in [0, T] \end{cases} \tag{3.3.3}$$

for every $x \in S_1$. Then from the definition $Q_1 x \in S_1$, and $Q_1 : S_1 \to S_1$ is a strict contraction on S_1. Indeed, let $x, y \in S_1$ with $x_0^n = y_0^n = \phi(0)$. We have

$$\|(Q_1 x)(t) - (Q_1 y)(t)\|_H = \|\phi(t) - \phi(t)\|_H = 0 \le \beta h \|x - y\|_T$$

for every $t \in [-r, 0]$, and

$$\begin{aligned} &\|(Q_1 x)(t) - (Q_1 y)(t)\|_H \\ &= \Big\| \Big(A_1(t_0^n) + \frac{1}{h} I\Big)^{-1} \Big(\frac{x_0^n}{h} + G(t_0^n, x_{t_0^n})\Big) \\ &\quad - \Big(A_1(t_0^n) + \frac{1}{h} I\Big)^{-1} \Big(\frac{y_0^n}{h} + G(t_0^n, y_{t_0^n})\Big) \Big\|_H \\ &\le h \|G(t_0^n, x_{t_0^n}) - G(t_0^n, y_{t_0^n})\|_H \\ &\le h\beta \|x_{t_0^n} - y_{t_0^n}\|_{PC} \\ &\le h\beta \|x - y\|_T \end{aligned}$$

for every $t \in [0, T]$. Thus

$$\|Q_1 x - Q_1 y\|_H \le \beta h \|x - y\|_T.$$

Since $\beta h < 1$, from Banach's contraction priciple, there exists $\overline{x_1^n} \in S_1$ uniquely such that $Q_1 \overline{x_1^n} = \overline{x_1^n}$. From (3.3.3)

$$\overline{x_1^n}(t) = \begin{cases} \phi(t), & t \in [-r, 0] \\ \Big(A_1(t_0^n) + \frac{1}{h} I\Big)^{-1} \Big(\frac{x_0^n}{h} + G(t_0^n, (\overline{x_1^n})_{t_0^n})\Big), & t \in [0, T]. \end{cases} \tag{3.3.4}$$

We put

$$x_1^n = \Big(A_1(t_0^n) + \frac{1}{h} I\Big)^{-1} \Big(\frac{x_0^n}{h} + G(t_0^n, (\overline{x_1^n})_{t_0^n})\Big)$$

in (3.3.4). Then

$$\overline{x_1^n}(t) = \begin{cases} \phi(t), & t \in [-r, 0] \\ x_1^n, & t \in [0, T] \end{cases}$$

and $\overline{x_1^n}(t) \in V \cap H$ for every $t \in [0, T]$. Also, $x_1^n \in D(A_1(t_0^n))$, $A_1(t_0^n) x_1^n = A(t_0^n) x_1^n$, and

$$\frac{x_1^n - x_0^n}{t_1^n - t_0^n} + A(t_0^n) x_1^n = G(t_0^n, (\overline{x_1^n})_{t_0^n}).$$

Proceeding in a similar manner we obtain the spaces S_i, $i = 2, 3, \cdots, n$, by

$$\begin{aligned} S_i = \{x(t) : [-r, T] \to H \mid x(t) = \overline{x_{i-1}^n}(t),\ t \in [-r, t_{i-1}^n] \\ x(t) \text{ is a constant on } (t_{i-1}^n, T]\}. \end{aligned}$$

Then each space S_i is a complete metric space with the norm in (3.3.2). Moreover, the strict contractions $Q_i : S_i \to S_i$ are defined by

$$(Q_i x)(t) = \begin{cases} \phi(t), & t \in [-r, 0] \\ x_j^n, & t \in (t_{j-1}^n, t_j^n],\ j = 1, 2, \cdots, i-1 \\ \left(A_1(t_{i-1}^n) + \frac{1}{h} I\right)^{-1} \left(\frac{x_{i-1}^n}{h} + G(t_{i-1}^n, (\overline{x_{i-1}^n})_{t_{i-1}^n})\right), & t \in (t_{i-1}^n, T]. \end{cases}$$

The unique fixed point of Q_i on S_i is denotes by $(\overline{x_{i-1}^n})_{t_0^n}$ and we put

$$x_i^n = \left(A_1(t_{i-1}^n) + \frac{1}{h} I\right)^{-1} \left(\frac{x_{i-1}^n}{h} + G(t_{i-1}^n, (\overline{x_i^n})_{t_{i-1}^n})\right) \tag{3.3.5}$$

for $i = 2, 3, \cdots, n$. Then

$$\overline{x_i^n}(t) = \begin{cases} \phi(t), & t \in [-r, 0] \\ x_j^n, & t \in (t_{j-1}^n, t_j^n],\ j = 1, 2, \cdot, i-1 \\ x_i^n, & t \in (t_{i-1}^n, T], \end{cases} \tag{3.3.6}$$

and $\overline{x_i^n}(t) \in V \cap H$ for every $t \in [0, T]$ and $i = 2, 3, \cdots, n$. Also, $x_i^n \in D(A_1(t_{i-1}^n))$, $A_1(t_{i-1}^n) x_i^n = A(t_{i-1}^n) x_i^n$, and from (3.3.5)

$$\frac{x_i^n - x_{i-1}^n}{t_i^n - t_{i-1}^n} + A(t_{i-1}^n) x_i^n = G(t_{i-1}^n, (\overline{x_i^n})_{t_{i-1}^n}) \tag{3.3.7}$$

for $i = 2, 3, \cdots, n$.

(1) By (3.3.7)

$$\left(\frac{x_i^n - x_{i-1}^n}{t_i^n - t_{i-1}^n}, x\right)_H + (A(t_{i-1}^n) x_i^n, x) = (G(t_{i-1}^n, (\overline{x_i^n})_{t_{i-1}^n}), x)_H \tag{3.3.8}$$

for every $x \in V \cap H$ and $i = 1, 2, \cdots, n$. Let $1 \leq k \leq n$. If we put $x = h x_i^n$ in (3.3.8) and sum through $i = 1, 2, \cdots, k$ we obtain

$$\sum_{i=1}^{k} (x_i^n - x_{i-1}^n, x_i^n)_H + \sum_{i=1}^{k} h(A(t_{i-1}^n) x_i^n, x_i^n) = \sum_{i=1}^{k} h(G(t_{i-1}^n, (\overline{x_i^n})_{t_{i-1}^n}), x_i^n)_H. \tag{3.3.9}$$

Now, since

$$\begin{aligned} |(G(t_{i-1}^n, (\overline{x_i^n})_{t_{i-1}^n}), x_i^n)_H| &\leq \|G(t_{i-1}^n, (\overline{x_i^n})_{t_{i-1}^n})\|_H \|x_i^n\|_H \\ &\leq \frac{1}{2}\|x_i^n\|_H^2 + \frac{1}{2}\|G(t_{i-1}^n, (\overline{x_i^n})_{t_{i-1}^n})\|_H^2 \end{aligned}$$

we have

$$\begin{aligned}
&\|G(t_{i-1}^n, (\overline{x_i^n})_{t_{i-1}^n})\|_H \\
&\quad \le \|G(t_{i-1}^n, (\overline{x_i^n})_{t_{i-1}^n}) - G(t_{i-1}^n, \bar{0})\|_H + \|G(t_{i-1}^n, \bar{0})\|_H \\
&\quad \le \beta \|(\overline{x_i^n})_{t_{i-1}^n}\|_{PC} + K_1, \qquad (3.3.10)
\end{aligned}$$

where $\bar{0}$ is the zero function in $PC([-r,0];H)$ and $K_1 = \sup_{t\in[0,T]} \|G(t,\bar{0})\|_H$. Since $A(t) : X \to X^*$ is coercive there exists a constant $K_2 > 0$ such that

$$(A(t_{i-1}^n)x_i^n, x_i^n) \ge -K_2 \qquad (3.3.11)$$

for $i = 1, 2, \cdots, n$ and $n = 1, 2, \cdots$. By (3.3.8)–(3.3.11) we obtain

$$\begin{aligned}
&2\sum_{i=1}^k (x_i^n - x_{i-1}^n, x_i^n)_H \\
&\quad = \sum_{i=1}^k \|x_i^n - x_{i-1}^n\|_H^2 + \|x_k^n\|_H^2 - \|x_0^n\|_H^2 \\
&\quad = -2\sum_{i=1}^k h(A(t_{i-1}^n)x_i^n, x_i^n) + 2\sum_{i=1}^k h(G(t_{i-1}^n(\overline{x_i^n})_{t_{i-1}^n}), x_i^n)_H \\
&\quad \le K_3 + \sum_{i=1}^k h\|x_i^n\|_H^2 + \sum_{i=1}^k h(K_1 + \beta\|(\overline{x_i^n})_{t_{i-1}^n}\|_{PC})^2,
\end{aligned}$$

where $K_3 = 2TK_2$. Thus

$$\begin{aligned}
\|x_k^n\|_H^2 \;\le\; & \|\phi(0)\|_H^2 + K_3 \qquad (3.3.12) \\
& + \sum_{i=1}^k h\|x_i^n\|_H^2 + \sum_{i=1}^k h(K_1 + \beta\|(\overline{x_i^n})_{t_{i-1}^n}\|_{PC})^2.
\end{aligned}$$

Since

$$\begin{aligned}
\|(\overline{x_i^n})_{t_{i-1}^n}\|_{PC} &= \sup_{\tau\in[-r,0]} \|\overline{x_i^n}(t_{i-1}^n + \tau)\|_H \\
&= \sup_{t\in[t_{i-1}^n - r, t_{i-1}^n]} \|\overline{x_i^n}(t)\|_H \\
&\le \sup_{t\in[-r,T]} \|\overline{x_i^n}(t)\|_H = \|\overline{x_i^n}\|_T
\end{aligned}$$

and $\|x_i^n\|_H \le \|\overline{x_i^n}\|_T$, from (3.3.12)

$$\|x_k^n\|_H^2 \;\le\; \|\phi(0)\|_H^2 + K_3 + \sum_{i=1}^k hK_1^2 +$$

$$\begin{aligned}
&+h(1+\beta^2)\sum_{i=1}^{k}\|\overline{x_i^n}\|_T^2 + 2K_1\beta\sum_{i=1}^{k}h\|\overline{x_i^n}\|_T \\
\leq\ & K_4 + \|\phi(0)\|_H^2 + hK_5\sum_{i=1}^{k}\|\overline{x_i^n}\|_T^2,
\end{aligned}$$

where $K_4 = K_3 + K_1T(K_1+\beta)$ and $K_5 = 1+\beta^2+K_1\beta$. Since

$$\begin{aligned}
\|x_j^n\|_H^2 &\leq K_4 + \|\phi(0)\|_H^2 + hK_5\sum_{i=1}^{j}\|\overline{x_i^n}\|_T^2 \\
&\leq K_4 + \|\phi\|_{PC}^2 + hK_5\sum_{i=1}^{k}\|\overline{x_i^n}\|_T^2
\end{aligned}$$

for every integer j with $1 \leq j \leq k$ we conclude that

$$\begin{aligned}
\|\overline{x_k^n}\|_T^2 &= \sup_{t\in[-r,T]}\|\overline{x_k^n}(t)\|_H^2 \\
&= \max\{\|\phi\|_{PC}^2, \|x_1^n\|_H^2, \cdots, \|x_k^n\|_H^2\} \\
&\leq K_6 + hK_5\sum_{i=1}^{k}\|\overline{x_k^n}\|_T^2, \qquad (3.3.13)
\end{aligned}$$

where $K_6 = K_4 + \|\phi\|_{PC}^2$. Let $h < \dfrac{1}{K_5}$.

For $k=1$ (3.3.13) gives

$$\|\overline{x_1^n}\|_T^2 \leq K_6 + hK_5\|\overline{x_1^n}\|_T^2,$$

and thus

$$\|\overline{x_1^n}\|_T^2 \leq \frac{K_6}{1-hK_5}. \qquad (3.3.14)$$

For $k=2$ also from (3.3.13) we obtain

$$\|\overline{x_2^n}\|_T^2 \leq K_6 + hK_5(\|\overline{x_1^n}\|_T^2 + \|\overline{x_2^n}\|_T^2),$$

from which by (3.3.14)

$$\begin{aligned}
(1-hK_5)\|\overline{x_2^n}\|_T^2 &\leq K_6 + hK_5\|\overline{x_1^n}\|_T^2 \\
&\leq K_6 + \frac{hK_5K_6}{1-hK_5},
\end{aligned}$$

and thus

$$\|\overline{x_2^n}\|_T^2 \leq \frac{K_6}{(1-hK_5)^2}.$$

Similarly we have from (3.3.13)

$$\|\overline{x_i^n}\|_T^2 \leq \frac{K_6}{(1-hK_5)^i}$$

for $i = 1, 2, \cdots, n$. Since

$$\frac{1}{(1-hK_5)^i} \leq \frac{1}{(1-hK_5)^n} = \frac{1}{\left(1-\frac{T}{n}K_5\right)^n} \to e^{TK_5}$$

as $n \to \infty$ there exist constants $h_0 \in (0, \min t\{1/\beta, 1/K_5\})$ and $K_7 > 0$ with $\frac{1}{(1-hK_5)^i} \leq K_7$ for $h \leq h_0$. Hence there exist constant $h_0 \in (0, 1/\beta)$ and $K > 0$ such that $\|x_i^n\|_T \leq K$ for $i = 0, 1, 2, \cdots, n$, $n = 1, 2, \cdots$ and $h \leq h_0$.

(2) Let $h \in (0, h_0]$. Letting $x = x_1^n - x_0^n$ and $i = 1$ in (3.3.8) we obtain

$$\begin{aligned}&\left(\frac{x_1^n - x_0^n}{h}, x_1^n - x_0^n\right)_H + (A(t_0^n)x_1^n, x_1^n - x_0^n)\\ &= (G(t_0^n, (\overline{x_1^n})_{t_0^n}), x_1^n - x_0^n)_H.\end{aligned}$$

Thus

$$\begin{aligned}&\frac{\|x_1^n - x_0^n\|_H^2}{h} + (A(0)x_1^n - A(0)x_0^n, x_1^n - x_0^n)\\ &= (G(0, (\overline{x_1^n})_{t_0^n}), x_1^n - x_0^n)_H - (A(0)x_0^n, x_1^n - x_0^n),\end{aligned}$$

which, from the assumptions of monotonicity of $A(0)$ and $A(0)\phi(0) \in H$, yields

$$\begin{aligned}&\frac{\|x_1^n - x_0^n\|_H^2}{h} \\ &\leq \|G(0, (\overline{x_1^n})_{t_0^n})\|_H \|x_1^n - x_0^n\|_H + \|A(0)x_0^n\|_H \|x_1^n - x_0^n\|_H.\end{aligned} \tag{3.3.15}$$

Since

$$(\overline{x_1^n})_{t_0^n}(\tau) = (\overline{x_1^n})_0(\tau) = \phi(\tau)$$

for every $\tau \in [-r, 0]$, from (3.3.15)

$$\frac{\|x_1^n - x_0^n\|_H}{h} \leq \|G(0, \phi)\|_H + \|A(0)\phi(0)\|_H = M_1. \tag{3.3.16}$$

In a similar way, putting $x = \dfrac{x_i^n - x_{i-1}^n}{h}$ and $i = 2, 3, \cdots, n$ in (3.3.8) we have

$$\frac{\|x_i^n - x_{i-1}^n\|_H^2}{h} =$$

$$\begin{aligned}
&= \Big(G(t_{i-1}^n, (\overline{x_i^n})_{t_{i-1}^n}), \frac{x_i^n - x_{i-1}^n}{h}\Big)_H \\
&\quad -\Big(A(t_{i-1}^n)x_i^n, \frac{x_i^n - x_{i-1}^n}{h}\Big) \\
&= \Big(G(t_{i-1}^n, (\overline{x_i^n})_{t_{i-1}^n}) - G(t_{i-1}^n, (\overline{x_i^n})_{t_{i-2}^n}), \frac{x_i^n - x_{i-1}^n}{h}\Big)_H \\
&\quad +\Big(G(t_{i-1}^n, (\overline{x_i^n})_{t_{i-2}^n}) - G(t_{i-2}^n, (\overline{x_i^n})_{t_{i-2}^n}), \frac{x_i^n - x_{i-1}^n}{h}\Big)_H \\
&\quad +\Big(G(t_{i-2}^n, (\overline{x_i^n})_{t_{i-2}^n}), \frac{x_i^n - x_{i-1}^n}{h}\Big)_H \\
&\quad -\Big(A(t_{i-2}^n)x_{i-1}^n, \frac{x_i^n - x_{i-1}^n}{h}\Big) \\
&\quad -\Big(A(t_{i-2}^n)x_i^n - A(t_{i-2}^n)x_{i-1}^n, \frac{x_i^n - x_{i-1}^n}{h}\Big) \\
&\quad -\Big(A(t_{i-1}^n)x_i^n - A(t_{i-2}^n)x_i^n, \frac{x_i^n - x_{i-1}^n}{h}\Big).
\end{aligned}$$

From $(A1)_3$, $(A2)_3$, $(G1)_3$, $(G2)_3$ and (3.3.7) we obtain

$$\begin{aligned}
&\frac{\|x_i^n - x_{i-1}^n\|_H^2}{h^2} \\
&\qquad \le \beta \|(\overline{x_i^n})_{t_{i-1}^n} - (\overline{x_i^n})_{t_{i-2}^n}\|_{PC} \frac{\|x_i^n - x_{i-1}^n\|_H}{h} \\
&\qquad\quad + hl(\|(\overline{x_i^n})_{t_{i-2}^n}\|_{PC}) \frac{\|x_i^n - x_{i-1}^n\|_H}{h} \\
&\qquad\quad + \frac{\|x_{i-1}^n - x_{i-2}^n\|_H}{h} \frac{\|x_i^n - x_{i-1}^n\|_H}{h} \\
&\qquad\quad + h(c_1 + c_2 L_1(\|x_i^n\|))\|x_i^n - x_{i-1}^n\|, \qquad (3.3.17)
\end{aligned}$$

where $L_1(s)$ is the function in Remark 3.3.2. For the last term in (3.3.17) we have

$$\begin{aligned}
&(c_1 + c_2 L_1(\|x_i^n\|))\|x_i^n - x_{i-1}^n\| \\
&\quad \le c_1(\|x_i^n\| + \|x_{i-1}^n\|) + c_2 L_1(\|x_i^n\|)(\|x_i^n\| + \|x_{i-1}^n\|) \\
&\quad \le 2c_1 \max_{1\le j\le n} \|x_j^n\| \\
&\qquad + 2c_2 \max_{1\le j\le n} L_1(\|x_j^n\|) \max_{1\le j\le n} \|x_j^n\| \\
&\quad \le 2c_1 \max_{1\le j\le n} L_1(\|x_j^n\|) \max_{1\le j\le n} \|x_j^n\| \\
&\qquad + 2c_2 \max_{1\le j\le n} L_1(\|x_j^n\|) \max_{1\le j\le n} \|x_j^n\| \\
&\quad = M_2 \max_{1\le j\le n} L_1(\|x_j^n\|)\|x_j^n\|, \qquad (3.3.18)
\end{aligned}$$

where $M_2 = 2(c_1 + c_2)$.

We now show that

$$\|(\overline{x_i^n})_{t_{i-1}^n} - (\overline{x_i^n})_{t_{i-2}^n}\|_{PC} \leq \max_{1\leq j\leq i} \|x_j^n - x_{j-1}^n\|_H + L_0 h. \tag{3.3.19}$$

Indeed, since

$$\|(\overline{x_i^n})_{t_{i-1}^n} - (\overline{x_i^n})_{t_{i-2}^n}\|_{PC} = \sup_{\tau\in[-r,0]} \|\overline{x_i^n}(t_{i-1}^n + \tau) - \overline{x_i^n}(t_{i-2}^n + \tau)\|_H$$

we distinguish three cases.

Case 1. $t_{i-2}^n \geq r$.

We have $t_{i-1}^n \geq r$ and

$$\begin{aligned}
&\|\overline{x_i^n}(t_{i-1}^n + \tau) - \overline{x_i^n}(t_{i-2}^n + \tau)\|_H \\
&\quad = \|\overline{x_i^n}(t_{i-1}^n + \tau) - \overline{x_{i-1}^n}(t_{i-2}^n + \tau)\|_H \\
&\quad \leq \max\nolimits_{1\leq j\leq i} \|x_j^n - x_{j-1}^n\|_H
\end{aligned}$$

for every $\tau \in [-r, 0]$.

Case 2. $t_{i-1}^n < r$.

We also have $t_{i-2}^n < r$.

(a) Let $\tau \in [-r, -t_{i-1}^n]$. Then we have $t_{i-1}^n + \tau \in [t_{i-1}^n - r, 0]$ and $t_{i-2}^n + \tau \in [t_{i-2}^n - r, 0]$. Thus

$$\begin{aligned}
&\|\overline{x_i^n}(t_{i-1}^n + \tau) - \overline{x_i^n}(t_{i-2}^n + \tau)\|_H \\
&\quad = \|\phi(t_{i-1}^n + \tau) - \phi(t_{i-2}^n + \tau)\|_H \\
&\quad \leq L_0 |t_{i-1}^n - t_{i-2}^n| \\
&\quad = M_0 h.
\end{aligned}$$

(b) Let $\tau \in (-t_{i-2}^n, 0]$. Then we have $t_{i-1}^n + \tau \in (t_{i-1}^n - t_{i-2}^n, t_{i-1}^n]$ and $t_{i-2}^n + \tau \in (0, t_{i-2}^n]$. Thus

$$\|\overline{x_i^n}(t_{i-1}^n + \tau) - \overline{x_i^n}(t_{i-2}^n + \tau)\|_H \leq \|\overline{x_i^n}(t_{i-1}^n + \tau) - \overline{x_i^n}(t_{i-2}^n + \tau)\|_H.$$

(c) Let $\tau \in (-t_{i-1}^n, -t_{i-2}^n]$. Then we have $t_{i-1}^n + \tau \in (0, h]$ and $t_{i-2}^n + \tau \in (-h, 0]$. Thus

$$\begin{aligned}
&\|\overline{x_i^n}(t_{i-1}^n + \tau) - \overline{x_i^n}(t_{i-2}^n + \tau)\|_H \\
&\quad \leq \|\overline{x_i^n}(t_{i-1}^n + \tau) - \overline{x_i^n}(0)\|_H + \|\overline{x_i^n}(0) - \overline{x_{i-1}^n}(0)\|_H \\
&\qquad + \|\overline{x_{i-1}^n}(0) - \overline{x_i^n}(t_{i-2}^n + \tau)\|_H \\
&\quad \leq \|x_1^n - x_0^n\|_H + \|\phi(0) - \phi(t_{i-2}^n + \tau)\|_H \\
&\quad \leq \|x_1^n - x_0^n\|_H + L_0 h,
\end{aligned}$$

where L_0 is the Lipschitz constant of ϕ.

Case 3. $t_{i-1}^n \ge r$ and $t_{i-2}^n \le r$.

(a) Let $\tau \in [-r, -t_{i-2}^n]$. Then $t_{i-1}^n + \tau \in [0, h]$ and $t_{i-2}^n + \tau \in [-h, 0]$. Thus we have as in (c) of Case 2 above

$$\|\overline{x_i^n}(t_{i-1}^n + \tau) - \overline{x_i^n}(t_{i-2}^n + \tau)\|_H \le \|x_1^n - x_0^n\|_H + L_0 h.$$

(b) Let $\tau \in (-t_{i-2}^n, 0]$. Then $t_{i-1}^n + \tau \in (t_{i-1}^n - t_{i-2}^n, t_{i-1}^n]$ and $t_{i-2}^n + \tau \in (0, t_{i-2}^n]$. Thus

$$\|\overline{x_i^n}(t_{i-1}^n + \tau) - \overline{x_i^n}(t_{i-2}^n + \tau)\|_H = \max_{1\le j\le i} \|x_j^n - x_{j-1}^n\|_H.$$

From the above three cases we see that

$$\|\overline{x_i^n}(t_{i-1}^n + \tau) - \overline{x_i^n}(t_{i-2}^n + \tau)\|_H \le \max_{1\le j\le i} \|x_j^n - x_{j-1}^n\|_H + L_0 h$$

for every $\tau \in [-r, 0]$ and $i = 2, 3, \cdots$. Hence

$$\begin{aligned} &\sup_{\tau\in[-r,0]} \|\overline{x_i^n}(t_{i-1}^n + \tau) - \overline{x_i^n}(t_{i-2}^n + \tau)\|_H \\ &\qquad \le \max_{1\le j\le i} \|x_j^n - x_{j-1}^n\|_H + L_0 h \end{aligned}$$

for $i = 2, 3, \cdots$. Therefore (3.3.19) holds. From (3.3.17)–(3.3.19),

$$\begin{aligned} &\frac{\|x_i^n - x_{i-1}^n\|_H^2}{h^2} \\ &\quad \le (\beta \max_{1\le j\le i} \|x_j^n - x_{j-1}^n\|_H + \beta L_0 h)\frac{\|x_i^n - x_{i-1}^n\|_H}{h} \\ &\quad\quad + hl(\|(\overline{x_i^n})_{t_{i-2}^n}\|_{PC})\frac{\|x_i^n - x_{i-1}^n\|_H}{h} \\ &\quad\quad + \|x_{i-1}^n - x_{i-2}^n\|_H \frac{\|x_i^n - x_{i-1}^n\|_H^2}{h^2} \\ &\quad\quad + hM_2 \max_{1\le j\le n} L_1(\|x_j^n\|)\|x_j^n\| \\ &\quad \le \frac{\beta}{h}\max_{1\le j\le i} \frac{\|x_i^n - x_{i-1}^n\|_H^2}{h^2} + h(l(K) + \beta L_0)\frac{\|x_i^n - x_{i-1}^n\|_H}{h} \\ &\quad\quad + \frac{1}{2}\frac{\|x_{i-1}^n - x_{i-2}^n\|_H^2}{h^2} + \frac{1}{2}\frac{\|x_i^n - x_{i-1}^n\|_H^2}{h^2} \\ &\quad\quad + hM_2 \max_{1\le j\le n} L_1(\|x_j^n\|)\|x_j^n\| \\ &\quad \le h\beta \max_{1\le j\le i} \frac{\|x_j^n - x_{j-1}^n\|_H^2}{h^2} + \frac{h}{2}M_3 + \frac{h}{2}\frac{\|x_i^n - x_{i-1}^n\|_H^2}{h^2} + \end{aligned}$$

$$+\frac{1}{2}\frac{\|x_i^n - x_{i-1}^n\|_H^2}{h^2} + \frac{1}{2}\frac{\|x_{i-1}^n - x_{i-2}^n\|_H^2}{h^2}$$
$$+hM_1 \max_{1\leq j\leq n} L_1(\|x_j^n\|)\|x_j^n\|,$$

where $M_3 = (l(K) + \beta L_0)^2$, from which we have

$$\left(\frac{1}{2} - \frac{h}{2} - h\beta\right) \max_{1\leq j\leq i} \frac{\|x_j^n - x_{j-1}^n\|_H^2}{h^2}$$
$$\leq hM_3 + \frac{1}{2} \max_{1\leq j\leq i} \frac{\|x_{j-1}^n - x_{j-2}^n\|_H^2}{h^2}$$
$$+hM_1 \max_{1\leq j\leq n} L_1(\|x_j^n\|)\|x_j^n\|$$
$$\leq hM_3 \sum_{s=0}^{n-1} \frac{1}{(2p)^s} + \frac{1}{(2p)^{n-1}}$$
$$+hM_2 \sum_{s=0}^{n-1} \frac{1}{(2p)^s} \max_{1\leq j\leq n} L_1(\|x_j^n\|)\|x_j^n\|,$$

where $p = \frac{1}{2}(1 - h - 2h\beta)$. If we further choose $h < \frac{1}{1+2\beta}$ then $p \in \left(0, \frac{1}{2}\right)$. Since by (3.3.16) $\frac{\|x_1^n - x_0^n\|_H}{h} \leq M_1$, and $\lim_{n\to\infty} \frac{1}{(2p)^n}$ exists there exist $h_1 < \min\left\{\frac{1}{1+2\beta}, h_0\right\}$ and constants $M_4 > 0$, $M_5 > 0$ such that

$$\max_{1\leq j\leq i} \frac{\|x_j^n - x_{j-1}^n\|_H^2}{h^2} \leq M_4 + M_5 \max_{1\leq j\leq n} L_1(\|x_j^n\|)\|x_j^n\| \qquad (3.3.20)$$

for every $h \leq h_1$ and $i = 1, 2, \cdots, n$. Assume that

$$\max_{1\leq j\leq n} L_1(\|x_j^n\|)\|x_j^n\| = L_1(\|x_m^n\|)\|x_m^n\|.$$

Then from (3.3.20) we obtain

$$\frac{\|x_m^n - x_{m-1}^n\|_H^2}{h^2} \leq M_4 + M_5 L_1(\|x_m^n\|)\|x_m^n\|$$

and thus from Young's inequality and (1) we obtain

$$\left(\frac{x_m^n - x_{m-1}^n}{h}, x_m^n\right)_H$$
$$\leq \frac{\|x_m^n - x_{m-1}^n\|_H}{h}\|x_m^n\|_H \leq$$

$$\leq \frac{1}{2\epsilon^2}\|x_m^n\|_H^2 + \frac{\epsilon^2}{2}\frac{\|x_m^n - x_{m-1}^n\|_H^2}{h^2}$$
$$\leq \frac{1}{2\epsilon^2}K^2 + \frac{\epsilon^2}{2}M_4 + \frac{\epsilon^2}{2}M_5 L(\|x_m^n\|)\|x_m^n\|$$

for every $\epsilon > 0$. We choose $\epsilon > 0$ such that $\epsilon^2 M_5 = 1$. Then from (3.3.20), (3.3.8), with $i = m$, $x = x_m^n$, (1) and (3.3.10) we have

$$\begin{aligned}
&(A(t_{m-1}^n)x_m^n, x_m^n)\\
&\leq M_6 + \frac{1}{2}L_1(\|x_m^n\|)\|x_m^n\| + \|G(t_{m-1}^n, (\overline{x_m^n})_{t_{m-1}^n})\|_H\|x_m^n\|_H\\
&\leq M_6 + \frac{1}{2}L_1(\|x_m^n\|)\|x_m^n\| + (\beta K + K_1)K,
\end{aligned}$$

where $M_6 = \frac{1}{2}(M_5K^2 + M_4/M_5)$.

Using the coerciveness of $A(t)$ we obtain

$$\begin{aligned}
L_1(\|x_m^n\|)\|x_m^n\| &\leq (A(t_{m-1}^n)x_m^n, x_m^n) + \mu\\
&\leq M_7 + \frac{1}{2}L_1(\|x_m^n\|)\|x_m^n\|,
\end{aligned}$$

which implies

$$L_1(\|x_m^n\|)\|x_m^n\| \leq 2M_7 \tag{3.3.21}$$

with $M_7 = M_6 + (\beta + K_1)K + \mu$. From (3.3.20) and (3.3.21) there exist constants $M_0 > 0$ and $M > 0$ such that

$$\|x_i^n\| \leq M_0 \quad \text{and} \quad \frac{\|x_i^n - x_{i-1}^n\|_H}{h} \leq M \tag{3.3.22}$$

for every $h \leq h_1$ and $i = 1, 2, \cdots, n$. Therefore the proof of (2) is completed.

We now consider Rothe's functions

$$x_{(n)}(t) = \begin{cases} \phi(t), & t \in [-r, 0] \\ x_{i-1}^n + (t - t_{i-1}^n)\dfrac{x_i^n - x_{i-1}^n}{h}, & t \in [t_{i-1}^n, t_i^n] \end{cases} \tag{3.3.23}$$

for $i = 1, 2, \cdots, n$. Then $x_{(n)}(t)$ is uniformly Lipschitz continuous in $t \in [-r, T]$ with Lipschitz constant $N_1 = \max\{M, L_0\}$. Let

$$x_n(t) = \overline{x_n^n}(t) \tag{3.3.24}$$

for every $t \in [-r, T]$, where $\overline{x_n^n}(t)$ is obtained from (3.3.6).

PROPOSITION 3.3.2 *Suppose that the assumptions in Proposition 3.3.1 are satisfied. Let* $x_{(n)}(t)$ *and* $x_n(t)$ *be defined in* (3.3.23) *and* (3.3.24), *respectively. Then*

$$\|x_n(t) - x_{(n)}(t)\|_H \leq 2M\frac{T}{n}$$

for every $t \in [-r, T]$ *and* $n = 1, 2, \cdots$.

Proof. Let $t \in (0, T]$. Then $t \in (t_{i-1}^n, t_i^n]$ for some $i = 1, 2, \cdots, n$. From the definition of $x_{(n)}(t)$ and $x_n(t)$ we have

$$\begin{aligned}
&\|x_n(t) - x_{(n)}(t)\|_H \\
&= \left\| x_i^n - x_{i-1}^n - (t - t_{i-1}^n)\frac{x_i^n - x_{i-1}^n}{h} \right\|_H \\
&= \left\| (h - (t - t_{i-1}^n))\frac{x_i^n - x_{i-1}^n}{h} \right\|_H \\
&\leq 2h\left\| \frac{x_i^n - x_{i-1}^n}{h} \right\|_H \\
&\leq 2Mh = 2M\frac{T}{n}.
\end{aligned}$$

Let $t \in [-r, 0]$. Then $x_n(t) = x_{(n)}(t) = \phi(t)$. Thus the proof is completed.

The main theorem is as follows:

THEOREM 3.3.1 *Assume that* $(A1)_3$, $(A2)_3$, $(G1)_3$, *and* $(G2)_3$ *are satisfied. Let* $\phi \in PC([-r, 0]; H)$ *is Lipschitz continuous with Lipschitz constant* $L_0 > 0$, *such that* $\phi(0) \in X \cap H$ *and* $A(0)\phi(0) \in H$. *Then there exists a unique solution* $x(t)$ *of* (3.3.1) *on* $[-r, T]$ *into* H *such that:*

(1) $x(t)$ *is Lipschitz continuous on* $[0, T]$;

(2) $x \in C_w([0, T]; V \cap H) \cap L^\infty(0, T; V \cap H)$;

(3) $A(\cdot)x(\cdot) \in C_w([0, T]; H)$;

(4) $\dfrac{dx}{dt} \subset L^\infty(0, T; H)$;

(5) Let $x_i(t)$ *be a solution of* (3.3.1) *on* $[-r, T]$ *corresponding to* $\phi_i \in PC([-r, 0]; H)$ *which is Lipschitz continuous with Lipschitz constant* $L_0 > 0$, $\phi_i(0) \in V \cap H$ *and* $A(0)\phi_i(0) \in H$, $i = 1, 2$. *Then*

$$\max_{t\in[0,T]} \|x_1(t) - x_2(t)\|_H \leq \|\phi_1 - \phi_2\|_{PC} + \int_0^T \|G(t, (x_1)_t) - G(t, (x_2)_t)\|_H dt.$$

Proof. We define the operators $A^n(t)$ and $G^n(t)$ by

$$\begin{aligned} A^n(0) &= A(0)\phi(0), \\ A^n(t) &= A(t_{i-1}^n)x_i^n, \quad t \in (t_{i-1}^n, t_i^n] \\ G^n(t) &= G(t_{i-1}^n, (\overline{x_i^n})_{t_{i-1}^n}), \quad t \in (t_{i-1}^n, t_i^n] \end{aligned}$$

for $i = 1, 2, \cdots, n$. Since $x_{(n)}(t)$ defined in (3.3.23) is strongly differentiable on $[0, T]$ except at a finite number of points at which the strong left derivative $\dfrac{d^- x_{(n)}}{dt}(t)$ exists and equals $\dfrac{x_i^n - x_{i-1}^n}{h}$ for every $t \in (t_{i-1}^n, t_i^n]$ we have from (3.3.7) and (3.3.23)

$$\frac{d^- x_{(n)}}{dt}(t) + A^n(t) = G^n(t), \quad t \in (t_{i-1}^n, t_i^n]$$

for $i = 1, 2, \cdots, n$ and thus

$$\left(\frac{d^- x_{(n)}}{dt}(t), y\right)_H + (A^n(t), y) = (G^n(t), y)_H \tag{3.3.25}$$

for every $y \in V \cap H$ and $t \in [0, T]$. From Proposition 3.3.1 and (3.3.10) we obtain

$$\begin{aligned} |(A^n(t), y)| &\leq \left|\left(\frac{d^- x_{(n)}}{dt}(t), y\right)_H\right| + |(G^n(t), y)_H| \qquad (3.3.26) \\ &\leq (M + \|G^n(t)\|_H)\|y\|_H \\ &\leq N_2\|y\|_H \end{aligned}$$

for every $y \in V \cap H$, $t \in [0, T]$ and sufficiently large n. Since $V \cap H$ is dense in H $A^n(t) \in H$ and

$$\|A^n(t)\|_H \leq N_2 \tag{3.3.27}$$

for every $t \in [0, T]$. Putting $y = y_1 - y_2$ in (3.3.26) for $y_i \in V \cap H$, $i = 1, 2$ we have

$$|(A^n(t), y_1 - y_2)| \leq N_2\|y_1 - y_2\|_H \tag{3.3.28}$$

for every $y_i \in V \cap H$, $i = 1, 2$, $t \in [0, T]$ and sufficiently large n.

Let $\{t_i^n\}$ and $\{t_j^m\}$ be two partitions of $[0, T]$, where $t_i^n = i\dfrac{T}{n}$ for $i = 0, 1, 2, \cdots, n$ and $t_j^m = j\dfrac{T}{m}$ for $j = 0, 1, 2, \cdots, m$. Let $t \in (t_{j-1}^m, t_j^m] \cap (t_{i-1}^n, t_i^n]$. By (3.3.25) and (12) of Lemma 1.1.2 we obtain

$$\frac{d^-}{dt}\|x_{(n)}(t) - x_{(m)}(t)\|_H^2 =$$

$$
\begin{aligned}
&= 2\Big(\frac{d^- x_{(n)}}{dt}(t) - \frac{d^- x_{(m)}}{dt}(t), x_{(n)}(t) - x_{(m)}(t)\Big)_H \\
&= 2(G^n(t) - G^m(t), x_{(n)}(t) - x_{(m)}(t))_H \\
&\quad -2(A^n(t) - A^m(t), x_{(n)}(t) - x_{(m)}(t)) \\
&\le 2\|G^n(t) - G^m(t)\|_H \|x_{(n)}(t) - x_{(m)}(t)\|_H \\
&\quad -2(A^n(t) - A^m(t), x_{(n)}(t) - x_{(m)}(t)). \qquad (3.3.29)
\end{aligned}
$$

But then we have

$$
\|G^n(t) - G^m(t)\|_H \le \epsilon_{nm} + \beta \sup_{s\in[-r,t]} \|x_{(n)}(s) - x_{(m)}(s)\|_H, \qquad (3.3.30)
$$

where $\lim_{n,m\to\infty} \epsilon_{nm} = 0$, actually ϵ_{nm} has the form $N_3\Big(\frac{1}{n} + \frac{1}{m}\Big)$ for some constant $N_3 > 0$.

In fact, we obtain

$$
\begin{aligned}
&\|G^n(t) - G^m(t)\|_H \\
&\quad = \|G(t^n_{i-1}, (\overline{x^n_i})_{t^n_{i-1}}) - G(t^m_{j-1}, (\overline{x^m_j})_{t^m_{j-1}})\|_H \\
&\quad \le \|G(t^n_{i-1}, (\overline{x^n_i})_{t^n_{i-1}}) - G(t^m_{j-1}, (\overline{x^n_i})_{t^n_{i-1}})\|_H \\
&\qquad + \|G(t^m_{j-1}, (\overline{x^n_i})_{t^n_{i-1}}) - G(t^m_{j-1}, (\overline{x^n_i})_{t^m_{j-1}})\|_H \\
&\qquad + \|G(t^m_{j-1}, (\overline{x^n_i})_{t^m_{j-1}}) - G(t^m_{j-1}, (\overline{x^m_j})_{t^m_{j-1}})\|_H \\
&\quad \le |t^n_{i-1} - t^m_{j-1}| l(\|(\overline{x^n_i})_{t^n_{i-1}}\|_{PC}) \\
&\qquad + \beta\|(\overline{x^n_i})_{t^n_{i-1}} - (\overline{x^n_i})_{t^m_{j-1}})\|_{PC} \\
&\qquad + \beta\|(\overline{x^n_i})_{t^m_{j-1}} - (\overline{x^m_j})_{t^m_{j-1}}\|_{PC} \qquad (3.3.31)
\end{aligned}
$$

and

$$
\begin{aligned}
&\|(\overline{x^n_i})_{t^m_{j-1}} - (\overline{x^m_j})_{t^m_{j-1}}\|_{PC} \\
&\quad = \sup_{\tau\in[-r,0]} \|(\overline{x^n_i})_{t^m_{j-1}}(\tau) - (\overline{x^m_j})_{t^m_{j-1}}(\tau)\|_H \\
&\quad = \sup_{\tau\in[-r,0]} \|(\overline{x^n_i})(t^m_{j-1} + \tau) - (\overline{x^m_j})(t^m_{j-1} + \tau)\|_H \\
&\quad = \sup_{s\in[t^m_{j-1}-r, t^m_{j-1}]} \|(\overline{x^n_i})(s) - (\overline{x^m_j})(s)\|_H \\
&\quad \le \sup_{s\in[-r,t]} \|(\overline{x^n_i})(s) - (\overline{x^m_j})(s)\|_H.
\end{aligned}
$$

If $t^m_{j-1} < t^n_{i-1}$ then

$$
\overline{x^n_i}(t^n_{i-1} + \tau) = \overline{x^n_n}(t^n_{i-1} + \tau) \text{ and } \overline{x^n_i}(t^m_{j-1} + \tau) = \overline{x^n_n}(t^m_{j-1} + \tau)
$$

for every $\tau \in [-r, 0]$. Also from Proposition 3.3.2 and the Lipschitz continuity of $x_{(n)}(t)$ on $[-r, T]$

$$\begin{aligned}
&\|(\overline{x_i^n})_{t_{i-1}^n} - (\overline{x_j^n})_{t_{j-1}^m}\|_{PC} \\
&\quad = \sup_{\tau\in[-r,0]} \|(\overline{x_i^n})_{t_{i-1}^n}(\tau) - (\overline{x_i^n})_{t_{j-1}^m}(\tau)\|_H \\
&\quad = \sup_{\tau\in[-r,0]} \|\overline{x_i^n}(t_{i-1}^n+\tau) - \overline{x_i^n}(t_{j-1}^m+\tau)\|_H \\
&\quad = \sup_{\tau\in[-r,0]} \|\overline{x_n^n}(t_{i-1}^n+\tau) - \overline{x_n^n}(t_{j-1}^m+\tau)\|_H \\
&\quad = \sup_{\tau\in[-r,0]} \|x_n(t_{i-1}^n+\tau) - x_n(t_{j-1}^m+\tau)\|_H \\
&\quad \le \sup_{\tau\in[-r,0]} \|x_{(n)}(t_{i-1}^n+\tau) - x_{(n)}(t_{j-1}^m+\tau)\|_H + 4M\frac{T}{n} \\
&\quad \le N_1|t_{i-1}^n - t_{j-1}^m| + 4M\frac{T}{n}.
\end{aligned}$$

Since

$$|t_{i-1}^n - t_{j-1}^m| = |t_{i-1}^n - t + t - t_{j-1}^m| \le \Big(\frac{1}{n} + \frac{1}{m}\Big)T$$

and

$$\sup_{s\in[-r,t]} \|\overline{x_i^n}(s) - \overline{x_j^m}(s)\|_H \le \sup_{s\in[-r,t]} \|x_{(n)}(s) - x_{(m)}(s)\|_H + 2M\Big(\frac{1}{n} + \frac{1}{m}\Big)T,$$

we obtain from (3.3.22) and (3.3.31)

$$\begin{aligned}
&\|G^n(t) - G^m(t)\|_H \le \\
&\quad \le l(K)\left(\frac{1}{n} + \frac{1}{m}\right)T + \beta \sup_{s\in[-r,t]} \|\overline{x_i^n}(s) - \overline{x_j^m}(s)\|_H \\
&\qquad + N_1\Big(\frac{1}{n} + \frac{1}{m}\Big)T + 4M\frac{T}{n} \\
&\quad \le l(K)\Big(\frac{1}{n} + \frac{1}{m}\Big)T + \beta\Big(\sup_{s\in[-r,t]} \|x_{(n)}(s) - x_{(m)}(s)\|_H + 2M\Big(\frac{1}{n} + \frac{1}{m}\Big)T\Big) \\
&\qquad + N_1\Big(\frac{1}{n} + \frac{1}{m}\Big)T + 4M\frac{T}{n} \\
&\quad \le \epsilon_{nm} + \beta \sup_{s\in[-r,t]} \|x_{(n)}(s) - x_{(m)}(s)\|_H,
\end{aligned}$$

where

$$\epsilon_{nm} = (l(K) + 2\beta M + N_1 + 4M)\Big(\frac{1}{n} + \frac{1}{m}\Big)T = N_3\Big(\frac{1}{n} + \frac{1}{m}\Big)$$

with $N_3 = (l(K) + 2\beta M + N_1 + 4M)T$ and $\lim_{n,m\to\infty} \epsilon_{nm} = 0$. Hence (3.3.30) holds.

We now have

$$\begin{aligned}\|\overline{x_i^n}(s) - \overline{x_j^m}(s)\|_H &= \|\overline{x_n^n}(s) - \overline{x_m^m}(s)\|_H = \|x_n(s) - x_m(s)\|_H \\ &\leq \|x_{(n)}(s) - x_{(m)}(s)\|_H + 2M\Big(\frac{1}{n} + \frac{1}{m}\Big)T\end{aligned}$$

for every $s \in [-r, t] \subset [-r, T]$. We write

$$\begin{aligned}&(A^n(t) - A^m(t), x_{(n)}(t) - x_{(m)}(t)) \\ &\quad = (A^n(t) - A^m(t), x_{(n)}(t) - x_{(m)}(t) - x_n(t) + x_m(t)) \\ &\qquad + (A^n(t) - A^m(t), x_n(t) - x_m(t))\end{aligned} \tag{3.3.32}$$

for every $t \in [0, T]$. From Proposition 3.3.2, (3.3.27), and (3.3.28) there exists a constant $N_4 > 0$ such that

$$(A^n(t) - A^m(t), x_{(n)}(t) - x_n(t) - x_{(m)}(t) + x_m(t)) \leq N_4\Big(\frac{1}{n} + \frac{1}{m}\Big).$$

for every $t \in [0, T]$. Also from the monotonicity of $A(t)$

$$\begin{aligned}&(A^n(t) - A^m(t), x_n(t) - x_m(t)) \\ &\quad = (A(t_{i-1}^n)x_n(t) - A(t_{j-1}^m)x_m(t), x_n(t) - x_m(t)) \\ &\quad = (A(t_{i-1}^n)x_n(t) - A(t_{i-1}^n)x_m(t), x_n(t) - x_m(t)) \\ &\qquad + (A(t_{i-1}^n)x_m(t) - A(t_{i-1}^m)x_m(t), x_n(t) - x_m(t)) \\ &\quad \geq (A(t_{i-1}^n)x_m(t) - A(t_{i-1}^m)x_m(t), x_n(t) - x_m(t)),\end{aligned}$$

and from $(A2)_3$

$$\begin{aligned}&|(A(t_{i-1}^n)x_m(t) - A(t_{j-1}^m)x_m(t), x_n(t) - x_m(t))| \\ &\quad \leq \|A(t_{i-1}^n)x_m(t) - A(t_{j-1}^m)x_m(t)\|_* \|x_n(t) - x_m(t)\| \\ &\quad \leq |t_{i-1}^n - t_{j-1}^m|\Big(\frac{1}{n} + \frac{1}{m}\Big)T(c_1 + c_2 L_1(\|x_m(t)\|)\end{aligned}$$

Hence, from (3.3.31)

$$\begin{aligned}&(A^n(t) - A^m(t), x_{(n)}(t) - x_{(m)}(t)) \\ &\quad \geq -N_4\Big(\frac{1}{n} + \frac{1}{m}\Big) - |t_{i-1}^n - t_{j-1}^m|(c_1 + c_2 L_1(\|x_m(t)\|) \\ &\quad \geq -N_4\Big(\frac{1}{n} + \frac{1}{m}\Big) - \Big(\frac{1}{n} + \frac{1}{m}\Big)T(c_1 + c_2 L_1(\|x_m(t)\|)\end{aligned}$$

From (3.3.7) with (3.3.29) and (3.3.30)

$$\begin{aligned}
&\frac{d^-}{dt}\|x_{(n)}(t) - x_{(m)}(t)\|_H^2 \\
&\quad \leq 2N_3\Big(\frac{1}{n} + \frac{1}{m}\Big) \\
&\quad\quad +\beta \sup_{s\in[-r,t]} \|x_{(n)}(s) - x_{(m)}(s)\|_H \|x_{(n)}(t) - x_{(m)}(t)\|_H \\
&\quad\quad +2N_4\Big(\frac{1}{n} + \frac{1}{m}\Big) + 2M_0T(c_1 + c_2L_1(M_0))\Big(\frac{1}{n} + \frac{1}{m}\Big) \\
&\quad = N_5\Big(\frac{1}{n} + \frac{1}{m}\Big) + 2\beta \sup_{s\in[-r,t]} \|x_{(n)}(s) - x_{(m)}(s)\|_H^2,
\end{aligned}$$

where

$$N_5 = 4N_3K + 3N_4 + 4M_0T(c_1 + c_2L_1(M_0)).$$

Integrating this inequality over $[0,t]$ we obtain

$$\begin{aligned}
&\|x_{(n)}(t) - x_{(m)}(t)\|_H^2 \\
&\quad \leq N_5\Big(\frac{1}{n} + \frac{1}{m}\Big)T + 2\beta \int_0^t \sup_{\sigma\in[-r,s]} \|x_{(n)}(\sigma) - x_{(m)}(\sigma)\|_H^2 ds,
\end{aligned}$$

where we have used $x_{(n)}(0) = x_{(m)}(0) = \phi(0)$. Since for every $t_1 \in [0,t]$ we have $t_1 \in (t^m_{j-1}, t^m_j] \cap (t^n_{i-1}, t^n_i]$ for some n, m, i, j we obtain

$$\begin{aligned}
&\|x_{(n)}(t_1) - x_{(m)}(t_1)\|_H^2 \\
&\quad \leq N_5\Big(\frac{1}{n} + \frac{1}{m}\Big)T + 2\beta \int_0^t \sup_{\sigma\in[-r,s]} \|x_{(n)}(\sigma) - x_{(m)}(\sigma)\|_H^2 ds.
\end{aligned}$$

Consequently,

$$\begin{aligned}
&\sup_{\sigma\in[-r,t]} \|x_{(n)}(\sigma) - x_{(m)}(\sigma)\|_H^2 \\
&\quad \leq N_5\left(\frac{1}{n} + \frac{1}{m}\right)T + 2\beta \int_0^t \sup_{\sigma\in[-r,s]} \|x_{(n)}(\sigma) - x_{(m)}(\sigma)\|_H^2 ds.
\end{aligned}$$

From Gronwall's inequality we have

$$\sup_{\sigma\in[-r,t]} \|x_{(n)}(\sigma) - x_{(m)}(\sigma)\|_H^2 \leq N_5Te^{2\beta T}\left(\frac{1}{n} + \frac{1}{m}\right),$$

which shows that the limit

$$\lim_{n,m\to\infty} \|x_{(n)}(t) - x_{(m)}(t)\|_H = 0$$

exists uniformly on $[-r, T]$. This implies that the limit

$$\lim_{n\to\infty} x_{(n)}(t) = x(t)$$

exists uniformly on $[-r, T]$, and thus by Proposition 3.3.2

$$\lim_{n\to\infty} x_n(t) = x(t)$$

uniformly on $[-r, T]$.

(1) Since $x_{(n)}(t)$ is Lipschitz continuous on $[-r, T]$ so is $x(t)$.

(2) First, $x(t) \in V \cap H$. Indeed, from (3.3.22) and (1) of Proposition 3.3.1,

$$\|x_n(t)\| + \|x_n(t)\|_H \leq M_0 + K$$

for every $t \in [0, T]$ and sufficiently large n, and hence by the reflexivity of $V \cap H$ there exist $\{x_{n_k}(t)\} \subset \{x_n(t)\}$ and $y^t \in V \cap H$ such that $w - \lim_{k\to\infty} x_{n_k}(t) = y^t$ in $V \cap H$ for every $t \in [0, T]$. Thus

$$\|y^t\| + \|y^t\|_H \leq M_0 + K$$

for every $t \in [0, T]$. On the other hand, $\lim_{n\to\infty} x_n(t) = x(t)$ in H. Then $x(t) = y^t$ for every $t \in [0, T]$. By the uniqueness of the limit we obtain $w - \lim_{n\to\infty} x_n(t) = x(t)$ in $V \cap H$ for every $t \in [0, T]$ and

$$\|x(t)\| + \|x(t)\|_H \leq M_0 + K \tag{3.3.33}$$

for every $t \in [0, T]$. Hence $x(t) \in V \cap H$ for every $t \in [0, T]$.

Next, we show that $x \in C_w([0, T]; V \cap H)$. Let $\{t_n\} \subset [0, T]$ and $t_0 \in [0, T]$ such that $\lim_{n\to\infty} t_n = t_0$. It follows from (3.3.32) that there exist $\{x(t_{n_k})\} \subset \{x(t_n)\}$ and $y \in V \cap H$ such that $w - \lim_{k\to\infty} = y$ in $V \cap H$. By the absolute continuity of $x(t)$, $\lim_{n\to\infty} x(t_{n_k}) = x(t_0)$ in H, and thus $x(t_0) = y$. Hence $w - \lim_{n\to\infty} = x(t_0)$ in $V \cap H$. That is, $x \in C_w([0, T]; V \cap H)$.

Finally, we prove that $x \in L^\infty(0, T; V \cap H)$. Since $x(t) \in V \cap H$ for every $t \in [0, T]$ with (3.3.32) it is sufficient to show that $x(t)$ is measurable. We prove that the set $\{x(t) \mid t \in [0, T]\}$ is separable in $V \cap H$ and $x(t)$ is weakly measurable. In fact, consider the countable set $M = \{x(r) \mid r \in [0, T]$ is a rational number$\}$. Let $L(M)$ be the smallest closed linear subspace of $V \cap H$ cotaining M. Then $L(M)$ is separable. Let $t \in [0, T]$. There exist rational numbers $r_n \in [0, T]$, $n = 1, 2, \cdots$ such that $\lim_{n\to\infty} r_n = t$. By the weak continuity of $x(t)$ in $V \cap H$, $w - \lim_{n\to\infty} x(r_n) = x(t)$ in $V \cap H$. Since $x(r_n) \in L(M)$ and $L(M)$ is weakly closed, $x(t) \in L(M)$ for every $t \in [0, T]$. Thus $\{x(t) \mid t \in [0, T]\}$ is separable in $V \cap H$. Since $x \in C_w([0, T]; V \cap H)$

$x(t)$ is weakly measurable. It follows from Pettis' Theorem [9] that $x(t)$ is measurable. Hence $x \in L^\infty(0,T;V \cap H)$.

Therefore $x \in C_w(0,T];V \cap H) \cap L^\infty(0,T;V \cap H)$.

(3) First, we show that

$$w - \lim_{n\to\infty} A^n(t) = A(t)x(t) \tag{3.3.34}$$

in H for every $t \in [0,T]$. In fact, from (3.3.27) there exist $\{n_k\} \subset \{n\}$ and $y^t \in H$ such that

$$w - \lim_{k\to\infty} A^{n_k}(t) = y^t \tag{3.3.35}$$

in H, also in $(V \cap H)^*$. By (3.3.27), (3.3.33), and the inequality

$$\begin{aligned}
&|(A^{n_k}(t), x_{n_k}(t)) - (y^t, x(t))| \\
&\quad \le |(A^{n_k}(t) - y^t, x(t))| + |(A^{n_k}(t), x_{n_k}(t) - x(t))| \\
&\quad \le |(A^{n_k}(t) - y^t, x(t))| + \|A^{n_k}(t)\|_H \|x_{n_k}(t) - x(t))\|_H,
\end{aligned}$$

we have

$$\lim_{k\to\infty} (A^{n_k}(t), x_{n_k}(t)) = (y^t, x(t))$$

for every $t \in [0,T]$. Using $(A1)_3$

$$(A(t^{n_k}_{i-1})y - A_{n_k}(t), y - x_{n_k}(t)) \ge 0 \tag{3.3.36}$$

for every $y \in V \cap H$, $t \in [0,T]$ and $i = 1, 2, \cdots$. From $(A2)_3$

$$\lim_{k\to\infty} A(t^{n_k}_{i-1})y = A(t)y$$

in $(V \cap H)^*$ for every $t \in [0,T]$. Since $w - \lim_{k\to\infty} x_{n_k}(t) = x(t)$ we have

$$\lim_{k\to\infty} (A(t^{n_k}_{i-1})y, y - x_{n_k}(t)) = (A(t)y, y - x(t))$$

for every $y \in V \cap H$, $t \in [0,T]$ and $i = 1, 2, \cdots$. By (3.3.35)

$$(A(t)y - y^t, y - x(t)) \ge 0 \tag{3.3.37}$$

for every $y \in V \cap H$ and $t \in [0,T]$. We put $y = x(t) + \lambda z$ in (3.3.36) with $z \in V \cap H$ and $\lambda > 0$. Then we obtain

$$(A(t)x(t) - y^t, z) = 0$$

for every $z \in V \cap H$ and $t \in [0,T]$ and thus $A(t)x(t) = y^t$ for every $t \in [0,T]$. Hence (3.3.34) holds.

Next, consider $(A(t_k)x(t_k), y)_H$ with $y \in V \cap H$ and $t_k \to t$ in $[0,T]$. From (3.3.27) and (3.3.33) $A(t_k)x(t_k)$ is bounded with respect to k. Thus there exist $\{A(t_{k_n})x(t_{k_n})\} \subset \{A(t_k)x(t_k)\}$ and $y_t \in H$ such that

$$w - \lim_{n\to\infty} A(t_{k_n})x(t_{k_n}) = y^t$$

in H. Using $(A1)_3$

$$(A(t_{k_n})y - A(t_{k_n})x(t_{k_n}), y - x(t_{k_n})) \geq 0$$

for every $y \in V \cap H$. From $\lim_{n\to\infty} A(t_{k_n})y = A(t)y$ in $(V \cap H)^*$ and the same argument as in the proof of (3.3.35), we conclude that $A(t)x(t) = y^t$ for every $t \in [0,T]$. Hence $A(\cdot)x(\cdot) \in C_w([0,T]; H)$.

(4) Integrating (3.3.25) over $[0,t]$ for $t \in [0,T]$ we have

$$\begin{aligned} &\int_0^t (A^n(\sigma), y)d\sigma + (x_{(n)}(t), y)_H \\ &\qquad = \int_0^t (G^n(\sigma), y)_H d\sigma + (\phi(0), y)_H \end{aligned} \tag{3.3.38}$$

for every $y \in V \cap H$ and $\in [0,T]$. By the same argument as that of the proof of (3.3.30) we obtain

$$\|G^n(t) - G(t, x_t)\|_H \leq \epsilon_n + \beta \sup_{s\in[-r,t]} \|x_{(n)}(s) - x(s)\|_H,$$

where $\epsilon_n > 0$, $\lim_{n\to\infty} \epsilon_n = 0$, and from $(G2)_3$ we obtain

$$\begin{aligned} \|G^n(t)\|_H &\leq \|G(t,\bar{0}) - G(0,\bar{0})\|_H + \|G(0,\bar{0})\|_H \\ &\leq tl(0) + \|G(0,\bar{0})\|_H \\ &\leq Tl(0) + \|G(0,\bar{0})\|_H. \end{aligned}$$

Using (3.3.26), (3.3.34) with this inequality and Lebesgue's dominated convergence theorem, as $n \to \infty$ in (3.3.38) we have

$$\begin{aligned} &\int_0^t (A(\sigma)x(s), y)d\sigma + (x(t), y)_H \\ &\qquad = \int_0^t (G(\sigma, x(\sigma), y)_H d\sigma + (\phi(0), y)_H \end{aligned} \tag{3.3.39}$$

for every $y \in V \cap H$ and $\in [0,T]$. Thus $\dfrac{dx}{dt}(t)$ is weakly continuous on $[0,T]$ in H and thus it is weakly measurable. By the same argument as that of the proof of (2) the set $\left\{\frac{dx}{dt}(t) \;\middle|\; t \in [0,T]\right\}$ is separable in H and hence $\frac{dx}{dt}(t)$ is measurable. By (3.3.39) we obtain

$$\left(\frac{dx}{dt}(t), y\right)_H = -(A(t)x(t), y) + (G(t, x_t), y)_H$$

for every $t \in [0,T]$ and $y \in V \cap H$, from which we have

$$\left|\left(\frac{dx}{dt}(t), y\right)_H\right| \leq C_0 \|y\|_H$$

for every $t \in [0,T]$ and $y \in V \cap H$ with some constant $C_0 > 0$. Hence

$$\left\|\left(\frac{dx}{dt}(t), y\right)\right\|_H \leq C_0$$

for every $t \in [0,T]$ and thus $\frac{dx}{dt} \in L^\infty(0,T;H)$.

(5) From (12) of Lemma 1.1.2 we obtain

$$\begin{aligned}
&\frac{d}{dt}\|x_1(t) - x_2(t)\|_H^2 \\
&\quad = 2\left(\frac{dx_1}{dt}(t) - \frac{dx_2}{dt}(t), x_1(t) - x_2(t)\right)_H \\
&\quad = 2(G(t,(x_1)_t) - G(t,x_2)_t), x_1(t) - x_2(t))_H \\
&\qquad -2(A(t)x_1(t) - A(t)x_2(t), x_1(t) - x_2(t)) \\
&\quad \leq 2\|G(t,(x_1)_t) - G(t,(x_2)_t)\|_H \|x_1(t) - x_2(t)\|_H
\end{aligned}$$

for every $t \in [0,T]$. Integrating this inequality over $[0,T]$ we deduce

$$\begin{aligned}
&\max_{t\in[0,T]} \|x_1(t) - x_2(t)\|_H^2 \\
&\quad \leq \|x_1(0) - x_2(0)\|_H^2 \\
&\qquad +2\max_{t\in[0,T]} \|x_1(t) - x_2(t)\|_H^2 \int_0^T \|G(t,(x_1)_t) - G(t,(x_2)_t)\|_H dt.
\end{aligned}$$

By this inequality we have

$$\max_{t\in[0,T]} \|x_1(t)-x_2(t)\|_H^2 \leq \|\phi_1-\phi_2\|_{PC}^2+2\int_0^T \|G(t,(x_1)_t)-G(t,(x_2)_t)\|_H dt.$$

From this inequality the uniqueness for the solution of (3.3.1) on $[-r,T]$ follows. Therefore the proof of Theorem is completed.

3.4. Methods of Lines

In this section we consider non-autonomous nonlinear functional evolutions of the type

$$\begin{cases} \dfrac{dx}{dt}(t) + A(t)x(t) \ni G(t, x_t), \quad 0 \le t \le T \\ x(t) = \phi(t), \quad -r \le t \le 0, \end{cases} \tag{3.4.1}$$

associated with an operator $A(t) : X \supset D(A(t)) \to 2^X$ for every $t \in [0, T]$, where $x(t) : [-r, T] \to X$ is an unknown function, $G : [0, T] \times PC([-r, 0]; X) \to X$ and $\phi \in PC([-r, 0]; X)$.

Actually, in Section 3.3 of this chapter the method of lines in $V \cap H$ was treated for (3.3.1), but we treated it for (3.4.1) by more general assumptions in Banach spaces.

Definition 3.4.1 *1. A function $x(t) : [-r, T] \to X$ is called a strong solution of* (3.4.1) *on $[-r, T]$ if:*

(1) $x(t) = \phi(t)$ for every $t \in [-r, 0]$;

(2) $x(t)$ is continuous on $[0, T]$;

(3) $x(t)$ is absolutely contionuous on every compact interval of $[0, T]$;

(4) $x(t)$ is differentiable a.e. $t \in (0, T)$;

(5) $x(t)$ satisfies

$$\frac{dx}{dt}(t) + A(t)x(t) = G(t, x_t)$$

a.e. $t \in (0, T)$.

2. A function $x(t) : [-r, T] \to X$ is called an integral solution of (3.4.1) *on $[-r, T]$ if:*

(1) $x(t) = \phi(t)$ for every $t \in [-r, 0]$;

(2) $x(t)$ is continuous on $[0, T]$;

(3) $x(t)$ satisfies

$$\|x(t) - x\| \le \|x(s) - x\| + \int_s^t ([G(\sigma, x_\sigma) - y, x(\sigma) - x]_+ + \theta(\sigma, \tau))d\sigma$$

for every $0 \le s \le t \le T$ and $[x, y] \in A(\tau)$ for some $\tau \in [0, T]$, where $\theta(\sigma, \tau)$ is a scalar function.

3. A function $x(t) : [-r, T] \to X$ is called a limit solution of (3.4.1) *on $[-r, T]$ if*

(1) $x(t) = \phi(t)$ *for every* $t \in [-r, 0]$;

(2) $x(t)$ *is continuous on* $[0, T]$;

(3) $x(t) = \lim_{n\to\infty} x_n(t)$ *uniformly on* $[0, T]$,

where $x_n(t) = \overline{x_n^n}(t)$ *for every* $t \in [-r, T]$ *with*

$$t_0^n = 0 < t_1^n < t_2^n < \cdots < t_n^n = T,$$
$$\lim_{n\to\infty} \max_{1\le i\le n}(t_i^n - t_{i-1}^n) = 0$$

and

$$\overline{x_i^n}(t) = \begin{cases} \phi(t), & t \in [-r, 0] \\ x_j^n, & t \in (t_{j-1}^n, t_j^n],\ j = 1, 2, \cdots, i-1 \\ x_i^n, & t \in (t_{i-1}^n, T] \end{cases}$$

satisfying

$$\frac{x_i^n - x_{i-1}^n}{t_i^n - t_{i-1}^n} + A(t^i)x_i^n \ni G(t_i^n, (\overline{x_i^n})_{t_i^n})$$

for $x_i^n \in D(A(t_i^n))$, $i = 1, 2, \cdots, n$; $n = 1, 2, \cdots$.

Also we call $\{\overline{x_i^n}(t)\}$, *or any such sequence valid for every large* n, *a method of lines for* (3.4.1), *and* $\{x_n(t)\}$ *a discrete scheme approximate solution of* (3.4.1).

We also assume the following conditions.

$(A1)_4$ $A(t) : X \supset D(A(t)) \to 2^X$ is m-accretive in X for every $t \in [0, T]$.

$(A2)_4$ There exists a non-decreasing continuous function $L : [0, \infty) \to [0, \infty)$ such that

$$\|A_\lambda(t_1)u - A_\lambda(t_2)u\| \le |t_1 - t_2| L(\|u\|)(1 + \|A_\lambda(t_2)u\|)$$

for every $u \in X$, $\lambda > 0$ and $t_i \in [0, T]$, $i = 1, 2$.

$(G1)_4$ There exists a constant $\beta > 0$ such that

$$\|G(t, \psi_1) - G(t, \psi_2)\| \le \beta\|\psi_1 - \psi_2\|_{PC}$$

for every $t \in [0, T]$ and $\psi_i \in PC([-r, 0] : X)$, $i = 1, 2$.

$(G2)_4$ There exists a non-decreasing function $l : [0, \infty) \to [0, \infty)$ such that

$$\|G(t_1, \psi) - G(t_2, \psi)\| \le |t_1 - t_2| l(\|\psi\|_{PC})$$

for every $\psi \in PC([-r, 0]; X)$ and $t_i \in [0, T]$, $i = 1, 2$.

Now, we give the existence of a method of lines for (3.4.1).

PROPOSITION 3.4.1 *Assume that* $(A1)_4$ *and* $(G1)_4$ *are satisfied. Let* $\phi \in C([-r,0];X)$. *Then there exists a method of lines for* (3.4.1).

Proof. Let $\{t_i^n\}$ be a partion of $[0,T]$, where $t_i^n = ih_n = i\frac{T}{n}$, $i = 1,2,\cdots,n$. Let $n > \beta T$. We define a space

$$S_1 = \{x : [-r,T] \to X \mid x(t) = \phi(t) \text{ for every } t \in [-r,0] \\ x(t) \text{ is constant on } (0,T]\}.$$

Then S_1 is a real Banach space with the norm $\|\cdot\|_T$ defined by

$$\|x\|_T = \sup_{t\in[-r,T]} \|x(t)\| \tag{3.4.2}$$

for every $x \in S_1$. We define a mapping Q_1 on S_1 by

$$(Q_1x)(t) = \begin{cases} \phi(t), & t \in [-r,0] \\ J_{h_n}(t_1^n)(x_0^n + h_nG(t_1^n, x_{t_1^n})), & t \in (0,T], \end{cases} \tag{3.4.3}$$

where $x_0^n = \phi(0)$. Since from (3.4.3) $Q_1x \in S_1$ for every $x \in S_1$ and by $(G1)_4$,

$$\begin{aligned} \|Q_1x - Q_1y\|_T &\leq h_n\|G(t_1^n, x_{t_1^n}) - G(t_1^n, y_{t_1^n})\| \\ &\leq h_n\beta\|x_{t_1^n} - y_{t_1^n}\|_{PC} \\ &\leq h_n\beta\|x-y\|_T \end{aligned}$$

for every x, $y \in S_1$ Q_1 is a strict contraction on S_1 into S_1. Using Banach's contraction principle there exists $\overline{x_1^n} \in S_1$ uniquely such that $Q_1\overline{x_1^n} = \overline{x_1^n}$. From (3.4.3)

$$\overline{x_1^n}(t) = \begin{cases} \phi(t), & t \in [-r,0] \\ J_{h_n}(t_1^n)(x_0^n + h_nG(t_1^n, (\overline{x_1^n})_{t_1^n})), & t \in (0,T]. \end{cases}$$

We also let

$$x_1^n = J_{h_n}(t_1^n)(x_0^n + h_nG(t_1^n, (\overline{x_1^n})_{t_1^n})).$$

Then

$$\overline{x_1^n}(t) = \begin{cases} \phi(t), & t \in [-r,0] \\ x_1^n, & t \in (0,T]. \end{cases}$$

Proceeding in a similar way we define a space S_i by

$$S_i = \{x : [-r,T] \to X \mid x(t) = \phi(t) \text{ for every } t \in [-r,0] \\ x(t) \text{ is constant on each interval} \\ (0,t_1^n],\ (t_1^n,t_2^n],\ \cdots,\ (t_{i-1}^n,T]\}.$$

for $i = 2, 3, \cdots, n$. Then each space S_i is a Banach space with norm (3.4.2). Moreover, the strict contractions $Q_i : S_i \to S_i$ are defined by

$$(Q_i x)(t) = \begin{cases} \phi(t), & t \in [-r, 0] \\ x_j^n, & t \in (t_{j-1}^n, t_j^n],\ j = 1, 2, \cdots, i-1 \\ J_{h_n}(t_i^n)(x_{i-1}^n + h_n G(t_i^n, x_{t_i^n})), & t \in (t_{i-1}^n, T]. \end{cases}$$

The unique fixed point of Q_i on S_i is denoted by $\overline{x_i^n}$ and we let

$$x_i^n = J_{h_n}(t_i^n)(x_{i-1}^n + h_n G(t_i^n, (\overline{x_i^n})_{t_i^n})$$

for $i = 2, 3, \cdots, n$. Then

$$\overline{x_i^n}(t) = \begin{cases} \phi(t), & t \in [-r, 0] \\ x_j^n, & t \in (t_{j-1}^n, t_j^n],\ j = 1, 2, \cdots, i-1 \\ J_{h_n}(t_i^n)(x_{i-1}^n + h_n G(t_i^n, (\overline{x_i^n})_{t_i^n})), & t \in (t_{i-1}^n, T], \end{cases}$$

and

$$\overline{x_i^n}(t) = \begin{cases} \phi(t), & t \in [-r, 0] \\ x_j^n, & t \in (t_{j-1}^n, t_j^n],\ j = 1, 2, \cdots, i-1 \\ x_i^n, & t \in (t_{i-1}^n, T], \end{cases}$$

that is, $x_i^n \in D(A(t_i^n))$, and thus

$$\frac{x_i^n - x_{i-1}^n}{t_i^n - t_{i-1}^n} + A(t^i)x_i^n \ni G(t_i^n, (\overline{x_i^n})_{t_i^n})$$

for $i = 2, 3, \cdots, n$. Hence we have

$$x_i^n = J_{h_n}(t_i^n)(x_{i-1}^n + h_n G(t_i^n, (\overline{x_i^n})_{t_i^n}) \tag{3.4.4}$$

with

$$\overline{x_i^n}(t) = \begin{cases} \phi(t), & t \in [-r, 0] \\ x_j^n, & t \in (t_{j-1}^n, t_j^n],\ j = 1, 2, \cdots, i-1 \\ x_i^n, & t \in (t_{i-1}^n, T], \end{cases}$$

and

$$\frac{x_i^n - x_{i-1}^n}{t_i^n - t_{i-1}^n} + A(t^i)x_i^n \ni G(t_i^n, (\overline{x_i^n})_{t_i^n}) \tag{3.4.5}$$

for $i = 1, 2, \cdots, n$; $n = 1, 2, \cdots$, and $\{\overline{x_i^n}(t)\}$ is a method of lines for (3.4.1)

Let $B_{\bar{r}}(0) = \{u \in X \mid \|u\| < \bar{r}\}$ for $\bar{r} > 0$. We consider the existence a method of lines for (3.4.1) with the boundary condition of the closure $\overline{B_{\bar{r}}(0)}$ of $B_{\bar{r}}(0)$. Moreover, we give the following conditions instead of $(A1)_4$ and $(G1)_4$:

$(A1)_D$ $A(t) : X \supset D \equiv D(A(t)) \to 2^X$ is m-accretive in X for every $t \in [0, T]$.

$(G1)_B$ There exists a constant $\beta > 0$ such that

$$\|G(t, \psi_1) - G(t, \psi_2)\| \leq \beta \|\psi_1 - \psi_2\|_{PC}$$

for every $t \in [0, T]$ and $\psi_i \in PC([-r, 0]; \overline{B_{\bar{r}}(0)})$, $i = 1, 2$. Here $\|\cdot\|_{PC}$ denotes the sup norm on $PC([-r, 0]; \overline{B_{\bar{r}}(0)})$.

$(G1)_D$ Condition $(G1)_B$ holds with D replaced by $B_{\bar{r}}(0)$.

Proposition 3.4.2 *Assume that $(A1)_D$ and $(G1)_B$ are satisfied. Let $\phi \in C([-r, 0]; \overline{B_{\bar{r}}(0)})$ with $\phi(0) \in D$. Assume furthermore that for every $\psi \in PC([-r, 0]; \overline{B_{\bar{r}}(0)})$ and $[u, v] \in A(t)$ with $\|u\| > \bar{r}$ there exists $u^* \in F(u)$ such that*

$$(v - G(t, \psi), u^*) \geq 0 \tag{3.4.6}$$

for every $t \in [0, T]$. Then there exists a method of lines for (3.4.1) *in $\overline{B_{\bar{r}}(0)}$.*

Proof. Let $\{t_i^n\}$ be a partition of $[0, T]$ with $t_0^n = 0$, $t_n^n = T$ and $t_i^n = ih_n = i\dfrac{T}{n}$ for $i = 0, 1, 2, \cdots, n$; $n = 1, 2, \cdots$. Let $x_0^n = \phi(0) \in D$. We put

$$\overline{x_0^n}(t) = \begin{cases} \phi(t), & t \in [-r, 0] \\ \phi(0), & t \in (0, T]. \end{cases}$$

We set

$$(g_1(u))(t) = \chi_{[-r,0]}(t)\overline{x_0^n}(t) + \chi_{(0,T]}(t)u$$

for every $t \in [-r, T]$ and $u \in X$, where χ_J denotes the characteristic function of the interval J, and

$$G_1(t)u = G(t, (g_1(u))_t)$$

for every $t \in [0, T]$ and $u \in \overline{B_{\bar{r}}(0)}$. Then it follows from $(G1)_B$ that the function $u \to G(t, (g_1(u)_t)$ is Lipschitz continuous on $\overline{B_{\bar{r}}(0)}$ with Lipschitz constant $\beta > 0$.

We now define a mapping $Q_1 : \overline{B_{\bar{r}}(0)} \to X$ by

$$Q_1 u = \Big(A(t_1^n) + \frac{1}{h_n} I\Big)^{-1} \Big(G_1(t_1^n)u + \frac{1}{h_n} x_0^n\Big) \tag{3.4.7}$$

for every $u \in \overline{B_{\bar{r}}(0)}$. Then from $(A1)_D$ and $(G1)_B$ we have

$$\begin{aligned} &\|Q_1 u_1 - Q_1 u_2\| \\ &\quad \le \|(h_n A(t_1^n) + I)^{-1}(h_n G_1(t_1^n)u_1 + x_0^n) \\ &\qquad -(h_n A(t_1^n) + I)^{-1}(h_n G_1(t_1^n)u_2 + x_0^n)\| \\ &\quad \le h_n \beta \|u_1 - u_2\| \end{aligned}$$

for every $u_i \in \overline{B_{\bar{r}}(0)}$, $i = 1, 2$.

We choose n sufficiently large that $h_n \beta < 1$ and show that $Q_1 u \in \overline{B_{\bar{r}}(0)}$ for every $u \in \overline{B_{\bar{r}}(0)}$. In fact, let $w = Q_1 u$ and $w \notin \overline{B_{\bar{r}}(0)}$. Then $w \in D$ and $\|w\| > \bar{r}$. From (3.4.7)

$$A(t_1^n)w + \frac{1}{h_n} w - G(t_1^n)w - \frac{1}{h_n} x_0^n \ni 0$$

and thus there exists $z \in A(t_1^n)w$ such that

$$z + \frac{1}{h_n} w - G(t_1^n)w - \frac{1}{h_n} x_0^n = 0.$$

By (3.4.6) there exists $w^* \in F(w)$ such that

$$\begin{aligned} 0 &= (z - G(t_1^n)w, w^*) + \frac{1}{h_n}(w - x_0^n, w^*) \\ &\ge \frac{1}{h_n}(\|w\|^2 - \|x_0^n\|\|w\|) \\ &\ge \frac{1}{h_n}(\|w\| - \bar{r})\|w\| > 0. \end{aligned}$$

Here we have used the fact that $G_1(t_1^n)u = G(t_1^n, (g_1(u))^{t_1^n})$ with $\|(g_1(u))_{t_1^n}\|_C \le \bar{r}$. The above contradition shows that $Q_1 : \overline{B_{\bar{r}}(0)} \to \overline{B_{\bar{r}}(0)}$. From Banach's contraction principle Q_1 has a unique fixed point $x_1^n \in \overline{B_{\bar{r}}(0)}$. Thus using (3.4.7) x_1^n is a solution of the equation

$$\Big(A(t_1^n) - G_1(t_1^n) + \frac{1}{h_n} I\Big) u \ni \frac{1}{h_n} x_0^n.$$

Moreover,

$$\frac{x_1^n - x_0^n}{h_n} + A(t_1^n)x_1^n \ni G(t_1^n, (\overline{x_1^n})_{t_1^n}). \tag{3.4.8}$$

We set

$$\overline{x_1^n}(t) = \begin{cases} \phi(t), & t \in [-r, 0] \\ x_1^n, & t \in (0, T]. \end{cases}$$

Then $\overline{x_1^n}(t) \in \overline{B_{\bar{r}}(0)}$ for every $t \in [-r, T]$. Similarly we define the functions g_i and G_i by

$$(g_i(u))(t) = \chi_{[-r, t_{i-1}^n]}(t)\overline{x_1^n}(t) + \chi_{(t_{i-1}^n, T]}(t)u$$

for every $t \in [-r, T]$ and $u \in X$, and

$$G_i(t)u = G(t, (g_i(u))_t)$$

for every $t \in [0, T]$, $u \in \overline{B_{\bar{r}}(0)}$ and for $i = 2, 3, \cdots, n$. Then the functions $G_i(t)u$ are Lipschitz continuous in $u \in \overline{B_{\bar{r}}(0)}$ with Lipschitz constant $\beta > 0$. Calling $x_i^n \in \overline{B_{\bar{r}}(0)}$ a solution of the equation

$$\Big(A(t_i^n) - G_i(t_i^n) + \frac{1}{h_n} I\Big)u \ni \frac{1}{h_n} x_{i-1}^n$$

we have constructed a method of lines $\{\overline{x_i^n}(t)\}$ for every $t \in [-r, T]$ with

$$\overline{x_i^n}(t) = \begin{cases} \phi(t), & t \in [-r, 0] \\ x_1^n, & t \in (0, t_1^n] \\ x_j^n, & t \in (t_{j-1}^n, t_j^n],\ j = 2, 3, \cdots, i-1 \\ x_i^n, & t \in (t_{i-1}^n, T] \end{cases}$$

for $i = 1, 2, \cdots, n$; $n = 1, 2, \cdots$. Moreover, with (3.4.8)

$$\frac{x_i^n - x_{i-1}^n}{h_n} + A(t_i^n)x_i^n \ni G(t_i^n, (\overline{x_i^n})_{t_i^n})$$

for $i = 1, 2, \cdots, n$; $n = 1, 2, \cdots$.

In the following we show the existence of a method of line for (3.4.1) using the homotopy type of result. First we need the next lemma for the proof of the homotopy type of lemma:

LEMMA 3.4.1 *Let V be an open subset of X with the closure $\overline{V}$ of V and the boundary ∂V of V. Let $Q : \overline{V} \to X$ be continuous and strongly accretive operator in X. Assume that there exists $v_0 \in V$ such that $\|Qv_0\| \leq \|Qv\|$ for every $v \in \partial V$. Then $0 \in Q(\overline{V})$.*

LEMMA 3.4.2 *Let V be an open subset of X with the closure $\overline{V}$ and boundary ∂V of V. Let $U : [0,1] \times \overline{V} \to X$ be a mapping satisfying*

(1) *$U(t,\cdot)$ is continuous and strongly accretive in X for every $t \in [0,1]$;*
(2) *$U(t,u)$ is continuous in $t \in [0,1]$ uniformly with respect to $u \in \overline{V}$;*
(3) *$U(t,v) \neq 0$ for every $t \in (0,1)$ and $v \in \partial V$;*
(4) *There exists $v_0 \in V$ such that*

$$\|U(0,v_0)\| \leq \|U(0,v)\|$$

for every $v \in \partial V$.

Then there exists $\bar{v} \in \overline{V}$ uniquely such that $U(1,\bar{v}) = 0$.

Proof. Suppose that $U(1,\bar{v}) = 0$ for some $\bar{v} \in \partial V$. Since $\bar{v} \in \overline{V}$ the conclusion of the lemma holds. Next, suppose that $U(1,v) \neq 0$ for every $v \in \partial V$. Let

$$\gamma = \inf\{\|U(t,v)\| \mid t \in [0,1],\ v \in \partial V\}.$$

We claim that $\gamma > 0$. Otherwise let $\gamma = 0$. Then there exist $\{t_k\} \subset [0,1]$ and $\{v_k\} \subset \partial V$ such that $\lim_{k\to\infty} U(t_k,v_k) = 0$. Let $\lim_{k\to\infty} t_k = t \in [0,1]$. From (2) $\lim_{k\to\infty} U(t,v_k) = 0$. Since by (1) $U(t,\cdot)$ is strongly accretive, $\lim_{k\to\infty} v_k = v \in \partial V$. Thus

$$\|U(t,v)\| \leq \|U(t,v) - U(t,v_k)\| + \|U(t,v_k)\| \to 0.$$

Thus $U(t,v) = 0$ for some $t \in [0,1]$ and $v \in \partial V$. But by (3) $t \notin (0,1)$, and $U(t,v) \neq 0$ for $v \in \partial V$. Suppose that $t = 0$. Then using (4) there exists $v_0 \in V$ such that

$$\|U(0,v_0)\| = \|U(0,v)\| = 0.$$

From the strong accretiveness of (1),

$$0 < \alpha\|v - v_0\| \leq \|U(0,v) - U(0,v_0)\| = 0,$$

where α is the strong accretiveness constant of U. Thus $\gamma > 0$. Let

$$E = \{t \in [0,1] \mid U(t,\cdot) \text{ has a zero in } \overline{V}\}.$$

From Lemma 3.4.1 $0 \in E$. We show that E is closed. Indeed, let $\{t_n\} \subset E$ converge to $t \in [0,1]$ as $n \to \infty$. Then there exists $\{v_n\} \subset \overline{V}$ such that $U(t_n,v_n) = 0$. By the above argument there exists $v \in \overline{V}$ such that $\lim_{n\to\infty} v_n = v$ and $U(t,v) = 0$. This implies $t \in E$, and thus E is closed.

Now we show that E is open. In fact, let $t_0 \in E$ and $v_0 \in \overline{V}$ such that $U(t_0, v_0) = 0$. Since $U(\cdot, v_0)$ is continuous there exists a neiborhood N of $t_0 \in [0,1]$ such that $\|U(t, v_0)\| < \gamma$ for every $t \in N$. Let $t \in N$. Then $v_0 \in \overline{V}$ and $\|U(t,v_0)\| < \gamma < \|U(t,v)\|$ for every $v \in \partial V$, from Lemma 3.4.1 there exists $v \in \overline{V}$ such that $U(t,v) = 0$. By the definition of E we have that $t \in E$. Thus E is open in $[0,1]$. It follows that $M = [0,1]$. Hence there exists $\bar{v} \in \overline{V}$ such that $U(1, \bar{v}) = 0$.

PROPOSITION 3.4.3 *Assume that* $(A1)_D$ *and* $(G1)_B$ *are satisfied with* $0 \in \bar{D}$. *Let* $\phi \in C([-r,0]; \overline{B_{\bar{r}}(0)})$ *with* $\phi(0) \in D$. *Assume furthermore that for every* $\psi \in PC([-r,0]; \overline{B_{\bar{r}}(0)})$ *and* $[u,v] \in A(t)$ *with* $\|u\| = \bar{r}$ *there exists* $u^* \in F(u)$ *such that*

$$(v - G(t,\psi), u^*) \geq 0 \tag{3.4.9}$$

for every $t \in [0,T]$. *Then there exists a method of lines for* (3.4.1) *in* $\overline{B_{\bar{r}}(0)}$.

Proof. In order to apply Lemma 3.4.2 we let $V = B_{\bar{r}}(0)$ and consider the homotopy equation

$$u = Q(t,u) \equiv \Big(tA(t_i^n) + \frac{1}{h_n} I\Big)^{-1} \Big(t\Big(G_i(t_i^n)u + \frac{1}{h_n} x_{i-1}^n\Big)\Big)$$

for every $t \in [0,1]$ $u \in \overline{B_{\bar{r}}(0)}$ and $i = 1, 2, \cdots, n$; $n = 1, 2, \cdots$, with the notations as in the proof of Proposition 3.4.2. Then since

$$\Big(tA(t_i^n) + \frac{1}{h_n} I\Big)^{-1} u = (h_n t A(t_i^n) + I)^{-1} h_n u$$

from $(A1)_D$ $Q(t,u)$ is Lipschitz continuous with respect to $u \in \overline{B_{\bar{r}}(0)}$ with Lipschitz constant $h_n\beta$. Let $U(t,u) = u - Q(t,u)$ and we show that $U(t,u)$ satisfies conditions of Lemma 3.4.2.

For sufficiently large n with $h_n\beta < 1$ we see that

$$(U(t,u_1) - U(t,u_2), u^*) \geq (1 - h_n\beta)\|u_1 - u_2\|^2$$

for every $u_i \in \overline{B_{\bar{r}}(0)}$, $i = 1, 2$, $t \in [0,1]$ and $u^* \in F(u_1 - u_2)$. Thus condition (1) of Lemma 3.4.2 is satisfied. For condition (2) of Lemma 3.4.2 we observe that

$$\begin{aligned} &\|Q(t,u) - Q(t_0,u)\| \\ &\quad = \|(h_n t A(t_i^n) + I)^{-1}(t(h_n G_i(t_i^n)u + x_{i-1}^n)) - \end{aligned}$$

$$
\begin{aligned}
&-(h_n t_0 A(t_i^n)+I)^{-1}(t_0(h_n G_i(t_i^n)u + x_{i-1}^n))\|\\
&\le \|(h_n t A(t_i^n)+I)^{-1}(t(h_n G_i(t_i^n)u + x_{i-1}^n))\\
&\quad -(h_n t_0 A(t_i^n)+I)^{-1}(t(h_n G_i(t_i^n)u + x_{i-1}^n))\|\\
&\quad +\|(h_n t_0 A(t_i^n)+I)^{-1}(t(h_n G_i(t_i^n)u + x_{i-1}^n))\\
&\quad -(h_n t_0 A(t_i^n)+I)^{-1}(t_0(h_n G_i(t_i^n)u + x_{i-1}^n))\|\\
&\le \frac{2|t-t_0|}{t_0}\|t(h_n G_i(t_i^n)u + x_{i-1}^n)\| + |t-t_0|\|h_n G_i(t_i^n)u + x_{i-1}^n\|
\end{aligned}
$$

for every t, $t_0 \in (0,1]$ and $u \in \overline{B_{\bar{r}}(0)}$. Thus $Q(t,u)$ is continuous at each $t_0 \in (0,1]$ uniformly with respect to $u \in \overline{B_{\bar{r}}(0)}$. To show this property at $t_0 = 0$ we have

$$
\begin{aligned}
&\left\|\left(tA(t_i^n)+\frac{1}{h_n}I\right)^{-1}\left(t\left(G_i(t_i^n)u+\frac{1}{h_n}x_{i-1}^n\right)\right)\right\|\\
&\quad\le \|(h_n t A(t_i^n)+I)^{-1}(t(h_n G_i(t_i^n)u + x_{i-1}^n) - (h_n t A(t_i^n)+I)^{-1}(0)\|\\
&\qquad +\|(h_n t A(t_i^n)+I)^{-1}(0)\|\\
&\quad\le \|t(h_n A(t_i^n)u + x_i^n\| + \|(h_n t A(t_i^n)+I)^{-1}(0)\|.
\end{aligned}
$$

Since $\lim_{t\to 0+}\|(h_n t A(t_i^n)+I)^{-1}(0)\| = 0$ $Q(t,u)$ is continuous at $t_0 = 0$ uniformly with respect to $u \in \overline{B_{\bar{r}}(0)}$. Thus $Q(t,u)$ is continuous in $t \in [0,1]$ uniformly with respect to $u \in \overline{B_{\bar{r}}(0)}$. Hence so is $U(t,u)$.

To show (3) of Lemma 3.4.2 let $t \in (0,1)$ and $u \in \partial B_{\bar{r}}(0)$ such that $U(t,u) = 0$. Then

$$
tA(t_i^n)u + \frac{1}{h_n}u \ni t\left(G_i(t_i^n)u + \frac{1}{h_n}x_{i-1}^n\right)
$$

or

$$
0 \in t(A(t_i^n)u - G_i(t_i^n)u) + \frac{1}{h_n}(u - t x_{i-1}^n).
$$

There exists $v \in A(t_i^n)u$ such that

$$
0 = t(v - G_i(t_i^n)u) + \frac{1}{h_n}(u - t x_{i-1}^n).
$$

Thus for an appropriate $u^* \in F(u)$

$$
\begin{aligned}
0 &= t(v - G_i(t_i^n)u, u^*) + \frac{1}{h_n}(u,u^*) - \frac{t}{h_n}(x_{i-1}^n, u^*)\\
&> \frac{1}{h_n}\|u\|^2 - \frac{1}{h_n}\bar{r}\|u\|\\
&= \frac{1}{h_n}\bar{r}^2 - \frac{1}{h_n}\bar{r}^2 = 0,
\end{aligned}
$$

where we have used (3.4.9).

Condition (4) of Lemma 3.4.2 is satisfied with $u_0 = 0$ because $U(0,0) = 0$. Thus, from Lemma 3.4.2 there exists $\bar{u} \in \overline{B_{\bar{r}}(0)}$ such that $U(1,\bar{u}) = 0$. This solution is the point x_i^n in the contruction of the method of lines as in the proof of Proposition 3.4.2.

PROPOSITION 3.4.4 *Assume $A(t) : D \equiv D(A(t)) \subset X \to X$ is accretive for every $t \in [0,T]$, $(G1)_B$ is satisfied and $\overline{B_{\bar{r}}(0)} \subset D$. Let $A(t)$ be continuous and bounded on $\overline{B_{\bar{r}}(0)} \subset D$. Let $\phi \in C([-r,0];\overline{B_{\bar{r}}(0)})$ with $\phi(0) \in D$. Assume, further, that for every $\psi \in PC([-r,0];\overline{B_{\bar{r}}(0)})$ and $u \in D$ with $\|u\| = \bar{r}$, there exists $u^* \in F(u)$ such that*

$$(A(t)u - G(t,\psi), u^*) \geq 0 \tag{3.4.10}$$

for every $t \in [0,T]$. Then there exists a method of lines for (3.4.1) *in* $\overline{B_{\bar{r}}(0)}$.

Proof. We set $V = B_{\bar{r}}(0)$ and consider the homotopy

$$U(t,u) = (1-t)u + tQu,$$

where

$$Qu = \Big(A(t_i^n) - G_i(t_i^n) + \frac{1}{h_n}I\Big)u - \frac{1}{h_n}x_{i-1}^n.$$

We show that the conditions in Lemma 3.4.2 are satisfied. We let sufficiently large n such that $\dfrac{1}{h_n} \geq 1 + \beta$ and have

$$\begin{aligned}
&(U(t,u_1) - U(t,u_2), u^*)\\
&\quad = (1-t)\|u_1 - u_2\|^2 + t(A(t_i^n)u_1 - A(t_i^n)u_2, u^*)\\
&\qquad - t(G_i(t_i^n)u_1 - G_i(t_i^n)u_2, u^*) + \frac{t}{h_n}\|u_1 - u_2\|^2\\
&\quad \geq (1-t)\|u_1 - u_2\|^2 - t\beta\|u_1 - u_2\|^2 + \frac{t}{h_n}\|u_1 - u_2\|^2\\
&\quad = (1-t)\|u_1 - u_2\|^2 + t\Big(\frac{t}{h_n} - \beta\Big)\|u_1 - u_2\|^2\\
&\quad = \Big(t\Big(\frac{t}{h_n} - (\beta+1\Big) + 1\Big)\|u_1 - u_2\|^2\\
&\quad \geq \|u_1 - u_2\|^2.
\end{aligned}$$

for every $u_i \in \overline{B_{\bar{r}}(0)}$, $i = 1,2$, $t \in [0,1]$, and an appropriate $u^* \in F(u_1 - u_2)$. Thus $U(t,u)$ is strongly accretive on $\overline{B_{\bar{r}}(0)})$ with repect to u. Since $A(t)$ is continuous condition (1) of Lemma 3.4.2 is satisfied.

Now we obtain

$$\begin{aligned}\|(\overline{x_i^n})_{t_i^n} - \phi(0)\|_{PC} &= \sup_{\tau\in[-r,0]} \|(\overline{x_i^n})(t_i^n+\tau) - \phi(0)\| \\ &= \sup_{\tau\in[t_i^n-r,0]} \|(\overline{x_i^n})(t) - \phi(0)\| \\ &\le \|\overline{x_i^n}(t) - \phi(0)\|_T.\end{aligned}$$

Thus we have that if $t \in (t_{i-1}^n, T]$ for $i = 1, 2, \cdots, n$ then

$$\|\overline{x_i^n}(t) - \phi(0)\| \le \|x_{i-1}^n - \phi(0)\|_T + h_n\beta\|\overline{x_i^n} - \phi(0)\|_T + h_nC_3,$$

where

$$\overline{x_0^n}(t) = \begin{cases} \phi(t), & t \in [-r, 0] \\ \phi(0), & t \in (0, T]. \end{cases}$$

If $t \in [-r, t_{i-1}^n]$ for $i = 1, 2, \cdots, n$ then

$$\|\overline{x_i^n}(t) - \phi(0)\| = \|\overline{x_{i-1}^n}(t) - \phi(0)\| \le \|\overline{x_i^n} - \phi(0)\|_T.$$

Hence

$$\|\overline{x_i^n} - \phi(0)\|_T \le \|x_{i-1}^n - \phi(0)\|_T + h_n\beta\|\overline{x_i^n} - \phi(0)\|_T + h_nC_3$$

for $i = 1, 2, \cdots, n$ which yields

$$(1 - \beta h_n)\|\overline{x_i^n} - \phi(0)\|_T \le \|x_{i-1}^n - \phi(0)\|_T + h_nC_3$$

for $i = 1, 2, \cdots, n$. Applying this inequality we obtain

$$\begin{aligned}&(1 - \beta h_n)\|\overline{x_i^n} - \phi(0)\|_T \\ &\le \frac{\|x_{i-2}^n - \phi(0)\|_T}{1 - \beta h_n} + \frac{h_nC_3}{1 - \beta h_n} + h_nC_3 \\ &\cdots \\ &\le \frac{\|x_0^n - \phi(0)\|_T}{(1 - \beta h_n)^{n-1}} + \sum_{j=0}^{n-1} \frac{h_nC_3}{(1 - \beta h_n)^j}.\end{aligned}$$

Thus

$$\begin{aligned}\|\overline{x_i^n} - \phi(0)\|_T &\le \frac{\|x_0^n - \phi(0)\|_T}{(1 - \beta h_n)^n} + \sum_{j=1}^{n} \frac{h_nC_3}{(1 - \beta h_n)^j} \\ &\le \frac{\|\phi - \phi(0)\|_{PC}}{(1 - \beta h_n)^n} + \sum_{j=1}^{n} \frac{h_nC_3}{(1 - \beta h_n)^n} \\ &\le \frac{\|\phi\|_{PC} + \|\phi(0)\|}{\left(1 - \beta\frac{T}{n}\right)^n} + \frac{TC_3}{\left(1 - \beta\frac{T}{n}\right)^n}.\end{aligned}$$

Since $\lim_{n\to\infty} 1\Big/\left(1-\beta\frac{T}{n}\right)^n = e^{\beta T}$ (3.4.11) holds.

Next we show that (3.4.12). Let n be sufficiently large such that $\beta h_n < 1$ and $h_n < 1$. From (3.4.4) for $i = 1$ we have

$$\begin{aligned}
\|x_1^n - x_0^n\| &= \|J_{h_n}(t_1^n)(x_0^n + h_n G(t_1^n, (\overline{x_1^n})_{t_1^n})) - x_0^n\| \\
&\le \|J_{h_n}(t_1^n)(x_0^n + h_n G(t_1^n, (\overline{x_1^n})_{t_1^n})) - J_{h_n}(t_1^n)x_0^n\| \\
&\quad + \|J_{h_n}(t_1^n)x_0^n - x_0^n\| \\
&\le h_n\|G(t_1^n, (\overline{x_1^n})_{t_1^n}) - G(t_1^n, \bar{0})\| + h_n\|G(t_1^n, \bar{0})\| \\
&\quad + h_n|A(t_1^n)\phi(0)| \\
&\le h_n\beta M_0 + h_n C_1 + h_n C_2 \\
&= h_n C_4,
\end{aligned}$$

where $C_4 = \beta M_0 + C_1 + C_2$.

By (3.4.4) for $i = 2, 3, \cdots, n$ we find

$$\begin{aligned}
\|x_i^n - x_{i-1}^n\| &= \|J_{h_n}(t_i^n)(x_{i-1}^n + h_n G(t_i^n, (\overline{x_i^n})_{t_i^n})) \\
&\qquad - J_{h_n}(t_{i-1}^n)(x_{i-2}^n + h_n G(t_{i-1}^n, (\overline{x_{i-1}^n})_{t_{i-1}^n}))\| \\
&\le \|J_{h_n}(t_i^n)(x_{i-1}^n + h_n G(t_i^n, (\overline{x_i^n})_{t_i^n})) \\
&\qquad - J_{h_n}(t_i^n))(x_{i-2}^n + h_n G(t_{i-1}^n, (\overline{x_{i-1}^n})_{t_{i-1}^n}))\| \\
&\quad + \|J_{h_n}(t_i^n))(x_{i-2}^n + h_n G(t_{i-1}^n, (\overline{x_{i-1}^n})_{t_{i-1}^n})) \\
&\qquad - J_{h_n}(t_{i-1}^n)(x_{i-2}^n + h_n G(t_{i-1}^n, (\overline{x_{i-1}^n})_{t_{i-1}^n}))\| \\
&\le \|x_{i-1}^n - x_{i-2}^n\| \\
&\quad + h_n\|G(t_i^n, (\overline{x_i^n})_{t_i^n}) - G(t_{i-1}^n, (\overline{x_{i-1}^n})_{t_{i-1}^n})\| \\
&\quad + h_n\|A_{h_n}(t_i^n))(x_{i-2}^n + h_n G(t_{i-1}^n, (\overline{x_{i-1}^n})_{t_{i-1}^n})) \\
&\qquad - A_{h_n}(t_{i-1}^n))(x_{i-2}^n + h_n G(t_{i-1}^n, (\overline{x_{i-1}^n})_{t_{i-1}^n}))\|.
\end{aligned}$$

Thus $(A2)_4$, $(G1)_4$, and $(G2)_4$ imply

$$\begin{aligned}
\|x_i^n - x_{i-1}^n\| &\le \|x_{i-1}^n - x_{i-2}^n\| + h_n\beta\|(\overline{x_i^n})_{t_i^n} - (\overline{x_{i-1}^n})_{t_{i-1}^n}\|_{PC} \\
&\quad + h_n|t_i^n - t_{i-1}^n| l(\|(\overline{x_{i-1}^n})_{t_{i-1}^n}\|_{PC}) \\
&\quad + h_n|t_i^n - t_{i-1}^n| L(\|x_{i-2}^n + h_n G(t_{i-1}^n, (\overline{x_{i-1}^n})_{t_{i-1}^n})\|) \\
&\qquad \times (1 + \|A_{h_n}(t_{i-1}^n)(x_{i-2}^n + h_n G(t_{i-1}^n, (\overline{x_{i-1}^n})_{t_{i-1}^n}))\|).
\end{aligned}$$

Since

$$\|G(t_{i-1}^n, (\overline{x_{i-1}^n})_{t_{i-1}^n})\| \le$$

$$\leq \|G(t_{i-1}^n, (\overline{x_{i-1}^n})_{t_{i-1}^n}) - G(t_{i-1}^n, \bar{0})\| - G(t_{i-1}^n, \bar{0})\|$$
$$\leq \beta M_0 + C_2$$

and

$$A_{h_n}(t_{i-1}^n)(x_{i-2}^n + h_n G(t_{i-1}^n, (\overline{x_{i-1}^n})_{t_{i-1}^n})) = G(t_{i-1}^n, (\overline{x_{i-1}^n})_{t_{i-1}^n}) - \frac{x_{i-1}^n - x_{i-2}^n}{h_n}$$

we have

$$\|x_i^n - x_{i-1}^n\| \leq \|x_{i-1}^n - x_{i-2}^n\| + h_n\beta\|(\overline{x_i^n})_{t_i^n} - (\overline{x_{i-1}^n})_{t_{i-1}^n}\|_{PC} + h_n^2 l(M_0) + h_n^2 L(M_0 + \beta M_0 + C_2)\Big(1 + \beta M_0 + C_2 + \frac{\|x_{i-1}^n - x_{i-2}^n\|}{h_n}\Big),$$

from which it follows that

$$\frac{\|x_i^n - x_{i-1}^n\|}{h_n} \leq \beta\|(\overline{x_i^n})_{t_i^n} - (\overline{x_{i-1}^n})_{t_{i-1}^n}\|_{PC} + h_n l(C_5) + h_n L(C_5)C_5 + L(C_5)\|x_{i-1}^n - x_{i-2}^n\| + \frac{\|x_{i-1}^n - x_{i-2}^n\|}{h_n}, \qquad (3.4.13)$$

where $C_5 = \max\{M_0(\beta + 1) + C_2, 1 + \beta M_0 + C_2\}$.

Moreover, by the definition of $\overline{x_i^n}$

$$(\overline{x_{i-1}^n})_{t_{i-1}^n}(\tau) = (\overline{x_i^n})_{t_{i-1}^n}(\tau)$$

for every $\tau \in [-r, 0]$, and

$$\|(\overline{x_i^n})_{t_i^n} - (\overline{x_{i-1}^n})_{t_{i-1}^n}\|_{PC} = \sup_{\tau\in[-r,0]} \|\overline{x_i^n}(t_i^n + \tau) - \overline{x_{i-1}^n}(t_{i-1}^n + \tau)\| \leq \max_{1\leq j\leq i} \|x_j^n - x_{j-1}^n\| + L_0 h_n.$$

Hence using (3.4.13)

$$\max_{1\leq j\leq i} \frac{\|x_j^n - x_{j-1}^n\|}{h_n} \leq \beta \max_{1\leq j\leq i} \|x_j^n - x_{j-1}^n\| + \beta L_0 h_n + h_n l(C_5) + h_n L(C_5)C_5 + L(C_5) \max_{1\leq j\leq i} \|x_j^n - x_{j-1}^n\| + \max_{1\leq j\leq i-1} \frac{\|x_j^n - x_{j-1}^n\|}{h_n}.$$

Choosing n sufficiently large such that $(\beta + L(C_5))h_n < 1$ and putting $p_n = 1 - (\beta + L(C_5))h_n \in (0, 1)$ we obtain

$$\begin{aligned} \frac{p_n}{h_n} \max_{1 \le j \le i} \frac{\|x_j^n - x_{j-1}^n\|}{h_n} \\ \le C_6 h_n + \max_{1 \le j \le i-1} \frac{\|x_j^n - x_{j-1}^n\|}{h_n} \end{aligned} \tag{3.4.14}$$

for $i = 2, 3, \cdots, n$, where $C_6 = l(C_5) + \beta L_0 L(C_5) C_5$. Applying (3.4.14) we have

$$\begin{aligned} &\frac{p_n}{h_n} \max_{1 \le j \le i} \frac{\|x_j^n - x_{j-1}^n\|}{h_n} \\ &\le h_n C_6 + \frac{h_n C_6}{p_n} + \frac{1}{p_n} \max_{1 \le j \le i-2} \frac{\|x_j^n - x_{j-1}^n\|}{h_n} \\ &\cdots \\ &\le h_n C_6 \sum_{k=0}^{n-2} \frac{1}{p_n^k} + \frac{1}{p_n^{n-2}} \frac{\|x_1^n - x_0^n\|}{h_n} \\ &\le h_n C_6 \sum_{k=0}^{n-1} \frac{1}{p_n^k} + \frac{1}{p_n^{n-1}} \frac{\|x_1^n - x_0^n\|}{h_n}. \end{aligned}$$

Therefore

$$\max_{1 \le j \le n} \frac{\|x_j^n - x_{j-1}^n\|}{h_n} \le h_n C_6 \sum_{k=1}^{n} \frac{1}{p_n^k} + \frac{C_4}{p_n^n}.$$

Since

$$h_n \sum_{k=1}^{n} \frac{1}{p_n^k} \le h_n \sum_{k=1}^{n} \frac{1}{p_n^n} = \frac{T}{\left(1 - \dfrac{(\beta + L(C_5))T}{n}\right)^n}$$

and

$$\lim_{n \to \infty} \frac{1}{\left(1 - \dfrac{(\beta + L(C_5))T}{n}\right)^n} = e^{(\beta + L(C_5))T}$$

(3.4.12) holds.

In the following we show that $\{x_n(t)\}$ is a uniform Cauchy sequence on $[-r, T]$.

LEMMA 3.4.3 *Under the assumptions of Proposition 3.4.6 let* $x_n(t) = \overline{x_n^n}(t)$ *for* $n \ge n_0$. *Then there exists a constant* $M = M(\phi) > 0$ *such that*

$$\|x_n(t) - x_n(s)\| \le M(|t - s| + h_n) \tag{3.4.15}$$

for every $t,\ s \in [-r, T]$ *and* $n \geq n_0$.

Proof. We consider Rothe's functions

$$x_{(n)}(t) = \begin{cases} \phi(t), & t \in [-r, 0] \\ x_{i-1}^n + (t - t_{i-1}^n)\dfrac{x_i^n - x_{i-1}^n}{h_n}, & t \in [t_{i-1}^n, t_i^n],\ i = 1, 2, \cdots, n. \end{cases}$$

Then $x_{(n)}(t)$ is Lipschitzian with Lipschitz constant $M_2 = \max\{M_1, L_0\}$, $t \in [-r, T]$.

Let $t \in (0, T]$ and $t \in (t_{i-1}^n, t_i^n]$ for some $i = 1, 2, \cdots, n$.

From the definition of $x_{(n)}(t)$ we have

$$\begin{aligned} &\|x_n(t) - x_{(n)}(t)\| \\ &= \left\| x_i^n - x_{i-1}^n - (t - t_{i-1}^n)\frac{x_i^n - x_{i-1}^n}{h_n} \right\| \\ &= \left\| (h_n - (t - t_{i-1}^n))\frac{x_i^n - x_{i-1}^n}{h_n} \right\| \\ &\leq (t_i^n - t)\frac{\|x_i^n - x_{i-1}^n\|}{h_n} \\ &\leq M_1 h_n \leq M_2 h_n. \end{aligned}$$

Since $x_n(t) = x_{(n)}(t) = \phi(t)$ for every $t \in [-r, 0]$ we have

$$\|x_n(t) - x_{(n)}(t)\| \leq M_2 h_n$$

for every $t \in [-r, T]$. By Lipschitz continuity of $x_{(n)}(t)$ we obtain

$$\begin{aligned} &\|x_n(t) - x_n(s)\| \\ &\quad \leq \|x_n(t) - x_{(n)}(t)\| + \|x_{(n)}(t) - x_{(n)}(s)\| + \|x_{(n)}(s) - x_n(s)\| \\ &\quad \leq M_2|t - s| + 2M_2 h_n \\ &\quad \leq 2M_2(|t - s| + h_n) \\ &\quad = M(|t - s| + h_n), \end{aligned}$$

where $M = 2M_2$.

LEMMA 3.4.4 *Under the assumptions of Proposition 3.4.6 let* $\{x_i^n\}$ *and* $\{x_j^m\}$ *be two methods of lined for* (3.4.1) *as in Proposition 3.4.1. Then* *there exists a constant $C_7 > 0$ such that*

$$
\begin{aligned}
&\|x_i^n - x_j^m\| \\
&\quad \le \frac{h_n}{h_n + h_m}\|x_i^n - x_{j-1}^m\| + \frac{h_n}{h_n + h_m}\|x_{i-1}^n - x_j^m\| \\
&\quad\quad + \frac{h_n h_m}{h_n + h_m}(C_7|t_i^n - t_j^m| \\
&\quad\quad\quad + [x_i^n - x_j^m, G(t_i^n, (\overline{x_i^n})_{t_i^n}) - G(t_j^m, (\overline{x_j^m})_{t_j^m})]_+).
\end{aligned}
$$

Proof. We put $h_{n,m} = \dfrac{h_n h_m}{h_n + h_m}$ and let $0 < \lambda < 1$. From (3.4.5) we have

$$A(t_i^n)x_i^n \ni \frac{x_{i-1}^n - x_i^n}{h_n} + G(t_i^n, (\overline{x_i^n})_{t_i^n})$$

and

$$A(t_j^m)x_j^m \ni \frac{x_{j-1}^m - x_j^m}{h_m} + G(t_j^m, (\overline{x_j^m})_{t_j^m}).$$

Thus

$$x_i^n = J_{\lambda h_{n,m}}(t_i^n)\Big(x_i^n + \lambda h_{n,m}\Big(\frac{x_{i-1}^n - x_i^n}{h_n} + G(t_i^n, (\overline{x_i^n})_{t_i^n})\Big)\Big)$$

and

$$x_j^m = J_{\lambda h_{n,m}}(t_j^m)\Big(x_j^m + \lambda h_{n,m}\Big(\frac{x_{j-1}^m - x_j^m}{h_m} + G(t_j^m, (\overline{x_j^m})_{t_j^m})\Big)\Big).$$

By $(A2)_4$

$$
\begin{aligned}
&\|x_i^n - x_j^m\| \\
&\quad = \Big\|J_{\lambda h_{n,m}}(t_i^n)\Big(x_i^n + \lambda h_{n,m}\Big(\frac{x_{i-1}^n - x_i^n}{h_n} + G(t_i^n, (\overline{x_i^n})_{t_i^n})\Big)\Big) \\
&\quad\quad - J_{\lambda h_{n,m}}(t_j^m)\Big(x_j^m + \lambda h_{n,m}\Big(\frac{x_{j-1}^m - x_j^m}{h_m} + G(t_j^m, (\overline{x_j^m})_{t_j^m})\Big)\Big)\Big\| \\
&\quad \le \Big\|J_{\lambda h_{n,m}}(t_i^n)\Big(x_i^n + \lambda h_{n,m}\Big(\frac{x_{i-1}^n - x_i^n}{h_n} + G(t_i^n, (\overline{x_i^n})_{t_i^n})\Big)\Big) \\
&\quad\quad - J_{\lambda h_{n,m}}(t_i^n)\Big(x_j^m + \lambda h_{n,m}\Big(\frac{x_{j-1}^m - x_j^m}{h_m} + G(t_j^m, (\overline{x_j^m})_{t_j^m})\Big)\Big)\Big\| \\
&\quad + \Big\|J_{\lambda h_{n,m}}(t_i^n)\Big(x_j^m + \lambda h_{n,m}\Big(\frac{x_{j-1}^m - x_j^m}{h_m} + G(t_j^m, (\overline{x_j^m})_{t_j^m})\Big)\Big) \\
&\quad\quad - J_{\lambda h_{n,m}}(t_j^m)\Big(x_j^m + \lambda h_{n,m}\Big(\frac{x_{j-1}^m - x_j^m}{h_m} + G(t_j^m, (\overline{x_j^m})_{t_j^m})\Big)\Big)\Big\| \le
\end{aligned}
$$

$$
\begin{aligned}
&\leq \Big\|(x_i^n - x_j^m) + \lambda h_{n,m}\Big(\frac{x_{i-1}^n - x_i^n}{h_n} + G(t_i^n, (\overline{x_i^n})_{t_i^n})\\
&\qquad -\frac{x_{j-1}^m - x_j^m}{h_m} - G(t_j^m, (\overline{x_j^m})_{t_j^m})\Big)\Big\|\\
&\quad +\lambda h_{n,m}|t_i^n - t_j^m|\\
&\quad \times L\Big(\Big\|x_j^m + \lambda h_{n,m}\Big(\frac{x_{j-1}^m - x_j^m}{h_m} + G(t_j^m, (\overline{x_j^m})_{t_j^m})\Big)\Big\|\Big)\\
&\qquad \times\Big(1 + \Big\|A_{\lambda h_{n,m}}(t_j^m)\\
&\qquad\qquad \times\Big(x_j^m + \lambda h_{n,m}\Big(\frac{x_{j-1}^m - x_j^m}{h_m} + G(t_j^m, (\overline{x_j^m})_{t_j^m})\Big)\Big)\Big\|\Big)\\
&= \Big\|(1-\lambda)(x_i^n - x_j^m) + \lambda(x_i^n - x_j^m) +\\
&\quad +\lambda h_{n,m}\Big(\frac{x_{i-1}^n - x_i^n}{h_n} + G(t_i^n, (\overline{x_i^n})_{t_i^n})\\
&\qquad -\frac{x_{j-1}^m - x_j^m}{h_m} - G(t_j^m, (\overline{x_j^m})_{t_j^m})\Big)\Big\| + U_\lambda\\
&\leq \|(1-\lambda)(x_i^n - x_j^m) + \lambda h_{n,m}(G(t_i^n, (\overline{x_i^n})_{t_i^n}) - G(t_j^m, (\overline{x_j^m})_{t_j^m}))\|\\
&\quad +\lambda\Big\|(x_i^n - x_j^m) + h_{n,m}\Big(\frac{x_{i-1}^n - x_i^n}{h_n} - \frac{x_{j-1}^m - x_j^m}{h_m}\Big)\Big\| + U_\lambda\\
&\leq \|(1-\lambda)(x_i^n - x_j^m) + \lambda h_{n,m}(G(t_i^n, (\overline{x_i^n})_{t_i^n}) - G(t_j^m, (\overline{x_j^m})_{t_j^m}))\|\\
&\quad +\frac{\lambda h_n}{h_n + h_m}\|x_i^n - x_{j-1}^m\| + \frac{\lambda h_m}{h_n + h_m}\|x_{i-1}^n - x_j^m\| + U_\lambda.
\end{aligned}
$$

Here

$$
\begin{aligned}
U_\lambda = {}& \lambda h_{n,m}|t_i^n - t_j^m| L\Big(\Big\|x_j^m + \lambda h_{n,m}\Big(\frac{x_{j-1}^m - x_j^m}{h_m} + G(t_j^m, (\overline{x_j^m})_{t_j^m})\Big)\Big\|\Big)\\
&\times\Big(1 + \Big\|A_{\lambda h_{n,m}}(t_j^m)\\
&\qquad \times\Big(x_j^m + \lambda h_{n,m}\Big(\frac{x_{j-1}^m - x_j^m}{h_m} + G(t_j^m, (\overline{x_j^m})_{t_j^m})\Big)\Big)\Big\|\Big).
\end{aligned}
$$

Thus

$$
\begin{aligned}
&\|x_i^n - x_j^m\|\\
&\quad \leq \frac{h_n}{h_n + h_m}\|x_i^n - x_{j-1}^m\| + \frac{h_m}{h_n + h_m}\|x_{i-1}^n - x_j^m\| + \frac{U_\lambda}{\lambda}\\
&\qquad +\frac{1}{\tau}(\|(x_i^n - x_j^m) + \tau h_{n,m}(G(t_i^n, (\overline{x_i^n})_{t_i^n}) - G(t_j^m, (\overline{x_j^m})_{t_j^m}))\| -
\end{aligned}
$$

$$
\begin{aligned}
&-\|x_i^n - x_j^m\|) \\
&= \frac{h_n}{h_n+h_m}\|x_i^n - x_{j-1}^m\| + \frac{h_m}{h_n+h_m}\|x_{i-1}^n - x_j^m\| + \frac{U_\lambda}{\lambda} \\
&+h_{n,m}[G(t_i^n, (\overline{x_i^n})_{t_i^n}) - G(t_j^m, (\overline{x_j^m})_{t_j^m}), x_i^n - x_j^m]_\tau. \qquad (3.4.16)
\end{aligned}
$$

On the other hand since, from Proposition 3.4.6 and its proof

$$
\begin{aligned}
&\left\|A_{\lambda h_{n,m}}(t_j^m)\Big(x_j^m + \lambda h_{n,m}\Big(\frac{x_{j-1}^m - x_j^m}{h_m} + G(t_j^m, (\overline{x_j^m})_{t_j^m})\Big)\Big)\right\| \\
&\quad = \left\|\frac{x_{j-1}^m - x_j^m}{h_m} + G(t_j^m, (\overline{x_j^m})_{t_j^m})\right\| \\
&\quad \le \left\|\frac{x_{j-1}^m - x_j^m}{h_m}\right\| + \|G(t_j^m, (\overline{x_j^m})_{t_j^m})\| \\
&\quad \le M_1 + \beta M_0 + C_2,
\end{aligned}
$$

and

$$
\frac{U_\lambda}{\lambda} \le |t_i^n - t_j^m| L(M_0 + \lambda h_{n,m}(M_1 + \beta M_0 + C_2))(1 + M_1 + \beta M_0 + C_2)
$$

we obtain

$$
\begin{aligned}
\lim_{\lambda \to 0+} \frac{U_\lambda}{\lambda} &\le |t_i^n - t_j^m| L(M_0)(1 + M_1 + \beta M_0 + C_2) \\
&= C_7 |t_i^n - t_j^m|,
\end{aligned}
$$

where $C_7 = L(M_0)(1 + M_1 + \beta M_0 + C_2)$. Letting $\lambda \to 0+$ and $\tau \to 0$ in (3.4.16) we obtain

$$
\begin{aligned}
&\|x_i^n - x_j^m\| \\
&\quad \le \frac{h_n}{h_n+h_m}\|x_i^n - x_{j-1}^m\| + \frac{h_n}{h_n+h_m}\|x_{i-1}^n - x_j^m\| \\
&\quad + \frac{h_n h_m}{h_n+h_m}(C_7|t_i^n - t_j^m| \\
&\qquad + [G(t_i^n, (\overline{x_i^n})_{t_i^n}) - G(t_j^m, (\overline{x_j^m})_{t_j^m}), x_i^n - x_j^m]_+),
\end{aligned}
$$

where we have used the continuity of $L : [0, \infty) \to [0, \infty)$.

PROPOSITION 3.4.7 *Under the assumptions of Proposition 3.4.6, let* $\{x_i^n\}$ *and* $\{x_j^m\}$ *be two methods of lined for* (3.4.1) *as in Proposition 3.4.1, and let* $x_n(t)$ *and* $x_m(t)$ *be the corresponding discrete scheme approximate solutions of* (3.4.1) *for* $n, m \ge n_0$*, respectively. Then there exist constants* $C = C(\phi)$ *and* $\epsilon_{n,m} > 0$ *with* $\lim_{n,m\to\infty} \epsilon_{n,m} = 0$ *such that*

$$
\|x_i^n - x_j^m\| \le CD_{i,j} + E_i + ih_n(CD_{i,j} + \epsilon_{n,m}) \qquad (3.4.17)
$$

for $i = 0, 1, 2, \cdots, n$ *and* $j = 0, 1, 2, \cdots, m$. *Here* $D_{i,j}$ *and* E_i *are defined by*

$$D_{i,j} = ((t_i^n - t_j^m)^2 + h_n t_i^n + h_m t_j^m)^{\frac{1}{2}}$$

and

$$E_i = \beta \sum_{k=1}^{i} \sup_{t \in [-r, t_k^n]} \|x_n(t) - x_m(t)\| h_n.$$

Proof. From $(G1)_4$ and $(G2)_4$ we have

$$\begin{aligned}
&\|G(t_i^n, (\overline{x_i^n})_{t_i^n}) - G(t_j^m, (\overline{x_j^m})_{t_j^m})\| \\
&\quad \leq \|G(t_i^n, (\overline{x_i^n})_{t_i^n}) - G(t_i^n, (\overline{x_j^m})_{t_j^m})\| \\
&\qquad + \|G(t_i^n, (\overline{x_j^m})_{t_j^m}) - G(t_j^m, (\overline{x_j^m})_{t_j^m})\| \\
&\quad \leq \beta \|(\overline{x_i^n})_{t_i^n} - (\overline{x_j^m})_{t_j^m}\|_{PC} + |t_i^n - t_j^m| l(M_0).
\end{aligned}$$

Now, by (3.4.15)

$$\begin{aligned}
&\|(\overline{x_i^n})_{t_i^n} - (\overline{x_j^m})_{t_j^m}\|_{PC} \\
&\quad = \|(\overline{x_n^n})_{t_i^n} - (\overline{x_m^m})_{t_j^m}\|_{PC} \\
&\quad \leq \|(\overline{x_n^n})_{t_i^n} - (\overline{x_m^m})_{t_i^n}\|_{PC} + \|(\overline{x_m^m})_{t_i^n} - (\overline{x_m^m})_{t_j^m}\|_{PC} \\
&\quad = \sup_{t \in [t_i^n - r, t_i^n]} \|x_n(t) - x_m(t)\| \\
&\qquad + \sup_{\tau \in [-r, 0]} \|x_m(t_i^n + \tau) - x_m(t_j^m + \tau)\| \\
&\quad \leq \sup_{t \in [-r, t_i^n]} \|x_n(t) - x_m(t)\| + M(|t_i^n - t_j^m| + h_m),
\end{aligned}$$

and

$$\begin{aligned}
|t_i^n - t_j^m| &\leq |(t_i^n - t_j^m) - h_n| + h_n \\
&\leq |t_{i-1}^n - t_j^m| + h_n \\
&\leq D_{i-1,j} + h_n.
\end{aligned}$$

Thus setting $A_{i,j} = \|x_i^n - x_j^m\|$ by Lemma 3.4.4 we obtain

$$\begin{aligned}
A_{i,j} \quad \leq \quad & \frac{h_n}{h_n + h_m} A_{i,j-1} + \frac{h_m}{h_n + h_m} A_{i-1,j} \\
& + \frac{h_n h_m}{h_n + h_m} \Big(\beta \sup_{t \in [-r, t_i^n]} \|x_n(t) - x_m(t)\| \\
& \qquad + C_8 D_{i-1,j} + \epsilon_{n,m} \Big), \qquad (3.4.18)
\end{aligned}$$

where $C_8 = \beta M + l(M_0) + C_7$ and

$$\epsilon_{n,m} = \beta M(h_n + h_m) + C_7 h_n + h_n l(M_0).$$

Using (3.4.4)

$$\begin{aligned}
&\|x_k^n - \phi(0)\| \\
&\quad = \|J_{h_n}(t_k^n)(x_{k-1}^n + h_n G(t_k^n, (\overline{x_k^n})_{t_k^n})) - \phi(0)\| \\
&\quad \le \|J_{h_n}(t_k^n)(x_{k-1}^n + h_n G(t_k^n, (\overline{x_k^n})_{t_k^n})) - J_{h_n}(t_k^n)\phi(0)\| \\
&\qquad + \|J_{h_n}(t_k^n)\phi(0) - \phi(0)\| \\
&\quad \le \|x_{k-1}^n - \phi(0)\| + h_n\|G(t_k^n, (\overline{x_k^n})_{t_k^n}))\| + h_n|A(t_k^n)\phi(0)| \\
&\quad \le \|x_{k-1}^n - \phi(0)\| + h_n C_4
\end{aligned}$$

for $k = 1, 2, \cdots, n$, where C_4 is as in the proof of Proposition 3.3.6.

Adding these inequalities for $k = 1, 2, \cdots, i$ we obtain

$$\|x_i^n - \phi(0)\| \le \sum_{k=1}^{i} h_n C_4 = t_i^n C_4 \le C_4 D_{i,0}.$$

Hence $A_{i,0} \le C D_{i,0}$ for $i = 1, 2, \cdots, n$, where $C = \max\{C_4, C_8\}$. So the inequality (3.4.17) claimed turnes out to hold for $0 \le i \le n$ and $j = 0$. In the same way (3.4.17) holds for $i = 0$ and $0 \le j \le m$. Next, let $1 \le i \le n$ and $1 \le j \le m$, and assume that (3.4.17) holds for the pairs $(i-1, j)$ and $(i, j-1)$.

From (3.4.17)

$$\begin{aligned}
A_{i,j} \;\le\; & \frac{h_n}{h_n + h_m}(C D_{i,j-1} + E_i + i h_n(C D_{i,j-1} + \epsilon_{n,m})) \\
& + \frac{h_m}{h_n + h_m}(C D_{i-1,j} + E_{i-1} + i h_n(C D_{i-1,j} + \epsilon_{n,m})) \\
& + \frac{h_n h_m}{h_n + h_m}(\beta \sup_{t \in [-r, t_i^n]} \|x_n(t) - x_m(t)\| + C D_{i-1,j} + \epsilon_{n,m}) \\
\le\; & C D_{i,j} + E_i + i h_n(C D_{i,j} + \epsilon_{n,m}).
\end{aligned}$$

Here we have used

$$\frac{h_n}{h_n + h_m} D_{i,j-1} + \frac{h_m}{h_n + h_m} D_{i-1,j} \le D_{i,j}.$$

Thus (3.4.17) holds for the pair (i, j). Therefore (3.4.17) holds for $0 \le i \le n$ and $0 \le j \le m$.

for every $t \in [0, T]$. By Gronwall's inequality we have

$$\sup_{s\in[-r,t]} \|x_n(s) - x_m(s)\| \leq \delta_{n,m} e^{\beta T}.$$

Thus $x_n(t)$ converges to a function $x(t)$ uniformly on $[-r, T]$.

Finally, from (3.4.15) we obtain

$$\|x(t) - x(s)\| \leq M|t - s| \tag{3.4.23}$$

for every $t, s \in [-r, T]$. This shows that $x(t)$ is absolutely continuous on $[-r, T]$. Therefore $x(t)$ is a limit solution of (3.4.1).

(2) From (1) let $x(t)$ and $x_n(t)$ be a limit solution and a discrete scheme approximate solution of (3.4.1) with $\lim_{n\to\infty} x_n(t) = x(t)$. Then there exists $[x_i^n, y_i^n] \in A(t_i^n)$ such that

$$(x_i^n - x) - (x_{i-1}^n - x) = h_n(G(t_i^n, (\overline{x_i^n})_{t_i^n} - y_i^n),$$

and thus we have

$$\begin{aligned} &\|x_i^n - x\| - \|x_{i-1}^n - x\| \\ &\quad \leq h_n[G(t_i^n, (\overline{x_i^n})_{t_i^n}) - y_i^n, x_i^n - x]_- \\ &\quad \leq h_n[y - y_i^n, x_i^n - x]_- \\ &\qquad + h_n[G(t_i^n, (\overline{x_i^n})_{t_i^n}) - y, x_i^n - x]_+ \end{aligned} \tag{3.4.24}$$

for every $[x, y] \in A(\sigma)$, $\sigma \in [0, T]$. Using $(A2)_4$ we obtain that

$$[y - y_i^n, x_i^n - x]_- \leq L(\|x\|)|t_i^n - \sigma|(1 + \|y\|)$$

for every $[x, y] \in A(\sigma)$, $\sigma \in [0, T]$. From (3.4.24)

$$\begin{aligned} &\|x_k^n - x\| - \|x_{k-1}^n - x\| \\ &\quad \leq h_n L(\|x\|)|t_k^n - \sigma|(1 + \|y\|) + h_n[G(t_k^n, (\overline{x_k^n})_{t_k^n}) - y, x_k^n - x]_+ \end{aligned}$$

for every $[x, y] \in A(\sigma)$, $\sigma \in [0, T]$ and $k = 1, 2, \cdots, n$. Adding these ineqaulites for $k = j + 1, \cdots, i$ we obtain

$$\begin{aligned} &\|x_i^n - x\| - \|x_j^n - x\| \\ &\quad \leq \sum_{k=j+1}^{i} h_n([G(t_k^n, (\overline{x_k^n})_{t_k^n}) - y, x_k^n - x]_+ \theta(t_k^n, \sigma)) \end{aligned} \tag{3.4.25}$$

for every $[x, y] \in A(\sigma)$, $\sigma \in [0, T]$, where $\theta(\tau, \sigma) = L(\|x\|)|\tau - \sigma|(1 + \|y\|)$. Let $s \in (t_{j-1}^n, t_j^n]$ and $t \in (t_{i-1}^n, t_i^n]$. Letting n tend to infinity in (3.4.25)

$$\begin{aligned} &\|x(t) - x\| - \|x(s) - x\| \\ &\quad \leq \int_s^t ([G(\tau, x_\tau) - y, x(\tau) - x]_+ + \theta(\tau, \sigma))d\tau \end{aligned} \tag{3.4.26}$$

for $0 \le t \le T$, $[x, y] \in A(\sigma)$, $\sigma \in [0, T]$. Since $x(t)$ is continuous $x(t)$ is an integral solution of (3.4.1).

(3) From (3.4.23) there exists an absolutely continuous limit solution $x(t)$ of (3.4.1) and thus, since X is reflexive $x(t)$ is differentiable a.e. $t \in (0, T)$.

Now let $x(t)$ be differentiable at $t = t_0$ and $h > 0$. Putting $s = \sigma = t_0$ and $t = t_0 + h$ in (3.4.26)

$$\begin{aligned}&\|x(t_0+h) - x\| - \|x(t_0) - x\| \\ &\quad \le \int_{t_0}^{t_0+h} ([G(\tau, x_\tau) - y, x(\tau) - x]_+ + \theta(\tau, t_0))d\tau\end{aligned}$$

for every $[x, y] \in A(t_0)$, where $\theta(\tau, t_0) = L(\|x\|)|\tau - t_0|(1 + \|y\|)$.

Dividing this inequality by h and letting $h \to 0+$ it follows that

$$\left[\frac{dx}{dt}(t_0), x(t_0) - x\right]_+ \le [G(t_0, x_{t_0}) - y, x(t_0) - x]_+$$

and thus

$$\left[-\frac{dx}{dt}(t_0) + G(t_0, x_{t_0}) - y, x(t_0) - x\right]_+ \ge 0$$

for every $[x, y] \in A(t_0)$. By the maximality of $A(t_0)$ we have

$$\frac{dx}{dt}(t_0) + A(t_0)x(t_0) \ni G(t_0, x_{t_0}).$$

Hence $x(t)$ is a strong solution of (3.4.1).

Finally, we prove the uniqueness of a strong solution. To this end let $x_i(t)$ be a strong solution of (3.4.1) for $i = 1, 2$. Then $\|x_1(t) - x_2(t)\|$ is absolutely continuous. Thus $\|x_1(t) - x_2(t)\|$ is differentiable a.e. $t \in (0, T)$ and

$$\begin{aligned}&\frac{d}{dt}\|x_1(t) - x_2(t)\| \\ &\quad = \left[\frac{dx_1}{dt}(t) - \frac{dx_2}{dt}(t), x_1(t) - x_2(t)\right]_- \\ &\quad \le [G(t, (x_1)_t) - G(t, (x_2)_t), x_1(t) - x_2(t)]_+ \\ &\quad\quad - \left[G(t, (x_1)_t) - \frac{dx_1}{dt}(t) - G(t, (x_2)_t) + \frac{dx_2}{dt}(t), x_1(t) - x_2(t)\right]_+\end{aligned}$$

for a.e. $t \in (0, T)$.

From $(A1)_4$ and $(G1)_4$ we obtain

$$\frac{d}{dt}\|x_1(t) - x_2(t)\| \le \beta\|(x_1)_t - (x_2)_t\|_{PC}$$

a.e. $t \in (0,T)$, which yields

$$\begin{aligned}\|x_1(t) - x_2(t)\| &\le \beta \int_0^t \|(x_1)_\tau - (x_2)_\tau\|_{PC} d\tau \\ &\le \beta \int_0^t \sup_{s\in[-r,\tau]} \|(x_1)_\tau - (x_2)_\tau\|_{PC} d\tau\end{aligned}$$

for every $t \in [0,T]$. Thus we have

$$\sup_{s\in[-r,t]} \|x_1(t) - x_2(t)\| \le \beta \int_0^t \sup_{s\in[-r,\tau]} \|(x_1)_\tau - (x_2)_\tau\|_{PC} d\tau$$

for every $t \in [0,T]$. From Gronwall's inequality

$$\sup_{s\in[-r,T]} \|x_1(t) - x_2(t)\| = 0.$$

Therefore $x_1 = x_2$.

Next, we show the following:

THEOREM 3.4.2 *Assume that* $(A1)_4$, $(A2)_4$, $(G1)_4$, *and* $(G2)_4$ *are satisfied. Let* ϕ, $\psi \in PC([-r,0];X)$ *be Lipschitzian with* $\phi(0)$, $\psi(0) \in \widehat{D}$. *If* $x(t)$ *and* $y(t)$ *are limit solutions of* (3.4.1) *corresponding to* ϕ *and* ψ, *respectively, then we have*

$$\begin{aligned}\|x(t) - y(t)\| &\le \|x(s) - y(s)\| \\ &\quad + \int_s^t [G(\sigma, x_\sigma) - G(\sigma, y_\sigma), x(\sigma) - y(\sigma)]_+ d\sigma\end{aligned} \tag{3.4.27}$$

for $0 \le s \le t \le T$.

Proof. From the definition of a limit solution of (3.4.1), there exists a discrete scheme approximate solution $x_n(t)$ of (3.4.1) such that

$$\frac{x_i^n - x_{i-1}^n}{h_n} + A(s_i^n)x_i^n \ni G(s_i^n, (\overline{x_i^n})_{s_i^n}),\ x_0^n = \phi(0),$$

where $h_n = s_i^n - s_{i-1}^n$ and

$$\lim_{n\to\infty} x_n(t) = x(t),\ x_n(t) = \overline{x_n^n}(t),$$

and also there exists a discrete sceme approximate solution $y_m(t)$ of (3.4.1) such that

$$\frac{y_j^m - y_{j-1}^m}{h_m} + A(t_j^m)y_j^m \ni G(t_j^m, (\overline{y_j^m})_{t_j^m}),\ y_0^m = \phi(0),$$

where $h_m = t_j^m - t_{j-1}^m$ and

$$\lim_{m\to\infty} y_m(t) = y(t), \quad y_m(t) = \overline{y_m^m}(t).$$

To end the proof of Theorem 3.4.2 we need the next lemma.

LEMMA 3.4.5 *Under the assumptions of Theorem 3.4.2, let $\delta \in (0, T/2)$. Assume that n and m are sufficiently large such that $\max\{h_n, h_m\} < \delta$ and $n, m \geq n_0$, where n_0 is as in (3.4.12). Then there exists a constant $K > 0$ such that for $p = 0, 1, 2, \cdots, n$ and $q = 0, 1, 2, \cdots, m$*

$$\begin{aligned} &\|x_i^n - y_j^m\| \\ &\quad \leq \|x_p^n - y_q^m\| + KD_{i,j} + \sum_{k=p}^{i} \delta_k^n h_n + \sum_{k=q}^{j} \epsilon_k^m h_m \\ &\quad\quad + ih_n\Big(\frac{1}{\delta}\rho(T) + K\Big)(D_{i,j} + |s_p^n - t_q^m|) \\ &\quad\quad\quad + \rho(2\delta) + Kh_n) \end{aligned} \qquad (3.4.28)$$

for $i = p, \cdots, n;\ j = q \cdots, m$, where

$$\delta_i^n = [G(s_i^n, (\overline{x_i^n})_{s_i^n}) - G(s_i^n, y_{s_i^n}), x_i^n - y(s_i^n)]_\lambda,$$

$$\epsilon_k^m = \|G(t_j^m, (\overline{y_j^m})_{t_j^m}) - G(t^m_, y_{t_j^m})\| + \frac{2}{\lambda}\|y_j^m - y(t_j^m)\|,$$

$$\rho(t) = \sup\Big\{\frac{2}{\lambda}\|y(s) - y(\sigma)\| + \|G(s, y_s) - G(\sigma, y_\sigma)\| \;\Big|\; |s - \sigma| \leq t\Big\},$$

and

$$D_{i,j} = (((s_i^n - s_p^n) - (t_j^m - t_q^m))^2 + (s_i^n - s_p^n)h_n + (t_j^m \quad t_q^m)h_m)^{\frac{1}{2}}.$$

Proof of Lemma 3.4.5 From Lemma 3.4.4 there exists a constant $K_1 > 0$ such that

$$\begin{aligned} &\|x_i^n - y_j^m\| \\ &\quad \leq \frac{h_n}{h_n + h_m}\|x_i^n - y_{j-1}^m\| + \frac{h_n}{h_n + h_m}\|x_{i-1}^n - y_j^m\| \\ &\quad\quad + \frac{h_n h_m}{h_n + h_m}(K_1|s_i^n - t_j^m| \\ &\quad\quad + [G(s_i^n, (\overline{x_i^n})_{s_i^n}) - G(t_j^m, (\overline{y_j^m})_{t_j^m}), x_i^n - y_j^m]_+). \end{aligned} \qquad (3.4.29)$$

Noting that $[y,z]_+ \le [y,z]_\lambda \le [u,z]_\lambda + \|z-v\| + \frac{2}{\lambda}\|y-u\|$ for every u, v, y, $z \in X$, we obtain

$$
\begin{aligned}
&[G(s_i^n,(\overline{x_i^n})_{s_i^n}) - G(t_j^m,(\overline{y_j^m})_{t_j^m}), x_i^n - y_j^m]_+ \\
&\quad \le [G(s_i^n,(\overline{x_i^n})_{s_i^n}) - G(s_i^n, y_{s_i^n}), x_i^n - y(s_i^n)]_\lambda \\
&\qquad + \|G(t_j^m,(\overline{y_j^m})_{t_j^m}) - G(t_j^m, y_{t_j^m})\| + \|G(t_j^m, y_{t_j^m}) - G(s_i^n, y_{s_i^n})\| \\
&\qquad + \frac{2}{\lambda}\|y_j^m - y(t_j^m)\| + \frac{2}{\lambda}\|y(t_j^m) - y(s_i^n)\|.
\end{aligned}
$$

Let $A_{i,j} = \|x_i^n - y_j^m\|$. Then (3.4.29) is written as follows.

$$
\begin{aligned}
A_{i,j} \;&\le\; \frac{h_n}{h_n+h_m}A_{i,j-1} + \frac{h_m}{h_n+h_m}A_{i-1,j} \\
&\quad + \frac{h_m}{h_n+h_m}(K_1 \mid s_i^n - t_j^m \mid + \delta_i^n + \epsilon_j^m + \rho(\mid s_i^n - t_j^m \mid)).
\end{aligned}
$$

But

$$
\begin{aligned}
|s_i^n - t_j^m| &\le |(s_i^n - t_j^m) - h_n| + h_n \\
&\le |(s_i^n - s_p^n) - (t_j^m - t_q^m) - h_n| + |s_p^n - t_q^m| + h_n \\
&\le D_{i-1,j} + |s_p^n - t_q^m| + h_n
\end{aligned}
$$

and

$$
\begin{aligned}
\rho(|s_i^n - t_j^m|) \;&\le\; \frac{1}{\delta}\rho(T)||s_i^n - t_j^m| - h_n| + \rho(2\delta) \\
&\le\; \frac{1}{\delta}\rho(T)(D_{i-1,j} + |s_p^n - t_q^m|) + \rho(2\delta).
\end{aligned}
$$

Thus

$$
\begin{aligned}
A_{i,j} \;&\le\; \frac{h_n}{h_n+h_m}A_{i,j-1} + \frac{h_m}{h_n+h_m}A_{i-1,j} \\
&\quad + \frac{h_m}{h_n+h_m}\Big(\Big(\frac{1}{\delta}\rho(T) + K_1\Big)(D_{i-1,j} + |s_p^n - t_q^m|) \\
&\qquad + k_1 h_n + \rho(2\delta) + \delta_i^n + \epsilon_j^m\Big). \qquad (3.4.30)
\end{aligned}
$$

Since

$$
A(s_i^n)x_i^n \ni G(s_i^n,(\overline{x_i^n})_{s_i^n}) - \frac{x_i^n - x_{i-1}^n}{h_n},
$$

we obtain that

$$
\begin{aligned}
|A(s_i^n)x_i^n| \;&\le\; \|G(s_i^n,(\overline{x_i^n})_{s_i^n})\| + \frac{\|x_i^n - x_{i-1}^n\|}{h_n} \\
&\le\; \beta M_0 + C_2 + M_1 \equiv K_2
\end{aligned}
$$

for $i = 1, 2, \cdots, n$. It follows from $(A2)_4$ and (1.4.2) that

$$|A(s_k^n)x_p^n| \leq |A(s_p^n)x_p^n| + |s_p^n - s_k^n|L(\|x_p^n\|)(1 + |A(s_p^n)x_p^n|) \\ \leq K_2 + TL(M_0)(1 + K_2) \equiv K_3.$$

By (3.4.4)

$$\begin{aligned} \|x_k^n - x_p^n\| &= \|J_{h_n}(s_k^n)(x_{k-1}^n + h_n G(s_k^n, (\overline{x_k^n})_{s_k^n})) - x_p^n\| \\ &\leq \|J_{h_n}(s_k^n)(x_{k-1}^n + h_n G(s_k^n, (\overline{x_k^n})_{s_k^n})) - J_{h_n}(s_k^n)x_p^n\| \\ &\quad + \|J_{h_n}(s_k^n)x_p^n - x_p^n\| \\ &\leq \|x_{k-1}^n - x_p^n\| + h_n\|G(s_k^n, (\overline{x_k^n})_{s_k^n}))\| + h_n|A(s_k^n)x_p^n| \\ &\leq \|x_{k-1}^n - x_p^n\| + h_n K \end{aligned}$$

for $k = p + 1, \cdots, n$, where

$$K = \max\{K_1, K_3 + \beta M_0 + C_2\}.$$

Adding these inequalities for $k = p + 1, \cdots, i$ we obtain

$$\|x_i^n - x_p^n\| \leq Kh_n(i - p) = K(s_i^n - s_p^n) \leq KD_{i,q}$$

which yields

$$\begin{aligned} \|x_i^n - y_q^m\| &\leq \|x_i^n - x_p^n\| + \|x_p^n - y_q^m\| \\ &\leq \|x_p^n - y_q^m\| + KD_{i,q} \end{aligned}$$

for $i = p, \cdots, n$. Thus (3.4.28) holds for $p \leq i \leq n$ and $j = p$. In the same way (3.4.28) also holds for $i = p$ and $q \leq j \leq m$.

Next, let $p + 1 \leq i \leq n$ and $q + 1 \leq j \leq m$, and assume that (3.4.28) holds for the pair $(i - 1, j)$ and $(i, j - 1)$. Using (3.4.28) and (3.4.30)

$$\begin{aligned} A_{i,j} \leq \; & \frac{h_n}{h_n + h_m}\Big(\|x_p^n - y_q^m\| + KD_{i,j-1} + \sum_{k=p}^{i} \delta_k^n h_n + \sum_{k=q}^{j-1} \epsilon_k^m h_m \\ & \quad + ih_n\Big(\Big(\frac{1}{\delta}\rho(T) + K\Big)(D_{i,j-1} + |s_p^n - t_q^m|) \\ & \quad + \rho(2\delta) + Kh_n\Big)\Big) \\ & + \frac{h_m}{h_n + h_m}\Big(\|x_p^n - y_q^m\| + KD_{i-1,j} + \sum_{k=p}^{i-1} \delta_k^n h_n + \sum_{k=q}^{j} \epsilon_k^m h_m \\ & \quad + (i - 1)h_n\Big(\Big(\frac{1}{\delta}\rho(T) + K\Big)(D_{i-1,j} + |s_p^n - t_q^m|) + \end{aligned}$$

$$
\begin{aligned}
&\qquad\qquad +\rho(2\delta) + Kh_n\Big)\Big) \\
&+\frac{h_n h_m}{h_n + h_m}\Big(\Big(\frac{1}{\delta}\rho(T) + K\Big)(D_{i-1,j} + |s_p^n - t_q^m|) \\
&\qquad +\rho(2\delta) + Kh_n + \delta_i^n + \epsilon_j^m\Big) \\
\leq\ & \|x_p^n - y_q^m\| + KD_{i,j} + \sum_{k=p}^{i} \delta_k^n h_n + \sum_{k=q}^{j} \epsilon_k^m h_m \\
&+ih_n\Big(\Big(\frac{1}{\delta}\rho(T) + K\Big)(D_{i,j} + |s_p^n - t_q^m|) + \rho(2\delta) + Kh_n\Big).
\end{aligned}
$$

Thus (3.4.28) holds for the pair (i, j). Hence (3.4.28) holds for $p \leq i \leq n$ and $q \leq j \leq m$.

Continued in Proof of Theorem 3.4.2 Let $s \in (s_{p-1}^n, s_p^n] \cap (t_{q-1}^m, t_q^m]$ and $t \in (s_{i-1}^n, s_i^n] \cap (t_{j-1}^m, t_j^m]$. Letting n, m tend to infinity in (3.4.28)

$$
\begin{aligned}
\|x(t) - y(t)\| \ \leq\ & \|x(s) - y(s)\| \\
&+\overline{\lim}_{n\to\infty} \sum_{k=p}^{i} \delta_k^n h_n + \overline{\lim}_{m\to\infty} \sum_{k=q}^{j} \epsilon_k^m h_m + T\rho(2\delta)
\end{aligned}
\tag{3.4.31}
$$

for every $\delta \in (0, \frac{T}{2})$. Since

$$
\lim_{n\to\infty} \sum_{k=p}^{i} \delta_k^n h_n = \int_s^t [G(\tau, x_\tau) - G(\tau, y_\tau), x(\tau) - y(\tau)]_\lambda d\tau
$$

and

$$
\lim_{m\to\infty} \sum_{k=q}^{j} \epsilon_k^m h_m = 0,
$$

let $\delta \to 0+$ in (3.4.31) to obtain

$$
\begin{aligned}
\|x(t) - y(t)\| \ \leq\ & \|x(s) - y(s)\| \\
&+ \int_s^t [G(\tau, x_\tau) - G(\tau, y_\tau), x(\tau) - y(\tau)]_\lambda d\tau.
\end{aligned}
$$

Letting $\lambda \to 0+$ here, we have (3.4.27).

We now show that the results of Theorem 3.4.1 remains true if $G(t, \phi)$ is defined only for every $\phi \in PC([-r, 0]; \overline{B_{\bar{r}}(0)})$ with Proposition 3.4.2, Proposition 3.4.3 and Proposition 3.4.5. The proof after the construction of the method of lines as in Proposition 3.4.2, Proposition 3.4.3 and Proposition 3.4.5, follows exactly as in Theorem 3.4.1.

THEOREM 3.4.3 *Let the assumptions of one of Proposition 3.4.2, Proposition 3.4.3 and Proposition 3.4.5 be satisfied. Furthermore, assume that* $(A2)_4$ *and* $(G2)_4$ *for every* $\phi \in PC([-r,0];\overline{B_{\bar{r}}(0)})$ *hold. If* $\phi : [-r,0] \to \overline{B_{\bar{r}}(0)}$ *is Lipschitz continuous then there exist a limit solution and an integral solution of* (3.4.1) *on* $[-r,T]$, *and moreover, when* X *is reflexive there exists a unique strong solution of* (3.4.1) $[-r,T]$.

3.5. Compactness Methods

In this section we consider non-autonomous nonlinear functional evolutions of the type

$$\begin{cases} \dfrac{dx}{dt}(t) + A(t)x(t) \ni G(t, x_t), & 0 \le t \le T \\ x(t) = \phi(t), & -r \le t \le 0 \end{cases} \tag{3.5.1}$$

associated with an operator $A(t) : X \supset D(A(t)) \to 2^X$ for every $t \in [0,T]$, where $T > 0$, $r > 0$, $x(t) : [-r,T] \to X$ is an unknown function for $r > 0$, $G : [0,T] \times C([-r,0];X) \to X$ and $\phi \in C([-r,0];X)$.

DEFINITION 3.5.1 *1. A function* $x(t) : [-r,T] \to X$ *is called a strong solution of* (3.4.1) *on* $[-r,T]$ *if:*

(1) $x(t) = \phi(t)$ *for every* $t \in [-r,0]$;

(2) $x(t)$ *is continuous on* $[0,T]$;

(3) $x(t)$ *is absolutely continuous on every compact interval of* $[0,T]$;

(4) $x(t)$ *is differentiable a.e.* $t \in (0,T)$;

(5) $x(t)$ *satisfies*

$$\frac{dx}{dt}(t) + A(t)x(t) - G(t, x_t)$$

a.e. $t \in (0,T)$.

2. A function $x(t) : [-r,T] \to X$ *is called an integral solution of* (3.4.1) *on* $[-r,T]$ *if:*

(1) $x(t) = \phi(t)$ *for every* $t \in [-r,0]$;

(2) $x(t)$ *is continuous on* $[0,T]$;

(3) $x(t)$ *satisfies*

$$\|x(t)-x\| \le \|x(s)-x\| + \int_s^t ([G(\sigma, x_\sigma) - y, x(\sigma) - x]_+ + \theta(\sigma,\tau))d\sigma$$

for every $0 \leq s \leq t \leq T$ *and* $[x,y] \in A(\tau)$ *for some* $\tau \in [0,T]$, *where* $\theta(\sigma,\tau)$ *is a scalar function.*

3. A function $x(t) : [-r,T] \to X$ *is called a limit solution of* (3.4.1) *on* $[-r,T]$ *if:*

(1) $x(t) = \phi(t),\ t \in [-r,0]$;

(2) $x(t)$ *is continuous on* $[0,T]$;

(3) $x(t) = \lim_{n\to\infty} x_n(t)$ *uniformly on* $[0,T]$,

where $x_n(t) = \overline{x_n^n}(t)$ *for every* $t \in [-r,T]$ *with*

$$\overline{x_i^n}(t) = \begin{cases} \phi(t), & t \in [-r,0] \\ x_j^n, & t \in (t_{j-1}^n, t_j^n],\ j = 1,2,\cdots,i-1 \\ x_i^n, & t \in (t_{i-1}^n, T] \end{cases}$$

satisfying

$$\frac{x_i^n - x_{i-1}^n}{t_i^n - t_{i-1}^n} + A(t_i^n)x_i^n \ni G(t_i^n, (\overline{x_i^n})_{t_i^n})$$

for $x_i^n \in D(A(t_i^n))$ *and a partition* $\{t_i^n\}$ *of* $[0,T]$ *such that*

$$0 = t_0^n < t_1^n < \cdots < t_n^n = T$$

and $\lim_{n\to\infty} \max_{1\leq i\leq n}(t_i^n - t_{i-1}^n) = 0$, $i = 1,2,\cdots,n$; $n = 1,2,\cdots$.

We also assume the following conditions.

$(A1)_5$ $A(t) : X \supset D(A(t)) \to 2^X$ is m-accretive in X for every $t \in [0,T]$;

$(A2)_5$ There exist two continuous function $q : [0,T] \to X$ and $L : [0,\infty) \to [0,\infty)$ such that

$$(v_1 - v_2, u_1 - u_2)_+ \geq -\|q(t) - q(s)\|\|u_1 - u_2\|L(\|u_2\|)$$

for every $t, s \in [0,T]$, $[u_1,v_1] \in A(t)$ and $[u_2,v_2] \in A(s)$;

$(A3)_5$ If $\lim_{n\to\infty} t_n = t$, $u_n \in D(A(t_n))$ for t, $t_n \in [0,T]$, $n = 1,2,\cdots$, and $\lim_{n\to\infty} u_n = u \in X$, then $u \in \overline{D(A(t))}$.

We note that if $(A1)_5$–$(A3)_5$ are satisfied then from Proposition 1.9.1 the set $\overline{D(A(t))}$ is independent of $t \in [0,T]$. This set will be denoted by $\overline{D}$.

$(G)_5$ $G : [0,T] \times C([-r,0];\overline{D}) \to X$ is continuous.

Let $\{U(t,s) \mid (t,s) \in \Delta\}$ be an evolution operator on $\overline{D}$ generated by $A(t)$. We show the existence of solutions of (3.5.1) under the assumptions $(A1)_5$–$(A3)_5$, $(G)_5$, and compactness of $\{U(t,s) \mid (t,s) \in \Delta\}$, where $\Delta = \{(t,s) \mid 0 \le s \le t \le T\}$.

First of all we deal with some basic compactness results for (1.8.1) and we use definitions in Section 2.4 in the following.

PROPOSITION 3.5.1 *Assume that $(A1)_5$–$(A3)_5$ are satisfied. Let $x_0 \in \overline{D}$. Assume that V is uniformly integrable subset of $L^1(0,T;X)$. Let*

$$W(V) = \{x_f \mid x_f \text{ is a limit solution of (1.9.1) for some } f \in V\}$$

and let

$$W(V)(t) = \{x_f(t) \mid x_f \in W(V)\}$$

for every $t \in [0,T]$. Then the following (1) and (2) are equivalent:

(1) $W(V)$ is relatively compact in $C([0,T];X)$;

(2) There exists a dense set $\Omega \subset [0,T]$ such that $W(V)(t)$ is relatively compact in X for evey $t \in \Omega$.

Proof (1)⇒(2) is obvious. Assume (2). In order to show (1) we start with

$$\begin{aligned} \|x_f(t) - x_0\| &\le \|x_f(t) - U(t,0)x_0\| + \|U(t,0)x_0 - x_0\| \\ &\le \int_0^t \|f(\sigma)\| d\sigma + \|U(t,0)x_0 - x_0\| \end{aligned} \tag{3.5.2}$$

for every $f \in V$ and $t \in [0,T]$. Here we have used (1.9.2) and that $U(t,0)x_0$ is a limit solution of (1.9.1).

Since $U(t,0)$ is continuous and V is uniformly integrable $W(V)$ is equicontinuous at $t=0$. Let $t \in (0,T]$ and $\epsilon > 0$. Choose $\lambda > 0$ such that $t - 2\lambda \in [0,T]$, $t - \lambda \in \Omega$ and

$$\int_\Omega \|f(s)\| ds < \frac{\epsilon}{16}$$

for every measurable set $\Omega \subset [0,T]$ with $m(\Omega) < 2\lambda$ and $f \in V$. Since $W(V)(t-\lambda)$ is relatively compact it is precompact . Thus there exists a family $\{f_1, f_2, \cdots, f_{N(\epsilon)}\} \subset V$ such that for every $f \in V$ there exists $1 \le i \le N(\epsilon)$ with

$$\|x_f(t-\lambda) - x_{f_i}(t-\lambda)\| \le \frac{\epsilon}{4}.$$

Since $\{f_1, f_2, \cdots, f_{N(\epsilon)}\}$ is obviously equicontinuous at $t \in (0,T]$ there exists $\delta(\epsilon) \in (0, \lambda)$ such that $\|x_{f_i}(t+h) - x_{f_i}(t)\| \le \frac{\epsilon}{4}$ for every $i = 1, 2, \cdots, N(\epsilon)$, $h \in \mathbf{R}$ with $|h| \le \delta(\epsilon)$ and $t + h \in [0,T]$. On the other hand

$$
\begin{aligned}
&\|x_f(t+h) - x_f(t)\| \\
&\quad \le \|x_f(t+h) - x_{f_i}(t+h)\| + \|x_{f_i}(t+h) - x_{f_i}(t)\| \\
&\qquad + \|x_{f_i}(t) - x_f(t)\| \\
&\quad \le \|x_f(t-\lambda) - x_{f_i}(t-\lambda)\| + \int_{t-\lambda}^{t+h} \|f(\sigma) - f_i(\sigma)\| d\sigma + \frac{\epsilon}{4} \\
&\qquad + \|x_f(t-\lambda) - x_{f_i}(t-\lambda)\| + \int_{t-\lambda}^{t} \|f(\sigma) - f_i(\sigma)\| d\sigma \\
&\quad \le 2\|x_f(t-\lambda) - x_{f_i}(t-\lambda)\| + \frac{\epsilon}{4} + \int_{t-\lambda}^{t} \|f(\sigma) - f_i(\sigma)\| d\sigma \\
&\qquad + \int_{t-\lambda}^{t+\lambda} \|f(\sigma) - f_i(\sigma)\| d\sigma \\
&\quad \le \frac{2\epsilon}{4} + \frac{\epsilon}{4} + \frac{\epsilon}{8} + \frac{\epsilon}{8} = \epsilon
\end{aligned}
$$

uniformly on V. Thus $W(V)$ is equicontinuous on $[0,T]$.

Since by the assumption $W(V)(t)$ is relatively compact for every $t \in \Omega$ we only have to show the relative compactness of $W(V)(t)$ for every $t \in [0,T] - \Omega$. Fix $t \in [0,T] - \Omega$ and choose a sequence $\{h_n\} \subset \mathbf{R}$ such that $t + h_n \in \Omega$. Then we define the operator $Q_n : W(V)(t) \to X$ by $Q_n x_f(t) = x_f(t+h_n)$ for $n = 1, 2, \cdots$. This implies that $Q_n(W(V)(t)) = W(V)(t+h_n)$ and thus $Q_n(W(V)(t))$ is relatively compact in X since $t + h_n \in \Omega$. Hence Q_n is a compact operator for every $n = 1, 2, \cdots$. Since $W(V)$ is equicontinuous on $[0,T]$ $\lim_{n\to\infty} Q_n = I$ uniformly on $W(V)(t)$. It follows that $W(V)(t)$ is bounded by the boundedness of $W(V)$ in $C([0,T];X)$, and thus $W(V)(t)$ is relatively compact in X.

LEMMA 3.5.1 *Let $f \in L^1(0,T;X)$ and let $x(t)$ be a limit solution of (1.9.1) on $[0,T]$ with $x_0 \in \overline{D}$. Then*

$$\|U(t, t-h)x(t-h) - x(t)\| \le \int_{t-\lambda}^{t} \|f(\sigma)\| d\sigma.$$

Proof. We define the function $v_h(\sigma) = U(\sigma, t-h)(x(t-h)$ for every $\sigma \in [t-h, t]$. Then v_h is a limit solution of (1.9.1) for $f = 0$ with initial condition $v_h(t-h) = x(t-h)$. Using (1.9.2) we have

$$\|v_h(t) - x(t)\| \le$$

$$\leq \|v_h(t-h) - x(t-h)\| + \int_{t-\lambda}^{t} \|f(\sigma)\| d\sigma$$
$$= \int_{t-\lambda}^{t} \|f(\sigma)\| d\sigma,$$

which yields our conclusion.

PROPOSITION 3.5.2 *Assume that* $(A1)_5$*–*$(A3)_5$ *are satisfied. Suppose that the evolution operator* $\{U(t,s) \mid (t,s) \in \Delta\}$ *generated by* $A(t)$ *for every* $t \in [0,T]$ *is compact. Then for every* $x_0 \in \overline{D}$ *and every uniformly integrable set* $V \subset L^1(0,T;X)$ *the set*

$$W(V) = \{x_f \mid x_f \text{ is a limit solution of (1.9.1) for some } f \in V\}$$

is relatively compact in $C([0,T];X)$.

Proof. From Proposition 3.5.1 it suffices to show that for every $t \in [0,T]$ the set $W(V)(t)$ is relatively compact in X. To this end fix $t \in [0,T]$ and let $h > 0$ be such that $t - h \in [0,T]$. Then from Lemma 3.5.1 we have

$$\|U(t,t-h)x_f(t-h) - x_f(t)\| \leq \int_{t-\lambda}^{t} \|f(\sigma)\| d\sigma \tag{3.5.3}$$

for $f \in V$. We define the operator $Q_h : W(V)(t) \to X$ by

$$Q_h(x_f(t)) = U(t,t-h)x_f(t-h)$$

for every $x_f(t) \in W(V)(t)$. Since $U(t,t-h)$ is compact and $W(V)(t-h)$ is bounded it follows that Q_h is compact operator for every $h > 0$ with $t - h \in [0,T]$. Since V is uniformly integrable (3.5.3) shows that $\lim_{h\to 0} Q_h = I$ uniformly on $W(V)(t)$. Thus the identity operator I is a compact operator. Hence $W(V)(t)$ is relatively compact in X for every $t \in [0,T]$. Therefore the proof is completed.

PROPOSITION 3.5.3 *Assume that* $(A1)_5$*–*$(A3)_5$ *are satisfied. Suppose that the evolution operator* $\{U(t,s) \mid (t,s) \in \Delta\}$ *generated by* $A(t)$ *for every* $t \in [0,T]$ *is equicontinuous. Then for every* $x_0 \in \overline{D}$ *and every uniformly integrable set* $V \subset L^1(0,T;X)$ *the set*

$$W(V) = \{x_f \mid x_f \text{ is a limit solution of (1.9.1) for some } f \in V\}$$

is bounded equicontinuous in $C([0,T];X)$.

Proof. We set

$$c_0 = \sup\{\|U(t,0)x_0\| \mid t \in [0,T]\} + \sup\left\{ \int_0^T \|f(t)\| dt \,\middle|\, f \in V\right\}.$$

Given $y \in K$, from Theorem 1.9.1 there exists a unique limit solution $x_y : [0, T_0] \to X$ of a non-autonomous nonlinear evolution

$$\begin{cases} \dfrac{dx}{dt}(t) + A(t)x(t) \in \widetilde{G}(t, (\hat{y})_t), & t \in [0, T_0] \\ x(0) = \phi(0) \end{cases} \tag{3.5.7}$$

with $x_y(t) \in \overline{D}$ for every $t \in [0, T_0]$.

We can thus define the solution operator $Q : D(Q) \subset K \to K_0$ by

$$\begin{aligned} D(Q) &= \{y \in K \mid x_y \in K_0\}, \\ Qy &= x_y, \quad y \in D(Q). \end{aligned}$$

Then $D(Q) = K$ and $Q(K) \subset K$.

In fact let $y \in K$; we have to show that $(\hat{x}_y)_t \in \overline{B}(\phi, h)$ for every $t \in [0, T_0]$. Given $t \in [0, T_0]$ and $\tau \in [-r, 0]$ assume that $t + \tau > 0$. Then we have using (3.5.4)–(3.5.6) and Theorem 1.9.2

$$\begin{aligned} &\|(\hat{x}_y)_t(\tau) - \phi(\tau)\| \\ &\quad = \|x_y(t+\tau) - \phi(\tau)\| \\ &\quad \le \|x_y(t+\tau) - U(t+\tau, 0)\phi(0)\| + \|U(t+\tau, 0)\phi(0) - \phi(0)\| \\ &\qquad + \|\phi(0) - \phi(\tau)\| \\ &\quad \le \int_0^t \|\widetilde{G}(\sigma, (\hat{y})_\sigma)\| d\sigma + \|U(t+\tau, 0)\phi(0) - \phi(0)\| + \frac{h}{3} \\ &\quad \le T_0 M + \|U(t+\tau, 0)\phi(0) - \phi(0)\| + \frac{h}{3} \le h. \end{aligned}$$

On the other hand if $t + \tau \le 0$ then from (3.5.5)

$$\|(\hat{x}_y)_t(\tau) - \phi(\tau)\| = \|\phi(t+\tau) - \phi(\tau)\| \le \frac{h}{3} \le h$$

so that the claim is proved.

We now show that $Q : K \to K$ is continuous. Let $y, y_n \in K$ for $n = 1, 2, \cdots$ such that $\lim_{n\to\infty} y_n = y$ in $C([0, T_0]; X)$. We put $x_n = Qy_n$ and $x = Qy$. Then by Theorem 1.9.2

$$\|x_n(t) - x(t)\| \le \int_0^{T_0} \|\widetilde{G}(\sigma, (\hat{y}_n)_\sigma) - \widetilde{G}(\sigma, (\hat{y})_\sigma)\| d\sigma.$$

But

$$\|(\widehat{y}_n)_\sigma(s) - (\widehat{y})_\sigma(s)\| = \begin{cases} \|y_n(\sigma+s) - y(\sigma+s)\|, & \sigma + s \ge 0 \\ 0, \quad \sigma + s \le 0 \end{cases}$$

converges to 0 uniformly for $s \in [-r, 0]$, and $\sigma \in [0, T_0]$, and

$$\|\widetilde{G}(\sigma, (\hat{y}_n)_\sigma) - \widetilde{G}(\sigma, (\hat{y})_\sigma)\| \leq 2M,$$

so that by Lebesgue's dominated convergence theorem

$$\lim_{n \to \infty} \|x_n(t) - x(t)\| = 0$$

uniformly over $[0, T_0]$. Thus $Q : K \to K$ is continuous.

Finally we show that $Q(K)$ is relatively compact in $C([0, T_0]; X)$.

The set

$$E = \{\{\sigma \to \widetilde{G}(\sigma, (\hat{y})_\sigma)\} \mid y \in K\} \tag{3.5.8}$$

is bounded in $L^\infty(0, T_0; X)$ and thus E is uniformly integrable in $L^1(0, T_0; X)$. Thus from Proposition 3.5.2 with compactness of the evolution operator $U(t, s)$, $Q(K)$ is relatively compact in $C([0, T_0]; X)$. Hence from Schauder's fixed point theorem there exists a fixed point $x \in K$ such that $Qx = x$ which, according to the definition of the operator Q, is a limit solution of (3.5.7). However by the definition of $\widetilde{G}$ and that $x(t) \in \overline{D}$ for every $t \in [0, T_0]$ x, in fact, is a local limit solution of (3.5.1) on $[-r, T]$. Therefore the proof is completed.

THEOREM 3.5.2 *Let the dual space X^* of X be uniformly convex. Assume that $(A1)_5$–$(A3)_5$ and $(G)_5$ are satisfied. Assume that*

(1) $G(t, \cdot) : C([-r, 0]; \overline{D}) \to X$ is compact for every $t \in [0, T]$;

(2) the evolution operator $\{U(t, s) \mid (t, s) \in \Delta\}$ generated by $A(t)$ for every $t \in [0, T]$ is equicontinuous.

Let $\phi \in C([-r, 0]; \overline{D})$. Then there exists a local limit solution $x(t)$ of (3.5.1) on $[-r, T]$ with $x(t) \in \overline{D}$ for every $t \in [0, T_0]$ for some $T_0 \in (0, T]$.

Proof. Using the notations and following the way as in Proof of Theorem 3.5.1 we only have to show that $Q(K)$ is relatively compact in $C([0, T_0]; X)$. Since E in (3.5.8) is uniformly integrable in $L^1(0, T_0; X)$ and $\{U(t, s) \mid (t, s) \in \Delta\}$ is equicontinuous from Proposition 3.5.2 $Q(K)$ is bounded and equicontinuous in $C([0, T_0]; X)$.

Now we show that $Q(K)(t)$ is relatively compact in X for every $t \in [0, T_0]$.

Let $t \in [0, T_0]$. Let $y_n \in K$ and we put $Qy_n = x_n$ for $n = 1, 2, \cdots$.

We first note that E is weakly relatively compact in $L^1(0, T_0; X)$. Since E is uniformly integrable it is enough to prove from Diestel [35] and [36] that $E(t)$ is weakly relatively compact in X for every $t \in [0, T_0]$.

We actually show that each of these sets is (norm-) relatively compact. In fact, let $y \in K$ and $(\hat{y})_t \in \overline{B}(\phi, h)$ for every $t \in [0, T_0]$. From the assumption that $G(t, \cdot)$ is compact $\widetilde{G}(t, \cdot)$ is also compact. Thus given $t \in [0, T_0]$ there exists a compact subset K_t of X such that

$$E(t) = \{\widetilde{G}(t, \hat{y}_t) \mid y \in K\} \subset \widetilde{G}(t, \overline{B}(\phi, h)) \subset K_t.$$

Thus $E(t)$ is relatively compact in X for every $t \in [0, T_0]$. Hence E is weakly relatively compact in $L^1(0, T_0; X)$.

So for the following we can assume that $f_n = \widetilde{G}(\cdot, (\hat{y}_n).)$ for $n = 1, 2, \cdots$, converges to some $f \in L^1(0, T_0; X)$ weakly in $L^1(0, T_0; X)$. We notice that, by assuming K_t for $t \in [0, T_0]$ to be convex we can further assume that $f(t) \in K_t$ for $t \in [0, T_0]$. We denote by x_f the unique limit solution of

$$\begin{cases} \dfrac{dx}{dt}(t) + A(t)x(t) \ni f(t), \quad t \in [0, T_0] \\ x(0) = \phi(0). \end{cases}$$

Since X^* is uniformly convex the duality mapping $F : X \to 2^{X^*}$ of X is single-valued and uniformly continuous on bounded sets of X. Thus we have

$$\|x_n(t) - x_f(t)\|^2 \leq 2 \int_0^t (f_n(\sigma) - f(\sigma), F(x_n(\sigma) - x_f(\sigma)))d\sigma \quad (3.5.9)$$

for $n = 1, 2, \cdots$, and the sequence $\{F(x_n(\cdot) - x_f(\cdot)\}$ is relatively sequentially compact in $C_{w^*}([0, T_0]; X^*)$. Hence we can assume that the sequence $\{F(x_n(\cdot) - x_f(\cdot)\}$ converges uniformly on $[0, T_0]$ weak* in X^* to some function $g \in C_{w^*}([0, T_0]; X^*)$. Clearly by Pettis' measurability theorem and the fact that X is reflexive we have $g \in L^1(0, T_0, X)^*$. We now return to (3.5.9) above and conclude that

$$\begin{aligned} &\|x_n(t) - x_f(t)\|^2 \qquad (3.5.10)\\ &\leq 2 \int_0^t (f_n(\sigma) - f(\sigma), F(x_n(\sigma) - x_f(\sigma)) - g(\sigma))d\sigma \\ &\quad +2 \int_0^t (f_n(\sigma) - f(\sigma), g(\sigma))d\sigma \\ &= I_n^1 + I_n^2. \end{aligned}$$

for every $t \in [0, T_0]$ and $n = 1, 2, \cdots$.

Since $\{f_n\}$ converges to f weakly in $L^1(0, T_0; X)$ and $g \in L^1(0, T_0; X)^*$ obviously $\lim_{n\to\infty} I_n^2 = 0$. As for I_n^1 let

$$g_n(\sigma) = (f_n(\sigma) - f(\sigma), F(x_n(\sigma) - x_f(\sigma)) - g(\sigma))$$

for every $\sigma \in [0, T_0]$. Then $\{g_n\}$ is bounded in $L^\infty(0, T_0)$ hence dominated in $L^1(0, T_0)$. Finally given $\sigma \in [0, T_0]$ $f_n(\sigma) - f(\sigma) \in K(\sigma)$ for some compact subset $K(\sigma)$ in X, and the sequence $\{F(x_n(\sigma) - x_f(\sigma)) - g(\sigma)\}$ converges weak* in X^* to zero hence uniformly on the compact subset $K(\sigma) \subset X$. Hence Lebesgue's dominated convergence theorem applies to show that $\lim_{n\to\infty} I_n^1 = 0$ as well. Therefore from (3.5.10) $Q(K)$ is relatively compact in $C([0, T_0]; X)$. The proof is completed as in Proof of Theorem 3.5.1.

3.6. L^p-Space Methods

In this section as in Section 2.4 we consider non-autonomous nonlinear functional evolutions of the type

$$\begin{cases} \dfrac{dx}{dt}(t) + A(t)x(t) \ni G(t, x_t), \quad 0 \le t \le T \\ x_0 = \phi \\ x(0) = h \end{cases} \tag{3.6.1}$$

associated with an operator $A(t) : X \supset D(A(t)) \to 2^X$ for every $t \in [0, T]$, where $T > 0$, $r > 0$, $x(t) : [-r, T] \to X$ is an unknown function, $G : [0, T] \times L^p(-r, 0; X) \to X$, $\phi \in L^p(-r, 0; X)$, $1 \le p < \infty$, and $h \in X$.

DEFINITION 3.6.1 *1. A function $x(t) : [-r, T] \to X$ is called a strong solution of* (3.6.1) *on $[-r, T]$ if:*

(1) $x_0 = \phi$;

(2) $x(0) = h$;

(3) $x(t)$ *is continuous on* $[0, T]$;

(4) $x(t)$ *is absolutely contionuous on every compact interval of* $[0, T]$;

(5) $x(t)$ *is differentiable a.e.* $t \in (0, T)$;

(6) $x(t)$ *satisfies*

$$\frac{dx}{dt}(t) + A(t)x(t) = G(t, x_t)$$

a.e. $t \in (0, T)$.

2. A function $x(t) : [-r, T] \to X$ is called an integral solution of (3.6.1) *on $[-r, T]$ if:*

(1) $x_0 = \phi$;

(2) $x(0) = h$;

(3) $x(t)$ *is continuous on* $[0, T]$;

(4) $x(t)$ satisfies

$$\|x(t) - x\| \\ \leq \|x(s) - x\| + \int_s^t ([G(\sigma, x_\sigma) - y, x(\sigma) - x]_+ + \theta(\sigma, \tau))d\sigma$$

for every $0 \leq s \leq t \leq T$ and $[x, y] \in A(\tau)$ for some $\tau \in [0, T]$, where $\theta(\sigma, \tau)$ is a scalar function.

To obtain the existence of solutions we assume the following conditions:

$(A1)_6$ Let $A(t) : X \supset D(A(t)) \equiv D_A \to 2^X$ be an operator for every $t \in [0, T]$ and let $\alpha \in \mathbf{R}$ such that $A(t) + \alpha I$ is accretive for every $t \in [0, T]$, and $R(I + \lambda A(t)) = X$ for sufficiently small $\lambda > 0$;

$(A1)'_6$ Let $A(t) : D_A \equiv D(A(t)) \subset X \to X$ be an operator for every $t \in [0, T]$ and let $\alpha \in \mathbf{R}$ such that $A(t) + \alpha I$ is accretive for every $t \in [0, T]$, and $R(I + \lambda A(t)) = X$ for sufficiently small $\lambda > 0$;

$(A2)_6$ There exists a continuous non-decreasing function $L : [0, \infty) \to [0, \infty)$ such that

$$\|A_\lambda(t_1)u - A_\lambda(t_2)u\| \leq |t_1 - t_2| L(\|u\|)(1 + \|A_\lambda(t_2)u\|)$$

for $t_i \in [0, T]$, $i = 1, 2$, $u \in X$ and sufficiently small $\lambda > 0$, where $A_\lambda(t)$ is the Yosida approximant of $A(t)$;

$(A2)'_6$ There exists a continuous non-decreasing function $L : [0, \infty) \to [0, \infty)$ such that

$$\|A(t_1)u - A(t_2)u\| \leq |t_1 - t_2| L(\|u\|)(1 + \|A(t_2)u\|)$$

for $t_i \in [0, T]$, $i = 1, 2$, and $u \in X$;

$(G1)_6$ There exists a constant $\beta > 0$ such that

$$\|G(t, \psi_1) - G(t, \psi_2)\| \leq \beta \|\psi_1 - \psi_2\|_p$$

for every $t \in [0, T]$ and $\psi_i \in L^p(-r, 0; X)$, $i = 1, 2$;

$(G2)_6$ There exists a continuous function $l : [0, \infty) \to [0, \infty)$ such that

$$\|G(t_1, \psi) - G(t_2, \psi)\| \leq |t_1 - t_2| l(\|\psi\|_p)$$

for every $t_i \in [0, T]$, $i = 1, 2$, and $\psi \in L^p(-r, 0; X)$.

We put $Y = L^p \times X$ for $1 \leq p < \infty$ with norm $\|\cdot\|_Y$ defined by

$$\|\{\phi, h\}\|_Y = \Big(\int_{-r}^0 \|\phi(\tau)\|^p + \|h\|^p\Big)^{\frac{1}{p}}$$

for every $\{\phi, h\} \in Y$. Then Y is a real Banach space with norm $\|\cdot\|_Y$ and

$$\|\phi\|_p \leq \|\{\phi, h\}\|_Y, \ \|h\| \leq \|\{\phi, h\}\|_Y \tag{3.6.2}$$

for every $\{\phi, h\} \in Y$.

We define as in (2.4.4) of Section 2.4 an operator $B(t) : Y \subset D(B(t)) \to 2^Y$ for every $t \in [0, T]$ by

$$\begin{aligned} D(B(t)) &= \Big\{\{\phi, h\} \in Y \ \Big| \ \phi : [-r, 0] \to X \text{ is absolutely continuous,} \\ &\qquad \frac{d\phi}{dt} \in L^p(-r, 0; X), \ h = \phi(0) \in D(A(t))\Big\}, \\ B(t)\{\phi, h\} &= \Big\{-\frac{d\phi}{dt}, A(t)h - G(t, \phi)\Big\}, \quad \{\phi, h\} \in D(B(t)). \end{aligned} \tag{3.6.3}$$

Then $D(B(t))$ is independent of $t \in [0, T]$. We denote it by D_B.

We first show that under the assumptions $(A1)_6$, $(A2)_6$,$(G1)_6$, and $G(2)_6$, $B(t)$ for every $t \in [0, T]$ satisfies conditions similar to $(A1)_6$ and $(A2)_6$.

PROPOSITION 3.6.1 *We have the following:*

(1) Assume that $(A1)_6$ and $(G1)_6$ are satisfied. Then, for every $t \in [0, T]$, $B(t) + \gamma I$ is accretive in Y and $R(I + \lambda B(t)) = Y$ for sufficiently small $\lambda > 0$, where $\gamma = \max\left\{0, \alpha + \frac{1}{p}\right\} + \beta$;

(2) Assume that $(A1)_6$, $(A2)_6$, $(G1)_6$, and $(G2)_6$ are satisfied. Then there exists a non-decreasing continuous function $L_1 : [0.\infty) \to [0, \infty)$ such that

$$\begin{aligned} &\|B_\lambda(t_1)\{\phi, h\} - B_\lambda(t_2)\{\phi, h\}\|_Y \\ &\quad \leq |t_1 - t_2| L_1(\|\{\phi, h\}\|_Y)(1 + \|B_\lambda(t)\{\phi, h\}\|_Y) \end{aligned}$$

for every $t_i \in [0, T]$ for $i = 1, 2$, $\{\phi, h\} \in Y$ and sufficiently small $\lambda > 0$.

Proof. (1) Let $t \in [0, T]$. Let $\{\phi_i, h_i\} \in D(B(t))$ and $y_i \in B(t)\{\phi_i, h_i\}$ and $\hat{h}_i \in A(t)h_i$ for $i = 1, 2$. We put $\phi = \phi_1 - \phi_2$ and $h = h_1 - h_2$. Let $\phi^* \in F_{L^p}(\phi)$, $h^* \in F(h)$ and $y^* \in F_Y(\{\phi, h\})$. Then from (2.4.5), $(A1)_6$, $(G1)_6$, (3.6.2), and (3.6.3) we obtain

$$(y_1 - y_2, y^*) =$$

$$
\begin{aligned}
&= \Big(\Big\{ -\frac{d\phi}{d\tau}, \hat{h}_1 - \hat{h}_2 - G(t,\phi_1) + G(t,\phi_2)\Big\}, y^*\Big) \\
&= \|\{\phi, h\}\|_Y^{2-p}\Big(\int_{-r}^{0} \Big(-\frac{d\phi}{d\tau}(\tau), \phi^*(\tau)\Big) \|\phi(\tau)\|^{p-2} d\tau \\
&\qquad + ((\hat{h}_1 - \hat{h}_2, h^*) - (G(t,\phi_1) - G(t,\phi_2), h^*))\|h\|^{p-2}\Big) \\
&\geq \|\{\phi, h\}\|_Y^{2-p}\Big(\int_{-r}^{0} \Big(-\frac{1}{p}\frac{d}{d\tau}\|\phi(\tau)\|^p\Big) d\tau - \alpha\|h\|^2\|h\|^{p-2} \\
&\qquad - \|G(t,\phi_1) - G(t,\phi_2)\|\|h\|\|h\|^{p-2}\Big) \\
&\geq \|\{\phi, h\}\|_Y^{2-p}\Big(\Big(-\frac{1}{p} - \alpha\Big)\|h\|^p \beta\|\phi\|_p\|h\|^{p-1}\Big) \\
&\geq \|\{\phi, h\}\|_Y^{2-p}(-\|\{\phi, h\}\|_Y^{2-p} - \alpha - \beta)\|\{\phi, h\}\|_Y^p \\
&\geq -\gamma\|\{\phi, h\}\|_Y^2.
\end{aligned}
$$

Thus $B(t) + \gamma I$ is accretive in Y for every $t \in [0, T]$.

Let $\{\psi, k\} \in Y$. We define a mapping $Q : X \to X$ by

$$
Q(h) = J_\lambda^A(t)\Big(k + \lambda G\Big(t, e^{\frac{\tau}{\lambda}} h + \frac{1}{\lambda} e^{\frac{\tau}{\lambda}} \int_{\tau}^{0} e^{-\frac{\sigma}{\lambda}} \psi(\sigma) d\sigma\Big)\Big)
$$

for every $h \in X$ and $\tau \in [-r, 0]$, where $J_\lambda^A(t) = (I + \lambda A(t))^{-1}$. Then for every $h_i \in X$ for $i = 1, 2$

$$
\begin{aligned}
\|Q(h_1) - Q(h_2)\| &\leq \frac{\lambda\beta}{1 - \lambda\alpha}\|e^{\frac{\tau}{\lambda}} h_1 - e^{\frac{\tau}{\lambda}} h_2\| \\
&\leq \frac{\lambda\beta}{1 - \lambda\alpha}\|h_1 - h_2\|.
\end{aligned}
$$

Since $\dfrac{\lambda\beta}{1 - \lambda\alpha} < 1$ for sufficiently small $\lambda > 0$ Q is a strict contraction. Thus from Banach's contraction principle Q has a unique fixed point $h \in X$. With this h we define $\phi : [-r, 0] \to X$ by

$$
\phi(\tau) = e^{\frac{\tau}{\lambda}} h + \frac{1}{\lambda} e^{\frac{\tau}{\lambda}} \int_{\tau}^{0} e^{-\frac{\sigma}{\lambda}} \psi(\sigma) d\sigma
$$

for every $\tau \in [-r, 0]$. Then ϕ is differentiable a.e. on $[-r, 0]$ and satisfied $\dfrac{d\phi}{d\tau} - \lambda\phi = \psi$ a.e. on $[-r, 0]$. Also since $h = J_\lambda^A(t)(k + \lambda G(t, \phi))$ we obtain $\phi(0) = h \in D(A(t))$ for every $t \in [0, T]$. Hence $\{\phi, h\} \in D(B(t))$

and from (3.6.3)

$$\begin{aligned}
&(I+\lambda B(t))\{\phi, h\} \\
&\quad = \{\phi, h\} + \lambda B(t)\{\phi, h\} \\
&\quad = \{\phi, h\} + \lambda\Big\{-\frac{d\phi}{dt}, A(t)h - G(t,\phi)\Big\} \\
&\quad = \Big\{\phi - \lambda\frac{d\phi}{d\tau}, h + \lambda A(t)h - \lambda G(t,\phi)\Big\} \\
&\quad = \{\psi, k\}
\end{aligned}$$

for every $t \in [0,T]$. Hence $R(I+\lambda B(t)) = Y$ for sufficiently small $\lambda > 0$.

(2) Let $\{\psi, k\} \in Y$ and let $t \in [0,T]$. From (1) there exists $\{\phi, h\} \in D(B(t))$ such that $(I+\lambda B(t))\{\phi, h\} \ni \{\psi, k\}$ for sufficiently small $\lambda > 0$. We denote $\{\phi, h\}$ by $\{\phi(t,\cdot), h_\phi(t)\}$. Then $J_\lambda^B\{\psi, k\} = \{\phi(t,\cdot), h_\phi(t)\}$, and thus $\{\phi(t,\cdot), h_\phi(t)\} \in D(B(t)$ and $\phi(t,0) = h_\phi(t) \in D(A(t))$, where $J_\lambda^B(t) = (I+\lambda B(t))^{-1}$, from which

$$\begin{aligned}
&\phi(t,\cdot) - \lambda\frac{d\phi(t,\cdot)}{d\tau} = \psi, \\
&\phi(t,0) + \lambda A(t)\phi(t,0) - \lambda G(t,\phi(t,\cdot)) \ni k
\end{aligned}$$

for every $t \in [0,T]$ and sufficiently small $\lambda > 0$. Hence

$$\phi(t,\tau) = e^{\frac{\tau}{\lambda}}\phi(t,0) + \frac{1}{\lambda}\int_\tau^0 e^{-\frac{\sigma-\tau}{\lambda}}\psi(\sigma)d\sigma$$

for every $\tau \in [-r,0]$, and

$$J_\lambda^A(t)(k + \lambda G(t,\phi(t,\cdot))) = \phi(t,0)$$

for every $t \in [0,T]$.

Now let $\{\psi, k\} \in Y$. Then we obtain

$$\begin{aligned}
&\|J_\lambda^B(t_1)\{\psi,k\} - J_\lambda^B(t_2)\{\psi,k\}\|_Y \\
&\quad = \|\{\phi(t_1,\cdot), \phi(t_1,0)\} - \{\phi(t_2,\cdot), \phi(t_2,0)\}\|_Y \\
&\quad = \|\{\phi(t_1,\cdot) - \phi(t_2,\cdot), \phi(t_1,0) - \phi(t_2,0)\}\|_Y \\
&\quad = \Big(\int_{-r}^0 \|\phi(t_1,\tau) - \phi(t_2,\tau)\|^p d\tau + \|\phi(t_1,0) - \phi(t_2,0)\|^p\Big)^{\frac{1}{p}} \\
&\quad = \Big(\int_{-r}^0 e^{\frac{p\tau}{\lambda}}\|\phi(t_1,0) - \phi(t_2,0)\|^p d\tau + \|\phi(t_1,0) - \phi(t_2,0)\|^p\Big)^{\frac{1}{p}} \\
&\quad = \|\phi(t_1,0) - \phi(t_2,0)\|\Big(\int_{-r}^0 e^{\frac{p\tau}{\lambda}} d\tau + 1\Big)^{\frac{1}{p}} \le
\end{aligned}$$

$$\leq \left(1+\frac{\lambda}{p}\right)^{\frac{1}{p}} \|\phi(t_1,0)-\phi(t_2,0)\|$$
$$\leq K_1\|\phi(t_1,0)-\phi(t_2,0)\| \tag{3.6.4}$$

for every $t_i \in [0,T]$, $i=1,2$, where K_1 is a constant with

$$0 < (1+\frac{\lambda}{p})^{\frac{1}{p}} < K_1$$

for sufficiently small $\lambda > 0$.

But from $(A1)_6$, $(A2)_6$, $(G1)_6$, and $(G2)_6$ we have

$$\begin{aligned}
&\|\phi(t_1,0)-\phi(t_2,0)\| \\
&\quad = \|J_\lambda^A(t_1)(k+\lambda G(t_1,\phi(t_1,\cdot))) - J_\lambda^A(t_2)(k+\lambda G(t_2,\phi(t_2,\cdot)))\| \\
&\quad \leq \|J_\lambda^A(t_1)(k+\lambda G(t_1,\phi(t_1,\cdot))) - J_\lambda^A(t_2)(k+\lambda G(t_1,\phi(t_1,\cdot)))\| \\
&\qquad +\|J_\lambda^A(t_2)(k+\lambda G(t_1,\phi(t_1,\cdot))) - J_\lambda^A(t_2)(k+\lambda G(t_2,\phi(t_2,\cdot)))\| \\
&\quad \leq \frac{\lambda}{1-\lambda\alpha}\|G(t_1,\phi(t_1,\cdot)) - G(t_2,\phi(t_2,\cdot))\| \\
&\qquad +\|A_\lambda(t_1)(k+\lambda G(t_2,\phi(t_2,\cdot))) - A_\lambda(t_2)(k+\lambda G(t_2,\phi(t_2,\cdot)))\| \\
&\quad \leq \frac{\lambda}{1-\lambda\alpha}(\|G(t_1,\phi(t_1,\cdot)) - G(t_1,\phi(t_2,\cdot))\| \\
&\qquad\qquad +\|G(t_1,\phi(t_2,\cdot)) - G(t_2,\phi(t_2,\cdot))\|) \\
&\qquad +\|A_\lambda(t_1)(k+\lambda G(t_2,\phi(t_2,\cdot))) - A_\lambda(t_2)(k+\lambda G(t_2,\phi(t_2,\cdot)))\| \\
&\quad \leq \frac{\lambda}{1-\lambda\alpha}(\beta\|\phi(t_1,\cdot)-\phi(t_2,\cdot)\|_p + |t_1-t_2|l(\|\phi(t_2,\cdot)\|_p)) \\
&\qquad +\lambda|t_1-t_2|L(\|k+\lambda G(t_2,\phi(t_2,\cdot))\|) \\
&\qquad\qquad \times(1+\|A_\lambda(t_2)(k+\lambda G(t_2,\phi(t_2,\cdot)))\|).
\end{aligned} \tag{3.6.5}$$

Since

$$\begin{aligned}
&\|A_\lambda(t_2)(k+\lambda G(t_2,\phi(t_2,\cdot)))\| \\
&\quad = \frac{1}{\lambda}\|k+\lambda G(t_2,\phi(t_2,\cdot)) - J_\lambda^A(t_2)(k+\lambda G(t_2,\phi(t_2,\cdot)))\| \\
&\quad \leq \|G(t_2,\phi(t_2,\cdot))\| + \frac{1}{\lambda}\|k - J_\lambda^A(t_2)(k+\lambda G(t_2,\phi(t_2,\cdot)))\| \\
&\quad = \|G(t_2,\phi(t_2,\cdot))\| + \frac{1}{\lambda}\|k-\phi(t_2,0)\| \\
&\quad \leq \|G(t_2,\phi(t_2,\cdot))\| + \|B_\lambda(t_2)\{\psi,k\}\|_Y,
\end{aligned}$$

and from (3.6.2) and (3.6.4)

$$\|\phi(t_1,\cdot)-\phi(t_2,\cdot)\|_p \leq K_1\|\phi(t_1,0)-\phi(t_2,0)\|,$$

we obtain from (3.6.5)

$$\begin{aligned}
&\|\phi(t_1,0)-\phi(t_2,0)\| \\
&\le \frac{\lambda}{1-\lambda\alpha}(\beta K_1\|\phi(t_1,0)-\phi(t_2,0)\| + |t_1-t_2| l(\|\phi(t_2,\cdot)\|_p)) \\
&\quad +\lambda|t_1-t_2|L(\|k\|+\lambda\|G(t_2,\phi(t_2,\cdot))\|) \\
&\qquad \times(1+\|G(t_2,\phi(t_2,\cdot))\|+\|B_\lambda(t_2)\{\psi,k\}\|_Y),
\end{aligned}$$

from which we obtain

$$\begin{aligned}
&\|\phi(t_1,0)-\phi(t_2,0)\| \\
&\quad \le \frac{\lambda}{1-\lambda\alpha-\lambda\beta K_1}|t_1-t_2| l(\|\phi(t_2,\cdot)\|_p) \\
&\qquad +\frac{\lambda(1-\lambda\alpha)}{1-\lambda\alpha-\lambda\beta K_1}|t_1-t_2|L(\|k\|+\lambda\|G(t_2,\phi(t_2,\cdot))\|) \\
&\qquad \times(1+\|G(t_2,\phi(t_2,\cdot))\|+\|B_\lambda(t_2)\{\psi,k\}\|_Y) \\
&\quad \le \lambda K_2|t_1-t_2|(l(\|\phi(t_2,\cdot)\|_p)+L(\|k\|+\lambda\|G(t_2,\phi(t_2,\cdot))\|) \\
&\qquad \times(1+\|G(t_2,\phi(t_2,\cdot))\|+\|B_\lambda(t_2)\{\psi,k\}\|_Y)) \\
&\quad \le \lambda K_2|t_1-t_2|(l(\|\phi(t_2,\cdot)\|_p)+(1+\|G(t_2,\phi(t_2,\cdot))\|) \\
&\qquad \times L(\|k\|+\lambda\|G(t_2,\phi(t_2,\cdot))\|) \\
&\qquad\quad \times(1+\|B_\lambda(t_2)\{\psi,k\}\|_Y)),
\end{aligned} \tag{3.6.6}$$

where K_2 is a constant with

$$0<\frac{1}{1-\lambda\alpha-\lambda\beta K_1},\quad \frac{1-\lambda\alpha}{1-\lambda\alpha-\lambda\beta K_1}<K_2$$

for sufficiently small $\lambda>0$.

To find the bound of $\|G(t_2,\phi(t_2,\cdot))\|$ let $\{\hat{\phi},\hat{h}\}\in D(B(0))$ be fixed. Then from $(G1)_6$ and $(G2)_6$

$$\begin{aligned}
&\|G(t_2,\phi(t_2,\cdot))\| \\
&\quad \le \|G(t_2,\phi(t_2,\cdot))-G(t_2,\hat{\phi})\|+\|G(t_2,\hat{\phi})-G(0,\hat{\phi})\|+\|G(0,\hat{\phi})\| \\
&\quad < \beta\|\phi(t_2,\cdot)-\hat{\phi}\|_p+|t_1-t_2|l(\|\hat{\phi}\|_p)+\|G(0,\hat{\phi})\| \\
&\quad \le \beta(\|\phi(t_2,\cdot)\|_p+\|\hat{\phi}\|_p)+|t_1-t_2|l(\|\hat{\phi}\|_p)+\|G(0,\hat{\phi})\| \\
&\quad \equiv L_2(\|\phi(t_2,\cdot)\|_p).
\end{aligned}$$

Thus from (3.6.6)

$$\begin{aligned}
&\|\phi(t_1,0)-\phi(t_2,0)\| \\
&\quad \le \lambda K_2|t_1-t_2|(l(\|\phi(t_2,\cdot)\|_p)+(1+L_2(\|\phi(t_2,\cdot)\|_p)) \\
&\qquad \times L(\|k\|+\lambda L_2(\|\phi(t_2,\cdot)\|_p))(1+\|B_\lambda(t_2)\{\psi,k\}\|_Y))
\end{aligned} \tag{3.6.7}$$

for sufficiently small $\lambda > 0$. We also estimate $\|\phi(t_2,\cdot))\|_p$. From (3.6.4) and (3.6.7)

$$\begin{aligned}
&\|\phi(t_2,\cdot)\|_p \\
&\quad \le \|\{\phi(t_2,\cdot),\phi(t_2,0)\}\|_Y \\
&\quad \le \|\{\phi(t_2,\cdot),\phi(t_2,0)\} - \{\phi(0,\cdot),\phi(0,0)\}\|_Y \\
&\quad\quad +\|\{\phi(0,\cdot),\phi(0,0)\}\|_Y \\
&\quad \le K_1\|\{\phi(t_2,0),\phi(0,0)\}\|_Y + \|\{\phi(0,\cdot),\phi(0,0)\}\|_Y \\
&\quad \le \lambda K_1K_2t_2(l(\|\phi(0,\cdot)\|_p) \\
&\quad\quad +(1+L_2(\|\phi(0,\cdot)\|_p)) \\
&\quad\quad\quad \times L(\|k\| + \lambda L_2(\|\phi(0,\cdot)\|_p))(1+\|B_\lambda(0)\{\psi,k\}\|_Y)) \\
&\quad\quad +\|\{\phi(0,\cdot),\phi(0,0)\}\|_Y \\
&\quad \le K_1K_2t_2(l(\|\phi(0,\cdot)\|_p) \\
&\quad\quad +(1+L_2(\|\phi(0,\cdot)\|_p)) \\
&\quad\quad\quad \times L(\|k\| + \lambda L_2(\|\phi(0,\cdot)\|_p))(1+\lambda\|B_\lambda(0)\{\psi,k\}\|_Y)) \\
&\quad\quad +\|\{\phi(0,\cdot),\phi(0,0)\}\|_Y
\end{aligned} \tag{3.6.8}$$

for sufficiently small $\lambda > 0$. Since

$$\begin{aligned}
1 + \lambda\|B_\lambda(0)\{\psi,k\}\|_Y \\
&= \|\{\psi,k\} - J_\lambda^B(0)\{\psi,k\}\|_Y \\
&\le \|\{\psi,k\}\|_Y + \|J_\lambda^B(0)\{\psi,k\}\|_Y \\
&= \|\{\psi,k\}\|_Y + \|\{\phi(0,\cdot),\phi(0,0)\}\|_Y,
\end{aligned}$$

we have from (3.6.8)

$$\begin{aligned}
&\|\phi(t_2,\cdot)\|_p \\
&\quad \le K_1K_2t_2(l(\|\phi(0,\cdot)\|_p) \\
&\quad\quad +(1+L_2(\|\phi(0,\cdot)\|_p)) \qquad\qquad (3.6.9)\\
&\quad\quad \times L(\|k\| + L_2(\|\phi(0,\cdot)\|_p))(1+\|\{\psi,k\}\|_Y + \|\{\phi(0,\cdot),\phi(0,0)\}\|_Y)) \\
&\quad\quad +\|\{\phi(0,\cdot),\phi(0,0)\}\|_Y.
\end{aligned}$$

But then

$$\begin{aligned}
&\|\phi(0,\cdot)\|_p \\
&\quad \le \|\{\phi(0,\cdot),\phi(0,0)\}\|_Y \\
&\quad = \|J_\lambda^B(0)\{\psi,k\}\|_Y \\
&\quad \le \|J_\lambda^B(0)\{\psi,k\} - \|J_\lambda^B(0)\{\hat{\phi},\hat{h}\}\|_Y \\
&\quad\quad +\|J_\lambda^B(0)\{\hat{\phi},\hat{h}\} - \{\hat{\phi},\hat{h}\}\|_Y + \|\{\hat{\phi},\hat{h}\}\|_Y \le
\end{aligned}$$

$$
\begin{aligned}
&\le \frac{1}{1-\lambda\gamma}\|\{\psi,k\}-\{\hat{\phi},\hat{h}\}\|_Y+\lambda|B(0)\{\hat{\phi},\hat{h}\}|_Y+\|\{\hat{\phi},\hat{h}\}\|_Y\\
&\le K_3(\|\{\psi,k\}\|_Y+\|\{\hat{\phi},\hat{h}\}\|_Y)+|B(0)\{\hat{\phi},\hat{h}\}|_Y+\|\{\hat{\phi},\hat{h}\}\|_Y\\
&\equiv L_3(\|\{\psi,k\}\|_Y)
\end{aligned}
$$

for sufficiently small $\lambda > 0$, where

$$
|B(0)\{\hat{\phi},\hat{h}\}|_Y=\lim_{\lambda\to 0+}\|B_\lambda(0)\{\hat{\phi},\hat{h}\}\|_Y.
$$

Since

$$
\begin{aligned}
&\|\phi(t_2,\cdot)\|_p\\
&\quad\le K_1K_2t_2(l(L_3(\|\{\psi,k\}\|_Y))\\
&\qquad+(1+L_2(L_3(\|\{\psi,k\}\|_Y)))\\
&\qquad\quad\times L(\|k\|+L_2(L_3(\|\{\psi,k\}\|_Y)))\\
&\qquad\qquad\times(1+\|\{\psi,k\}\|_Y+L_3(\|\{\psi,k\}\|_Y)))\\
&\qquad+L_3(\|\{\psi,k\}\|_Y)\\
&\quad\equiv L_4(\|\{\psi,k\}\|_Y)
\end{aligned}
$$

for sufficiently small $\lambda > 0$ from (3.6.7)

$$
\begin{aligned}
&\|\phi(t_1,0)-\phi(t_2,0)\|\\
&\quad\le \lambda K_2|t_1-t_2|(l(L_4(\|\{\psi,k\|_Y))\\
&\qquad+(1+L_2(L_4(\|\{\psi,k\}\|_Y)))\\
&\qquad\quad\times L(\|k\|+\lambda L_2(L_4(\{\psi,k\}\|_Y)))(1+\|B_\lambda(t_2)\{\psi,k\}\|_Y))
\end{aligned}
$$

for sufficiently small $\lambda > 0$ and from (3.6.4)

$$
\begin{aligned}
&\|J_\lambda^B(t_1)\{\psi,k\}-J_\lambda^B(t_2)\{\psi,k\}\|_Y\\
&\quad\le \lambda K_1K_2|t_1-t_2|(l(L_4(\|\{\psi,k\}\|_Y))\\
&\qquad+(1+L_2(L_4(\|\{\psi,k\}\|_Y)))\\
&\qquad\quad\times L(\|k\|+L_2(L_4(\{\psi,k\}\|_Y))))(1+\|B_\lambda(t_2)\{\psi,k\}\|_Y)\\
&\quad=\lambda|t_1-t_2|L_1(\|\{\psi,k\}\|_Y)(1+\|B_\lambda(t_2)\{\psi,k\}\|_Y)
\end{aligned}
$$

for sufficiently small $\lambda > 0$, where $L_1 : [0,\infty) \to [0,\infty)$ is a continuous nondecreasing function defined by

$$
L_1(c)\equiv K_1K_2(l(L_4(c))+(1+L_2(L_4(c)))L(\|k\|+L_2(L_4(c)))
$$

for every $c \in [0, \infty)$. Therefore

$$\|B_\lambda(t_1)\{\psi, k\} - B_\lambda(t_2)\{\psi, k\}\|_Y$$
$$\leq |t_1 - t_2| L_1(\|\{\psi, k\}\|_Y)(1 + \|B_\lambda(t_2)\{\psi, k\}\|_Y)$$

for sufficiently small $\lambda > 0$.

Now we consider the existence for solutions of (3.6.1).
We have the main theorem for $1 < p < \infty$.

THEOREM 3.6.1 *Let X be a real Hilbert space. Assume that $(A1)_6$, $(A2)_6$, $(G1)_6$, and $(G2)_6$ are satisfied. Let $\{U(t,s) \mid (t,s) \in \Delta\}$ be the evolution operator in Y of the type γ generated by $B(t)$ as in* (3.6.3) *for every $t \in [0,T]$. Let $1 < p < \infty$. Let $\{\phi, h\} \in D_B$. Then there exists a strong solution of* (3.6.1) *on $[-r, T]$.*

Proof. Let $\{\phi, h\} \in D_B$. Since Y is reflexive for $1 < p < \infty$, from Theorem 1.7.1, $u(t) \equiv U(t, 0)\{\phi, h\}$ is a unique strong solution of

$$\begin{cases} \dfrac{du}{dt}(t) + B(t)u(t) \ni 0, & t \in [0,T] \\ u(0) = \{\phi, h\}. \end{cases} \tag{3.6.10}$$

That is, $u(t) : [0,T] \to X$ is absolutely continuous, differentiable a.e. on $(0,T)$, and satisfies

$$\frac{d}{dt}U(t,0) + B(t)U(t,0) \ni 0 \tag{3.6.11}$$

a.e. on (0,T) with $U(0,0) = \{\phi, h\}$. We define $x(t) : [-r, T] \to X$ by

$$x(t) = \begin{cases} \phi(t), & t \in [-r, 0] \\ \pi_2 U(t,0)\{\phi, h\}, & t \in [0,T]. \end{cases} \tag{3.6.12}$$

Then we can show that

$$x_t = \pi_1 U(t,0)\{\phi, h\} \tag{3.6.13}$$

for every $t \in [0,T]$ as in Proof of Proposition 3.2.3. Since X is reflexive and $x(t)$ is absolutely continuous on $[0,T]$, $x(t)$ is differentiable a.e. on $(0,T)$. If we project (3.6.10) into X, then from (3.6.3)

$$\frac{d}{dt}\pi_2 U(t,0)\{\phi, h\} + A(t)\pi_2 U(t,0)\{\phi, h\} \ni G(t, \pi_1 U(t,0)\{\phi, h\})$$

a.e. $t \in (0,T)$. Thus by (3.6.11) and (3.6.12) we obtain

$$\frac{dx}{dt}(t) + A(t)x(t) \in G(t, x_t)$$

a.e. $t \in (0, T)$,

$$x(0) = \pi_2 U(0,0)\{\phi, h\} = h \text{ and } x_0 = \pi_1 U(0,0)\{\phi, h\} = \phi.$$

Therefore $x(t)$ is a strong solution of (3.6.1) on $[-r, T]$.

We obtain the main theorem for $p = 1$.

THEOREM 3.6.2 *Let X^* be uniformly convex. Suppose that $A(t) : D_A \equiv D(A(t)) \subset X \to X$ is single-valued. Assume that $(A1)'_6$, $(A2)'_6$, $(G1)_6$, and $(G2)_6$ are satisfied. Let $p = 1$. Let $\{\phi, h\} \in D_B$. Then there exists an integral solution of* (3.6.1) *on* $[-r, T]$.

Proof. By Proposition 1.4.4 $(A2)_6$ holds. Thus from Proposition 3.6.1 and Theorem 1.8.1 there exists a limit solution $u(t)$ of (3.6.9) on $[0, T]$. Then there exist a partition $0 = t_0^n < t_1^n < t_2^n, \cdots < t_{N(n)}^n = T$ of $[0, T]$ and a sequence $\{u_i^n\} \in Y$ for $i = 0, 1, 2, \cdots, N(n)$, $n = 1, 2, \cdots$ such that

$$\lim_{n\to\infty} u_n(t) = u(t),$$

where

$$u_n(t) = \begin{cases} u_i^n, & t \in (t_{i-1}^n, t_i^n],\ i = 1, 2, \cdots, N(n) \\ u_0^n, & t = 0. \end{cases}$$

Here,

$$\frac{u_i^n - u_{i-1}^n}{t_i^n - t_{i-1}^n} + B(t_i^n)u_i^n = 0 \tag{3.6.14}$$

with $u_i^n = \{\phi_i^n, h_i^n\}$ for $i = 1, 2, \cdots, N(n)$, $n = 1, 2, \cdots$

$$u_0^n = \{\phi_0^n, h_0^n\} = \{\phi, h\} \in D(B(0)),$$

and

$$\lim_{n\to\infty} \max_{1\le i\le N(n)} (t_i^n - t_{i-1}^n) = 0.$$

If we project (3.6.13) onto X we obtain

$$\frac{h_i^n - h_{i-1}^n}{t_i^n - t_{i-1}^n} + A(t_i^n)h_i^n = G(t_i^n, \phi_i^n) \tag{3.6.15}$$

for $i = 0, 1, 2, \cdots, N(n)$, $n = 1, 2, \cdots$. We define

$$x_n(t) = \begin{cases} h_0^n, & t = 0 \\ h_i^n, & t \in (t_{i-1}^n, t_i^n],\ i = 1, 2, \cdots, N(n) \end{cases}$$

for $n = 1, 2, \cdots$. Since

$$\lim_{n\to\infty} u_n(t) = u(t) \text{ and } h_i^n = \pi_2(u_n(t))$$

we have $\lim_{n\to\infty} x_n(t) = x(t)$, where

$$x(t) = \begin{cases} \phi(t), & t \in [-r, 0] \\ \pi_2(u(t)), & t \in [0, T] \end{cases} \tag{3.6.16}$$

and then $\pi_1(u(t)) = x_t$ for every $t \in [0, T]$.

Since $u(t)$ is continuous $x(t)$ is also continuous on $[0, T]$.

Let $x \in D_A$ and $\sigma \in [0, T]$. For $x^* \in F(h_i^n - x)$ since

$$\begin{aligned} \|h_i^n - x\|^2 &= (h_i^n - x, x^*) \\ &= (h_i^n - h_{i-1}^n, x^*) + (h_{i-1}^n - x, x^*) \\ &\leq (h_i^n - h_{i-1}^n, x^*) + \|h_{i-1}^n - x\|\|h_i^n - x\|, \end{aligned}$$

we obtain by (3.6.14)

$$\begin{aligned} &\|h_i^n - x\|^2 - \|h_i^n - x\|\|h_{i-1}^n - x\| \\ &\quad\leq (h_i^n - h_{i-1}^n, x^*) \\ &\quad= (t_i^n - t_{i-1}^n)(-A(t_i^n)h_i^n + G(t_i^n, \phi_i^n), x^*) \\ &\quad= -(t_i^n - t_{i-1}^n)(A(t_i^n)h_i^n - A(t_i^n)x, x^*) \\ &\qquad+(t_i^n - t_{i-1}^n)(A(\sigma)x - A(t_i^n)x, x^*) \\ &\qquad+(t_i^n - t_{i-1}^n)(-A(\sigma)x + G(t_i^n, \phi_i^n), x^*) \\ &\quad\leq (t_i^n - t_{i-1}^n)\alpha\|h_i^n - x\|^2 \\ &\qquad+(t_i^n - t_{i-1}^n)\|A(t_i^n)x - A(\sigma)x\|\|h_i^n - x\| \\ &\qquad+[-A(\sigma)x + G(t_i^n, \phi_i^n), h_i^n - x]_+ \end{aligned}$$

for $i = 1, 2, \cdots$ and $n = 1, 2, \cdots$. Thus from our assumptions and Proposition 1.4.4 there exists a continuous nondecreasing function $\widetilde{L}_1 : [0, \infty) \to [0, \infty)$ such that

$$\|A(t_1)u - A(t_2)u\| \leq |t_1 - t_2|\widetilde{L}_1(\|u\|)(1 + \|A(t_2)u\|) \tag{3.6.17}$$

for every $t_i \in [0, T]$, $i = 1, 2$, and $u \in D_A$. From (3.6.16) and (3.6.17) we have

$$\begin{aligned} &\|h_i^n - x\| - \|h_{i-1}^n - x\| \\ &\quad\leq (t_i^n - t_{i-1}^n)\alpha\|h_i^n - x\| \\ &\qquad+(t_i^n - t_{i-1}^n)|t_i^n - \sigma|\widetilde{L}_1(\|x\|)(1 + \|A(\sigma)x\|) \\ &\qquad+(t_i^n - t_{i-1}^n)[-A(\sigma)x + G(t_i^n, \phi_i^n), h_i^n - x]_+. \end{aligned}$$

for $i = 1, 2, \cdots, N(n)$, $n = 1, 2, \cdots$ and $x \in D_A$. Adding these inequalities for $i = j+1, \cdots, k$ we obtain

$$\begin{aligned}\|h_k^n - x\| &- \|h_j^n - x\| \\ &\leq \sum_{i=j+1}^{k} (\alpha \|h_i^n - x\| + C|t_i^n - \sigma| \\ &\qquad + [-A(\sigma)x + G(t_i^n, \phi_i^n), h_i^n - x]_+)\end{aligned}$$

for $n = 1, 2, \cdots$ and $x \in D_A$, where

$$C = \widetilde{L}_1(\|x\|)(1 + \|A(\sigma)x\|).$$

Let $t \in (t_{k-1}^n, t_k^n]$ and $s \in (t_{j-1}^n - t_j^n]$. Set $a_n(\tau) = t_i^n$ for $\tau(t_{i-1}^n, t_i^n]$. Then

$$\begin{aligned}\|\pi_2(u_n(t)) - x\| &- \|\pi_2(u_n(s)) - x\| \\ \leq \int_{t_j^n}^{t_k^n} &([-A(\sigma)x + G(a_n(\tau), \pi_1(u_n(\tau))), \pi_2(u_n(\tau)) - x]_+ \\ &+ \alpha \|\pi_2(u_n(\tau)) - x\| + C \mid a_n(\tau) - \sigma \mid) d\tau\end{aligned}$$

for $n = 1, 2, \cdots$. Since $\lim_{n \to \infty} a_n(\tau) = \tau$ passing to the limit as $n \to \infty$ we obtain

$$\begin{aligned}\|\pi_2(u(t)) - x\| &- \|\pi_2(u(s)) - x\| \\ \leq \int_s^t &([-A(\sigma)x + G(\tau, \pi_1(u(\tau))), \pi_2(u(\tau)) - x]_+ \\ &+ \alpha \|\pi_2(u(\tau)) - x\| + C \mid \tau - \sigma \mid) d\tau.\end{aligned}$$

From (3.6.15)

$$\begin{aligned}\|x(t) - x\| &- \|x(s) - x\| \\ &\leq \int_s^t ([-A(\sigma)x + G(\tau, x_\tau), x(\tau) - x]_+ \\ &\qquad + \alpha \|x(\tau) - x\| + C|\tau - \sigma|) d\tau.\end{aligned}$$

It implies that $x(t)$ is an integral solution of (3.6.1) on $[-r, T]$.

3.7. Examples and Applications

In this section we give some examples which the results of previous sections are applied.

EXAMPLE 3.7.1 (Kartsatos–Parrott [84], [87]) *We consider the following non-autonomous nonlinear functional heat equation*

$$\begin{cases} \dfrac{\partial x}{\partial t}(t,u) - a(t)k\Big(\dfrac{\partial x}{\partial u}(t,u)\Big)\dfrac{\partial^2 x}{\partial u^2}(t,u) = f(t, x(t-r,u)), \\ \qquad\qquad\qquad\qquad\qquad\qquad 0 \le t \le T, u \in (0,1) \\ x(t,u) = \phi(t,u), \quad t \in [-r,0], u \in (0,1) \\ x(t,0) = \alpha \dfrac{\partial x}{\partial t}(t,0), \quad 0 \le t \le T \\ x(t,1) = -\gamma \dfrac{\partial x}{\partial t}(t,1), \ 0 \le t \le T. \end{cases} \tag{3.7.1}$$

Here $x(t,u) : [-r,T] \times (0,1) \to \mathbf{R}$, $T > 0$, $r > 0$, $a(t) : [0,T] \to [m,\infty)$ for some constant $m > 0$, $k : \mathbf{R} \to \mathbf{R}$, $f : \mathbf{R} \to \mathbf{R}$, $r > 0$, a, $b \ge 0$ and $\phi : [-r,0] \times (0,1) \to \mathbf{R}$.

THEOREM 3.7.1 *Let $a(t) : [0,T] \to [m,\infty)$ be Lipschitzian, i.e.,*

$$|a(t_1) - a(t_2)| \le B|t_1 - t_2|$$

for every $t_i \in [0,T]$, $i = 1,2$, and for some constant $B > 0$. Let $f : [0,T] \times \boldsymbol{R} \to \boldsymbol{R}$ be Lipschitzian jointly in its two variables, i.e.,

$$|f(t_1,v_1) - f(t_2,v_2)| \le \beta(|t_1 - t_2| + |v_1 - v_2|) \tag{3.7.2}$$

for every $t_i \in [0,T]$, $v_i \in \boldsymbol{R}$, $i = 1,2$ and some $\beta > 0$. Let $\phi \in C^1([-r,0]; L^p(0,1))$ with $\phi(0) \in D$ for $1 < p < \infty$. Then (3.7.1) *has at most one solution $x(t) \equiv x(t,u;\phi)$ on $[-r,T]$ a.e. $u \in (0,1)$, where D is the set of* (3.7.3) *below.*

Proof. Let $X = L^p(0,1)$ for $1 < p < \infty$. Now we define an operator $\widetilde{A}(t) : D \equiv D(\widetilde{A}(t)) \subset X \to X$ for every $t \in [0,T]$ by

$$\begin{aligned} D &= \Big\{x \in X \;\Big|\; x \in C^2([0,1]),\ x(0) = \alpha \frac{dx}{du}(0), \\ &\qquad\qquad x(1) = -\gamma \frac{dx}{du}(1)\Big\}, \qquad (3.7.3) \\ (\widetilde{A}(t)x)(u) &= -a(t)k\Big(\frac{dx}{du}(u)\Big)\frac{d^2x}{du^2}(u), \quad x \in D. \end{aligned}$$

Let $A(t)$ be the closure of $\widetilde{A}(t)$. We define $G : [0,T] \times C([-r,0];X) \to X$ by

$$G(t,\psi) = f(t,\psi(-r))$$

for every $t \in [0,T]$ and $\psi \in C([-r,0];X)$. Then we have

$$G(t,x_t) = f(t,x_t(-r)) = f(t,x(t-r))$$

for every $x \in C([-r,0];X)$ and $t \in [0,T]$. Thus (3.7.1) now takes the abstract form of the type

$$\begin{cases} \dfrac{dx}{dt}(t) + A(t)x(t) = G(t,x_t), & 0 \le t \le T \\ x(t) = \phi(t), & -r \le t \le 0. \end{cases}$$

We show that $(A1)_1$, $(A2)_1$, $(G1)_1$, and $(G2)_1$ are satisfied. By Burch–Goldstein [18,Theorem 2.2], $A(t)$ is m-accretive in X with $a(t) \equiv 1$ for every $1 \le p < \infty$. Since $a(t) > 0$ $A(t)$ is also m-accretive in X.

Moreover

$$\begin{aligned} &\|\widetilde{A}(t_1)x - \widetilde{A}(t_2)x\|_p \\ &\quad = |a(t_1) - a(t_2)| \Big\| k\Big(\frac{dx}{du}\Big)\frac{d^2x}{du^2}\Big\|_p \\ &\quad \le B|t_1 - t_2| \frac{\|\widetilde{A}(t)x\|_p}{a(t_1)} \\ &\quad \le \frac{B}{a}|t_1 - t_2| \|\widetilde{A}(t)x\|_p \\ &\quad \le \frac{B}{a}|t_1 - t_2|(1 + \|\widetilde{A}(t)x\|_p) \end{aligned}$$

for every $t_i \in [0,T]$, $i = 1,2$ and $x \in D$. Thus we see that $(A1)_1$ and $(A2)_I$ are satisfied.

For $(G1)_1$ and $(G2)_2$ from (3.7.2) we obtain

$$\begin{aligned} &\|G(t,\psi_1) - G(t,\psi_2)\|_p \\ &\quad = \Big(\int_0^1 |f(t,\psi_1(-r,u)) - f(t,\psi_2(-r,u))|^p du\Big)^{\frac{1}{p}} \\ &\quad \le \beta\Big(\int_0^1 |\psi_1(-r,u) - \psi_2(-r,u)|^p du\Big)^{\frac{1}{p}} \\ &\quad = \beta\|\psi_1 - \psi_2\|_{C([0,1];L^p(0,1))} \end{aligned}$$

for every $\psi_i \in C([0,1];L^p(0,1))$, $i = 1,2$, and

$$\begin{aligned} &\|G(t_1,\psi) - G(t_2,\psi)\|_p \\ &\quad = \Big(\int_0^1 |f(t_1,\psi(-r,u)) - f(t_2,\psi(-r,u))|^p du\Big)^{\frac{1}{p}} \le \end{aligned}$$

$$\begin{aligned}&\leq \beta\Big(\int_0^1 |t_1 - t_2|^p du\Big)^{\frac{1}{p}}\\ &= \beta|t_1 - t_2|\end{aligned}$$

for every $t_i \in [0,T]$, $i = 1,2$. Since $X = L^p(0,1)$ is uniformly convex for $1 < p < \infty$, it follows from Theorem 3.1.1 that there exists a unique strong solutuon $x(t) \equiv x(t;\phi)$ of (3.1.1) on $[-r,T]$. Hence $x(t,u) = (x(t))(u)$ satisfies (3.7.1) for every $t \in [-r,T]$ and a.e. $u \in (0,1)$.

EXAMPLE 3.7.2 *We consider the first order non-autonomous nonlinear hyperbolic functional equation*

$$\begin{cases} \dfrac{\partial x}{\partial t}(t,u) + b(t)g(x(t,u)) = f(t, x(t-r,u)), \\ \qquad\qquad\qquad\qquad t \in [0,T],\ u \in (0,1) \\ x(t,u) = \phi(t,u), \quad t \in [-r,0],\ u \in (0,1), \end{cases} \tag{3.7.4}$$

where $x(t,u) : [-r,T] \times (0,1) \to \boldsymbol{R}$, $T > 0$, $r > 0$, $g : \boldsymbol{R} \to \boldsymbol{R}$, $b(t) : [0,T] \to [b,\infty)$ *for some constant* $b \geq 0$, $f : [0,T] \times \boldsymbol{R} \to \boldsymbol{R}$ *and* $\phi : [-r,0] \times (0,1) \to \boldsymbol{R}$.

THEOREM 3.7.2 $g : \boldsymbol{R} \to \boldsymbol{R}$ *is continuous and strictly monotone,* $g(\boldsymbol{R}) = \boldsymbol{R}$, $b(t) \geq b$ *for some* $b > 0$, $b(t)$ *is Lipschitzian, i.e.,*

$$|b(t_1) - b(t_2)| \leq B_1|t_1 - t_2|$$

for every $t_i \in [0,T]$, $i = 1,2$ *and some constant* $B_1 > 0$. *Assume that* $f : [0,T] \times \boldsymbol{R} \to \boldsymbol{R}$ *satisfies*

$$|f(t_1,v_1) - f(t_2,v_2)| \leq \beta(|t_1 - t_2| + |v_1 - v_2|)$$

for every $t_i \in [0,T]$, $v_i \in \boldsymbol{R}$, $i = 1,2$ *and some constant* $\beta > 0$. *Let* $\phi \in C([-r,0]; L^1(0,1))$ *be Lipschitz continuous in the first variable and* $\phi(0,\cdot) \in L^1(0,1) \cap L^\infty(0,1)$ *with* $\dfrac{\partial g(\phi(0,u))}{\partial u} \in L^1(0,1)$. *Then* (3.7.2) *has a limit solution and an integral solution* $x(t,u)$ *on* $[-r,T]$ *and a.e. on* $(0,1)$.

Proof. Let $X = L^1(0,1)$. We define an operator $A(t) : D \equiv D(A(t)) \subset X \to X$ for every $t \in [0,T]$, by

$$\begin{aligned} D &= \Big\{x \in L^1(0,1) \cap L^\infty(0,1) \;\Big|\; \frac{\partial g(x(u))}{\partial u} \in L^1(0,1)\Big\}, \\ (A(t)x)(u) &= b(t)\frac{\partial g(x(u))}{\partial u}, \quad x \in D. \end{aligned}$$

We also define $G : [0,T] \times C([-r,0];X) \to X$ by

$$G(t,\psi) = f(t,\psi(-r))$$

for every $t \in [0,T]$ and $\psi \in C([-r,0];X)$ Then we have

$$G(t,x_t) = f(t,x_t(-r)) = f(t,x(t-r)) \tag{3.7.5}$$

for every $x \in D$ and $t \in [0,T]$. Thus (3.7.3) can be formulated in the type of (3.2.1)

$$\begin{cases} \dfrac{dx}{dt}(t) + A(t)x(t) = G(t,x_t), & 0 \le t \le T \\ x(t) = \phi(t), & -r \le t \le 0. \end{cases}$$

We show that $(A1)_2$, $(A2)_2$, $(G1)_2$, and $(G2)_2$ are satisfied.

By Crandall [23, Theorem 3.1], $A(t)$ is m-accretive in X, which is condition $(A1)_2$. In order to verify condition $(A2)_2$ we see that

$$\begin{aligned} &\|A(t_1)x - A(t_2)x\|_1 \\ &\quad = \int_0^1 |(A(t_1)x)(u) - (A(t_2)x)(u)|du \\ &\quad = |b(t_1) - b(t_2)| \int_0^1 \left|\frac{\partial g(x(u))}{\partial u}\right| du \\ &\quad \le B_1|t_1 - t_2| \frac{\|A(t_2)x\|_1}{b(t_2)} \\ &\quad \le \frac{B_1}{b}|t_1 - t_2|\|A(t_2)x\|_1 \\ &\quad \le \frac{B_1}{b}|t_1 - t_2|(1 + \|A(t_2)x\|_1) \end{aligned}$$

for every $t_i \in [0,T]$, $i = 1,2$ and $x \in D$. For $(G1)_1$ and $(G2)_2$ from (3.7.4) we obtain

$$\begin{aligned} &\|G(t,\psi_1) - G(t,\psi_2)\|_1 \\ &\quad = \int_0^1 |f(t,\psi_1(-r,u)) - f(t,\psi_2(-r,u))|du \\ &\quad \le \beta \int_0^1 |\psi_1(-r,u) - \psi_2(-r,u)|du \\ &\quad = \beta\|\psi_1 - \psi_2\|_{C([-r,0];X)} \end{aligned}$$

for every $\psi_i \in C([-r,0];X)$, $i = 1,2$, and

$$\|G(t_1,\psi) - G(t_2,\psi)\|_1 =$$

$$\begin{aligned}
&= \int_0^1 |f(t_1, \psi(u,-r)) - f(t_2, \psi(u,-r))| du \\
&\leq \beta \int_0^1 |t_1 - t_2| du \\
&= \beta |t_1 - t_2|
\end{aligned}$$

for every $t_i \in [0,T]$, $i = 1,2$. From (1) and (2) of Theorem 4.3.1, there exists a limit solution solution and an integral solution , $x(t) \equiv x(t;\phi)$ of (3.2.1) on $[-r,T]$. Hence $x(u,t) = (x(t))(u)$ satisfies (3.7.1) as a limit solution and an integral solution for every $t \in [-r,T]$ and a.e. $u \in (0,1)$.

3.8. Notes for References

The methods in Section 3.1 follows those in Kato [95]. The results containing Theorem 3.1.1 are taken from Kartsatos–Parrott [86]. The technique for proving the existence for solutions of (3.1.1) does not involve the existence of a nonlinear evolution operator. We show in a straightforward manner that the unique strong solution $x(t)$ for (3.1.1) exists as the uniform limit of a sequence $\{x_n(t)\}$ for (classical) solutions of the approximate evolutions of the type

$$\begin{cases} \dfrac{dx_n}{dt}(t) + A_n(t)x_n(t) = G(t,(x_n)_t), \quad 0 \leq t \leq T \\ x_n(t) = \phi(t), \quad t \in [-r,0]. \end{cases}$$

For non-functional results in this direction, Kartsatos–Zigler [94] and Kartsatos [79] are referred.

Proposition 3.2.1 is due to Dyson–Villella–Bressan [39], Ha–Shin–Jin [62] and Tanaka [130]. Proposition 3.2.2 is from Crandall–Liggett [26], while Proposition 3.2.3 is taken from Dyson–Villella–Bressan [38]. Theorem 3.2.1 is owed to Ruess [122].

Dyson–Villella–Bressan [40], [41], [45], [46], Kartsatos [77], Kartsatos–Parrott [87], [88], [91] and Parrott [115] are referred.

The method in Section 3.3 of proving the existence of a unique solution of (3.3.1) in the sense of Definition 3.3.1 is based on Rothe [121], a method of lines for (3.3.1). Theorem 3.3.1 is owed to Kartsatos–Parrott [90] refering with the proof in Kačur [71] and Kartsatos–Parrott [89]. Also Kačur [72], Kartsatos–Zigler [94] and Necǎs [110] are referred.

The results in Section 3.4 are taken from Kartsatos [81] and Tanaka [131]. Proposition 3.4.1 is taken from Tanaka [131] and Proposition 3.4.2 is due to Kartsatos [81]. The boundary condition (3.4.6) is needed to show that the method of lines is in the closed ball. We also use the homotopy type of lemma to prove the existence of methods of lines.

Theorem 3.4.1 is taken from Tanaka [131] and Kartsatos [81]. Kartsatos [78] and Kartsatos–Parrott [87] considered the existence for a method of lines and solutions in the case of uniformly convex dual.

For operator equations Kartsatos–Shin [93] is referred.

Compactness results are from Vrabii [135]. The results in Section 3.5 with Theorem 3.5.1 and 3.5.2 are taken from Ruess [123]. While Theorem 3.5.1 obviously extends the results of Mitidieri–Vrabie [105] to the nonautonomous case, it also slightly extends the corresponding results of Theorem 5 and Theorem 6 of Kartsatos–Shin [92] in various respects: the restriction of Theorem 5 and Theorem 6 of Kartsatos–Shin [92] to $\phi \in C([-r,0];\overline{D})$ with the special properties of Lemma 8 of Kartsatos–Shin [92] can be dispensed with; in Theorem 3.5.2, the extra assumptions of the corresponding result of Theorem 6 of Kartsatos-Shin [92] that also the space X itself be uniformly convex, and the compactness of $G : [0,T] \times C([-r,0];\overline{D}) \to X$ are not needed. Note that the assumption that X be uniformly convex in Kartsatos-Shin [92] is used only to ensure convexity of the set $\bar{D}$, and that we have circumvented this problem by suitably extending $G(\cdot,\cdot)$ in the proof of Theorem 3.5.1.

Theorem 3.6.1 and Theorem 3.6.2 are owed to Ha [60] and Ha [61], respectively. One can refer to Dyson–Villella–Bressan [39], Ha [59] and Webb–Badii [139].

Example 3.7.1 is taken from Kartsatos-Parrott [86], [89] and Example 3.7.2 is owed to Parrott [114].

for every $u \in D$ and $t_i \in [0, T_1]$, $i = 1, 2$, where $\widetilde{L} : [0, \infty) \to [0, \infty)$ is defined by

$$\widetilde{L}(a) = (1 + C_1)L(2\|\phi\|_C + c, 2\|\phi\|_C + c, a)$$

for every $a \in [0, \infty)$, which is (4.1.4). From Theorem 3.1.1 with $G \equiv 0$, (4.1.2) has a unique strong solution $x^v(t)$ on $[-r, T_1]$. $x^v(t)$ is weakly continuously differentiable on $[0, T_1]$ and such that $A(t, v_t)x^v(t)$ is weakly continuous in t. Furthermore, $x^v(t)$ satisfies (4.1.2) everywhere on $[0, T_1]$ if $\dfrac{dx}{dt}(t)$ denotes now the weak derivative of $x(t)$.

We show that the operator $Q : v \to x^v$ is a strict contraction on S if T_1 is chosen sufficiently small. To this end fix $v \in S$ and consider the approximating equations of the type

$$\begin{cases} \dfrac{dx_n}{dt}(t) + A_n(t)x_n(t) = 0, \quad 0 \le t \le T_1 \\ x_n(t) = \phi(t), \quad -r \le t \le 0, \end{cases} \tag{4.1.6}$$

where $A_n(t) = A_n(t, v_t)\Big(I + \dfrac{1}{n}A(t, v_t)\Big)^{-1}$ is the Yosida approximant of $A(t, v_t)$ for $n = 1, 2, \cdots$. The operators $A_n(t)$ are defined and Lipschitz continuous on X with Lipschitz constants less than $2n$. Moreover, the operator $J_n(t) = \Big(I + \dfrac{1}{n}A(t, v_t)\Big)^{-1} : X \to D$ is also Lipschitz continuous on X with Lipschitz constant less than 1 for $n = 1, 2, \cdots$. Since $A^v(t)$ is m-accretive for every $t \in [0, T_1]$, so is the operator $A_n(t)$ for $n = 1, 2, \cdots$. From Proposition 1.4.4 and $(A2)_1$ we obtain

$$\|A_n(t_1)u - A_n(t_2)u\| \le \widetilde{L}(\|J_n(t_2)u\|)|t_1 - t_2|(1 + \|A_n(t_2)u\|)$$

for every $u \in X$, $t_i \in [0, T_1]$, $i = 1, 2$, and $n = 1, 2, \cdots$. Since, using (4.1.5)

$$\begin{aligned} &\frac{1}{n}\|A_n(t_2)u\| \\ &\quad \le \frac{1}{n}(\|A_n(t_2)u - A_n(t_2)\phi(0)\| + \|A_n(t_2)\phi(0)\|) \\ &\quad \le 2\|u - \phi(0)\| + \frac{1}{n}\|A_n(t_2)\phi(0)\| \\ &\quad \le \|A_n(t_2, v_{t_2})\phi(0)\| + 2(\|u\| + \|\phi(0)\|) \\ &\quad \le \|A(0, \phi)\phi(0)\| + \widetilde{L}(\|\phi(0)\|)\frac{c}{C_1}(1 + \|A(0, \phi)\phi(0)\|) \\ &\quad\quad +2(\|u\| + \|\phi(0)\|) = C_4 + 2\|u\|, \end{aligned}$$

where

$$C_4 = \|A_n(0,\phi)\phi(0)\| + \widetilde{L}(\|\phi(0)\|)\frac{c}{C_1}(1 + \|A(0,\phi)\phi(0)\|) + 2\|\phi(0)\|$$

and

$$\|J_n(t_2)u\| \le \|u\| + \frac{1}{n}\|A_n(t_2)u\|,$$

we finally arrive at

$$\begin{aligned}
&\|A_n(t_1, v_{t_1})u - A_n(t_1, v_{t_1})u\| \\
&\quad = \|A_n(t_1)u - A_n(t_2)u\| \\
&\quad \le L_2(\|u\|)|t_1 - t_2|(1 + \|A_n(t_2)u\|) \qquad (4.1.7)
\end{aligned}$$

for every $u \in D$, $t_i \in [0, T_1]$, $i = 1, 2$ and $n = 1, 2, \cdots$, where $L_2 : [0,\infty) \to [0,\infty)$ is defined by

$$L_2(a) = L_1(3a) + C_4$$

for every $a \in [0,\infty)$. From Theorem 1.6.1 (4.1.6) has a unique (classical) solution $x_n(t)$ on $[0, T_1]$ and such that $\lim_{n\to\infty} x_n(t) = x^v(t)$ uniformly on $[0, T_1]$.

We show next that $x_n(t)$ is uniformly bounded and uniformly Lipschitz continuous on $[0, T_1]$ independently of $v \in S$. To this end, from (12) of Lemma 1.1.2 with the accretiveness of $A_n(t)$ and (4.1.7) we have

$$\begin{aligned}
&\|x_n(t) - \phi(0)\|\frac{d}{dt}\|x_n(t) - \phi(0)\| \\
&\quad = \frac{1}{2}\frac{d}{dt}\|x_n(t) - \phi(0)\|^2 \\
&\quad = \left(\frac{dx_n}{dt}(t), F(x_n(t) - \phi(0))\right) \\
&\quad = -(A_n(t)x_n(t) - A_n(t)\phi(0), F(x_n(t) - \phi(0))) \\
&\qquad -(A_n(t)\phi(0), F(x_n(t) - \phi(0))) \\
&\quad \le \|A_n(t)\phi(0)\|\|x_n(t) - \phi(0)\| \\
&\quad \le (\|A_n(t, v_t)\phi(0) - A_n(0,\phi)\phi(0)\| \\
&\qquad +\|A_n(0,\phi)\phi(0)\|)\|x_n(t) - \phi(0)\| \\
&\quad \le (L_2(\|\phi(0)\|)T_1(1 + \|A_n(0,\phi)\phi(0)\|) \\
&\qquad +\|A_n(0,\phi)\phi(0)\|)\|x_n(t) - \phi(0)\| \\
&\quad \le (L_2(\|\phi(0)\|)\frac{c}{C_1}(1 + \|A(0,\phi)\phi(0)\|) \\
&\qquad +\|A(0,\phi)\phi(0)\|)\|x_n(t) - \phi(0)\|
\end{aligned}$$

for $n = 1, 2, \cdots$ and a.e. $t \in (0, T_1)$. Dividing this inequality by $\|x_n(t) - \phi(0)\|$ and integrating from 0 to $t \leq T_1$ we obtain

$$\|x_n(t) - \phi(0)\| \leq C_5 T_1$$

and thus

$$\|x_n(t)\| \leq \|\phi(0)\| + C_5 T_1, \tag{4.1.8}$$

where

$$C_5 = L_2(\|\phi(0)\|)\frac{c}{C_1}(1 + \|A(0,\phi)\phi(0)\|) + \|A(0,\phi)\phi(0)\|$$

is independent of $T_1 \leq T$, $v \in S$.

In order to find a uniform upper bound for $\frac{dx_n}{dt}(t)$, let $0 \leq t, t+h < T_1$. Using again (12) of Lemma 1.1.2, the accretiveness of $A_n(t+h)$, (4.1.7), and (4.1.8) we have

$$\begin{aligned}
&\|x_n(t+h) - x_n(t)\|\frac{d}{dt}\|x_n(t+h) - x_n(t)\| \\
&= \frac{1}{2}\frac{d}{dt}\|x_n(t+h) - x_n(t)\|^2 \\
&= \Big(\frac{d}{dt}(x_n(t+h) - x_n(t)), F(x_n(t+h) - x_n(t))\Big) \\
&= -(A_n(t+h)x_n(t+h) - A_n(t)x_n(t), F(x_n(t+h) - x_n(t))) \\
&= -(A_n(t+h)x_n(t+h) - A_n(t+h)x_n(t), F(x_n(t+h) - x_n(t))) \\
&\quad -(A_n(t+h)x_n(t) - A_n(t)x_n(t), F(x_n(t+h) - x_n(t))) \\
&\leq \|A_n(t+h)x_n(t+h) - A_n(t+h)x_n(t)\|\|x_n(t+h) - x_n(t)\| \\
&\leq |h|L_2(\|x_n(t)\|)(1 + \|A_n(t)x_n(t)\|)\|\|x_n(t+h) - x_n(t)\| \\
&\leq |h|L_2(\|\phi(0)\| + C_5T_1)\Big(1 + \Big\|\frac{dx_n}{dt}(t)\Big\|\Big)\|x_n(t+h) - x_n(t)\|
\end{aligned}$$

for $n = 1, 2, \cdots$. Dividing this inequality by $|h|\|x_n(t+h) - x_n(t)\|$, integrating it and then passing to the limit as $h \to 0$, we get

$$\begin{aligned}
\Big\|\frac{dx_n}{dt}(t)\Big\| &= \Big\|\frac{dx_n}{dt}(0)\Big\| \\
&\quad + \int_0^t L_2(\|\phi(0)\| + C_5T_1)\Big\|\frac{dx_n}{dt}(\sigma)\Big\|d\sigma \\
&\quad + L_2(\|\phi(0)\| + C_5T_1)T_1
\end{aligned}$$

for $n = 1, 2, \cdots$. Since

$$\Big\|\frac{dx_n}{dt}(0)\Big\| = \|A_n(0,\phi)\phi(0)\| \leq \|A(0,\phi)\phi(0)\|$$

and $T_1 \le \dfrac{c}{C_1}$, applying Gronwall's inequality, we find

$$\begin{aligned}\left\|\frac{dx_n}{dt}(t)\right\| &\le \left(\left\|\frac{dx_n}{dt}(0)\right\| + C_6T_1\right)e^{C_6T_1} \\ &\le (C_7 + C_6T_1)e^{C_6T_1}\end{aligned} \tag{4.1.9}$$

for $n = 1, 2, \cdots$, where $C_6 = L_2\left(\|\phi(0)\| + C_5\dfrac{c}{C_1}\right)$ is independent of T_1, $u \in S$ and $n = 1, 2, \cdots$, and $C_7 = \|A(0, \phi)\phi(0)\|$. From (4.1.9) we conclude

$$\|x^v(t_1) - x^v(t_2)\| \le C_8|t_1 - t_2|$$

for every $t_i \in [0, T_1]$, $i = 1, 2$, where $C_8 = (C_7 + C_6T_1)e^{C_6T_1}$.

Now let $v_i \in S$ be given and let $x_i(t) = x^{v_i}$ be the corresponding strong solutions of (4.1.2) on $[-r, T_1]$, $i = 1, 2$. Then, from (12) of Lemma 1.1.2, $(A1)_1$ and $(A2)_1$

$$\begin{aligned}&\|x_1(t) - x_2(t)\|\frac{d}{dt}\|x_1(t) - x_2(t)\| \\ &= \frac{1}{2}\frac{d}{dt}\|x_1(t) - x_2(t)\|^2 \\ &= -(A(t, (v_1)_t)x_1(t) - A(t, (v_2)_t)x_2(t), F(x_1(t) - x_2(t))) \\ &= -(A(t, (v_1)_t)x_1(t) - A(t, (v_1)_t)x_2(t), F(x_1(t) - x_2(t))) \\ &\quad -(A(t, (v_1)_t)x_2(t) - A(t, (v_2)_t)x_2(t), F(x_1(t) - x_2(t))) \\ &\le \|A(t, (v_1)_t)x_2(t) - A(t, (v_2)_t)x_2(t)\|\|x_1(t) - x_2(t)\| \\ &\le L(\|(v_1)_t\|_C, \|(v_2)_t\|_C, \|x_2(t)\|)\|(v_1)_t - (v_2)_t\|_C\|x_1 - x_2\|\end{aligned}$$

from which, dividing this by $\|x_1(t) - x_2(t)\|$ and then integrating it on $[0, t]$, we arrive at

$$\begin{aligned}&\|x_1(t) - x_2(t)\| \\ &\le \int_0^t L(\|(v_1)_\sigma\|_C, \|(v_2)_\sigma\|_C, \|x^{v_2}(\sigma)\|)\|(v_1)_\sigma - (v_2)_\sigma\|_C d\sigma \\ &\le \int_0^t L\left(2\|\phi\|_C, 2\|\phi\|_C, \|\phi(0)\| + C_5\frac{c}{C_1}\right) \sup_{\upsilon\in[0,T_1]} \|(v_1)_\sigma - (v_2)_\sigma\|_C d\sigma \\ &= T_1L\left(2\|\phi\|_C + c, 2\|\phi\|_C + c, \|\phi(0)\| + C_5\frac{c}{C_1}\right) \\ &\qquad \times \sup_{\sigma\in[0,T_1]} \|(v_1)_\sigma - (v_2)_\sigma\|_C \\ &= C_9 \sup_{t\in[0,T_1]} \|(v_1)_t - (v_2)_t\|_C \\ &\le C_9 \sup_{t\in[0,T_1]} \|v_1(t) - v_2(t)\|,\end{aligned}$$

where

$$C_9 = T_1 L\Big(2\|\phi\|_C + c, 2\|\phi\|_C + c, \|\phi(0)\| + C_5 \frac{c}{C_1}\Big).$$

Therefore we obtain

$$\sup_{t\in[0,T_1]} \|x_1(t) - x_2(t)\| \le C_9 \sup_{t\in[0,T_1]} \|v_1(t) - v_2(t)\|. \tag{4.1.10}$$

We now choose T_1 sufficiently small that $C_8 \le C_1$ and $C_9 < 1$. Then the operator $Q : v \to x^v$ is a strict contraction on S. From Banach's contraction principle, Q has a unique fixed point $x(t)$ for every $t \in [-r, T_1]$. $x(t)$ is a local strong solution of (4.2.1) on $[-r, T]$. Its uniqueness follows from (4.1.10) by replacing v_i by x_i, $i = 1, 2$, respectively.

4.2. Types with Integral Operators

In this section we consider quasi-nonlinear functional evolutions with an integral operator, of the type

$$\begin{cases} \dfrac{dx}{dt}(t) + A(t, x_t)x(t) \ni G\Big(t, x_t, \displaystyle\int_0^t k(t, \sigma, x_\sigma) d\sigma\Big), & 0 \le t \le T \\ x(t) = \phi(t), \quad -r \le t \le 0 \end{cases} \tag{4.2.1}$$

associated with a multi-valued operator $A(t, \psi) : X \supset D(A(t, \psi)) \to 2^X$ for every $t \in [0, T]$ and $\psi \in PC([-r, 0]; X)$, where $T > 0$, $r > 0$, $x(t) : [-r, T] \to X$ is an unknown function, $G : [0, T] \times PC([-r, 0]; X) \times X \to X$, $k : [0, T] \times [0, T] \times PC([-r, 0]; X) \to X$ and $\phi \in PC([-r, 0]; X)$.

Now, we are in the position to give definitions for solutions of (4.2.1).

DEFINITION 4.2.1 *1. A function $x(t) : [-r, T] \to X$ is called a strong solution of* (4.2.1) *on* $[-r, T]$ *if:*

(1) $x(t) = \phi(t)$ for every $t \in [-r, 0]$;

(2) $x(t)$ is continuous on $[0, T]$;

(3) $x(t)$ is absolutely continuous on every compact interval of $[0, T]$;

(4) $x(t)$ is differentiable a.e. $t \in (0, T)$;

(5) $x(t)$ satisfies

$$\frac{dx}{dt}(t) + A(t, x_t)x(t) \ni G\Big(t, x_t, \int_0^t k(t, \sigma, x_\sigma) d\sigma\Big)$$

a.e. $t \in (0, T)$.

2. A function $x(t) : [-r,T] \to X$ is called an integral solution of (4.2.1) on $[-r,T]$ if:

(1) $x(t) = \phi(t),\ t \in [-r,0]$;

(2) $x(t)$ is continuous on $[0,T]$;

(3) $x(t)$ satisfies the inequality

$$\begin{aligned}&\|x(t)-x\| - \|x(\tau)-x\| \\ &\quad \le \int_\tau^t \Big(\Big[G\Big(\sigma, x_\sigma, \int_0^\sigma k(\sigma,s,x_s)ds\Big) - y, x(\sigma)-x\Big]_+ \\ &\qquad + C|\sigma-\bar\sigma|\Big)d\sigma\end{aligned}$$

for every $[x,y] \in A(\bar\sigma, x_{\bar\sigma})$, $\bar\sigma \in [0,T]$, t, $\tau \in [0,T]$ and a constant $C \ge 0$.

3. A function $x(t) : [-r,T] \to X$ is called a local strong solution (resp. a local integral solution) of (4.2.1) on $[-r,T]$ if there exists $T_1 \in (0,T)$ such that the restriction $x_{|T_1}(t) : [-r,T_1] \to X$ of $x(t)$ on $[-r,T_1]$ is a strong solution (resp. an integral solution) of

$$\begin{cases} \dfrac{dx}{dt}(t) + A(t,x_t)x(t) \ni G\Big(t, x_t, \displaystyle\int_0^t k(t,\sigma,x_\sigma)d\sigma\Big), & 0 \le t \le T_1 \\ x(t) = \phi(t), \quad -r \le t \le 0 \end{cases}$$

on $[-r,T_1]$.

To show the existence for local integral solutions and local strong solutions, we impose the following conditions.

$(A1)_2$ $A(t,\psi) : X \supset D(A(t,\psi)) \to 2^X$ is m-accretive in X for every $t \in [0,T]$ and $\psi \in PC([-r,0];X)$;

$(A2)_2$ There exists a increasing continuous function $L : [0,\infty) \times [0,\infty) \times [0,\infty) \to [0,\infty)$ such that

$$\begin{aligned}&\|A_\lambda(t_1,\psi_1)u - A_\lambda(t_2,\psi_2)u\| \\ &\quad \le L(\|\psi_1\|_C, \|\psi_2\|_C, \|u\|)(|t_1-t_2|(1+\|A_\lambda(t_2,\psi_2)u\|) \\ &\qquad + \|\psi_1-\psi_2\|_{PC})\end{aligned}$$

for every $u \in X$, $t_i \in [0,T]$, $\psi_i \in PC([-r,0];X)$, $i = 1,2$ and $\lambda > 0$;

$(k1)_2$ There exists a continuous function $a_1 : [0,T] \times [0,T] \to [0,\infty)$ such that

$$\|k(t_1,t_2,\phi_1) - k(t_1,t_2,\phi_2)\| \le a_1(t_1,t_2)\|\psi_1-\psi_2\|_{PC}$$

for every $t_i \in [0,T]$ and $\psi_i \in PC([-r,0];X)$, $i = 1,2$;

$(k2)_2$ There exists a nondecreasing function $l_1; [0,\infty) \to [0,\infty)$ such that

$$\|k(t_1,t_2,\psi) - k(\bar{t}_1,\bar{t}_2,\psi)\| \le (|t_1 - \bar{t}_1| + |t_2 - \bar{t}_2|)l_1(\|\psi\|_{PC})$$

for every t_i, $\bar{t}_i \in [0,T]$, $i = 1,2$, and $\psi \in PC([-r,0];X)$;

$(G1)_2$ There exists a continuous function $a_2 : [0,T] \to [0,\infty)$ such that

$$\|G(t,\psi_1,u_1) - G(t,\psi_2,u_2)\| \le a_2(t)(\|\psi_1 - \psi_2\|_{PC} + \|u_1 - u_2\|)$$

for every $t \in [0,T]$, $\psi_i \in PC([-r,0];X)$ and $u_i \in X$, $i = 1,2$;

$(G2)_2$ There exists a nondecreasing function $l_2 : [0,\infty) \times [0,\infty) \to [0,\infty)$ such that

$$\|G(t_1,\psi,u) - G(t_2,\psi,u)\| \le |t_1 - t_2|l_2(\|\psi\|_{PC}, \|u\|)$$

for every $t_i \in [0,T]$, $i = 1,2$, $\psi \in PC([-r,0];X)$ and $u \in X$.

We recall the generalized domain of $A(t,\psi)$ for every $t \in [0,T]$ and $\psi \in PC([-r,0];X)$ as

$$\widehat{D}(A(t,\psi)) = \{u \in X \mid |A(t,\psi)u| = \lim_{\lambda \to 0+} \|A_\lambda(t,\psi)u\| < \infty\|\},$$

where $A_\lambda(t,\psi)$ for every $\lambda > 0$ is the Yosida approximant of $A(t,\psi)$ for every $t \in [0,T]$ and $\psi \in PC([-r,0];X)$.

LEMMA 4.2.1 *Assume that* $(A1)_2$ *and* $(A2)_2$ *are satisfied. Then* $\widehat{D}(A(t,\psi))$ *is independent of* $t \in [0,T]$ *and* $\psi \in PC([-r,0];X)$.

Proof. Let $v \in \widehat{D}(A(t_2,\psi_2))$. From $(A1)_2$ and $(A2)_2$ we observe that

$$\begin{aligned}
&\|A_\lambda(t_1,\psi_1)v\| \\
&\quad \le \|A_\lambda(t_2,\psi_2)v\| \\
&\qquad + L(\|\psi_1\|_C, \|\psi_2\|_C, \|v\|)(|t_1 - t_2|(1 + \|A_\lambda(t_2,\psi_2)v\|) \\
&\qquad\qquad + \|\psi_1 - \psi_2\|_{PC}))
\end{aligned}$$

for every $t_i \in [0,T]$, $\psi_i \in PC([-r,0];X)$, $i = 1,2$ and $\lambda > 0$. Letting $\lambda \to 0+$, we get

$$\begin{aligned}
&|A_\lambda(t_1,\psi_1)u| \\
&\quad \le |A_\lambda(t_2,\psi_2)v| \\
&\qquad + L(\|\psi_1\|_C, \|\psi_2\|_C, \|v\|)(|t_1 - t_2|(1 + |A_\lambda(t_2,\psi_2)v|) \\
&\qquad\qquad + \|\psi_1 - \psi_2\|_{PC})
\end{aligned}$$

for every $t_i \in [0,T]$ and $\psi_i \in PC([-r,0];X)$, $i = 1,2$. Hence $v \in \widehat{D}(A(t_1,\psi_1))$, which says that $\widehat{D}(A(t_2,\psi_2) \subset \widehat{D}(A(t_1,\psi_1))$. Similarly, $\widehat{D}(A(t_1,\psi_1)) \subset \widehat{D}(A(t_2,\psi_2))$ for every $t_i \in [0,T]$ and $\psi_i \in PC([-r,0];X)$, $i = 1,2$. Consequently $\widehat{D}(A(t_1,\psi_1)) = \widehat{D}(A(t_2,\psi_2))$ for every $t_i \in [0,T]$ and $\psi_i \in PC([-r,0];X)$, $i = 1,2$. Therefore $\widehat{D}(A(t,\psi)$ is independent of $t \in [0,T]$ and $\psi \in PC([-r,0];X)$.

We denote $\widehat{D} = \widehat{D}(A(t,\psi))$ for every $t \in [0,T]$ and $\psi \in PC([-r,0];X)$.

In order to establish the existence of a local integral solution and local strong solution of (4.2.1) on $[-r,T]$, we first ensure the existence of a limit solution of

$$\begin{cases} \dfrac{dx}{dt}(t) + A(t,v_t)x(t) \ni G\Big(t, x_t, \displaystyle\int_0^t k(t,\sigma,x_\sigma)d\sigma\Big), \quad 0 \le t \le T \\ x(t) = \phi(t), \quad -r \le t \le 0 \end{cases} \tag{4.2.2}$$

on $[-r,T]$, which is the intermediate step to arrive at (4.2.1), where v is a fixed element of a suitable metric space of continuous functions similar as in Section 4.1 and a Lipschitz continuous function $\phi \in PC([-r,0];X)$ with Lipschitz constant $c > 0$,

Let $a_1(t_1,t_2) \le C_1$ and $a_2(t) \le C_2$ for t, $t_i \in [0,T_1]$, $i = 1,2$ with $0 < T_1 < T$. Let

$$M_1 = 3(1 + |A(0,\phi)\phi(0)| + \|G(0,\bar{0},0)\| + 7C_2\|\phi\|_{PC})$$

and we put $M = \max\{c, M_1\}$. We define a set S by

$$\begin{aligned} S = \{v : [-r,T_1] \to X \mid\ & v(t) \text{ is continuous on } [0,T_1], \\ & v(t) = \phi(t),\ t \in [-r,0] \\ & \text{and } \|v(t_1) - v(t_2)\| \le M|t_1 - t_2|, \\ & t_i \in [0,T_1],\ i = 1,2\}. \end{aligned}$$

Then $S \neq \emptyset$ because the function $v(t) : [-r,T_1] \to X$ defined by

$$v(t) = \begin{cases} \phi(t), & t \in [-r,0] \\ \phi(0), & t \in [0,T_1], \end{cases}$$

belongs to S, and S is a real Banach space with norm $\|\cdot\|_{T_1}$ defined by

$$\|x\|_{T_1} = \sup_{t\in[-r,T_1]} \|x(t)\| \tag{4.2.3}$$

for every $x \in S$.

PROPOSITION 4.2.1 *Assume that* $(A1)_2$, $(A2)_2$, $(k1)_2$, $(k2)_2$, $(G1)_2$, *and* $(G2)_2$ *are satisfied. Let* $\{t_i^n\}$ *be a partition of* $[0,T_1]$ *with* $t_i^n =$

$ih_n = i\frac{T_1}{n}$, $i = 0, 1, 2, \cdots, n$. Let $\phi : [-r, 0] \to X$ be a Lipschitz continuous function with Lipschitz constant $c > 0$ and $\phi(0) \in \widehat{D}$. Then there exists $\{x_i^n\}$ such that

$$\frac{x_i^n - x_{i-1}^n}{t_i^n - t_{i-1}^n} + A(t_i^n, v_{t_i^n})x_i^n \ni G\Big(t_i^n, (\overline{x_i^n})_{t_i^n}, \int_0^{t_i^n} k(t_i^n, \sigma, (\overline{x_i^n})_\sigma)d\sigma\Big) \tag{4.2.4}$$

and

$$x_i^n = J_{h_n}(t_i^n, v_{t_i^n}) \times \Big(x_{i-1}^n + h_n G\Big(t_i^n, (\overline{x_i^n})_{t_i^n}, \int_0^{t_i^n} k(t_i^n, \sigma, (\overline{x_i^n})_\sigma)d\sigma\Big)\Big) \tag{4.2.5}$$

with

$$\overline{x_i^n}(t) = \begin{cases} \phi(t), & t \in [-r, 0] \\ x_i^n, & t \in (t_{j-1}^n, t_j^n],\ j = 1, 2, \cdots, i-1 \\ x_i^n, & t \in (t_{i-1}^n, T_1] \end{cases}$$

for $i = 1, 2, \cdots, n$, satisfying the following properties

(1) $\{x_i^n\}$ is uniformly bounded;

(2) $\left\{\frac{x_i^n - x_{i-1}^n}{t_i^n - t_{i-1}^n}\right\}$ is uniformly bounded.

Proof. Let n be sufficiently large so that $h_n(1 + C_1T_1)C_2 < 1$. We also define the space

$$S_1 = \{x : [-r, T_1] \to X \mid x(t) = \phi(t),\ t \in [-r, 0] \text{ and } x(t) \text{ is a constant on } [0, T_1]\}.$$

Then $S_1 \neq \emptyset$ because the function $x(t) : [-r, T_1] \to X$ defined by

$$x(t) = \begin{cases} \phi(t), & t \in [-r, 0] \\ \phi(0), & t \in [0, T_1], \end{cases}$$

belongs to S_1 and S_1 is a real Banach space with norm $\|\cdot\|_{T_1}$ defined in (4.2.3).

We define a mapping $Q_1 : S_1 \to S_1$ by

$$(Q_1x)(t) = \begin{cases} \phi(t), & t \in [-r, 0] \\ J_{h_n}(t_1^n, v_{t_1^n})\Big(x_0^n + h_n G\Big(t_1^n, x_{t_1^n}, \int_0^{t_1^n} k(t_1^n, \sigma, x_\sigma)d\sigma\Big)\Big), & t \in (0, T_1] \end{cases}$$

for every $x \in S_1$, where $x_0^n = \phi(0)$ and

$$J_{h_n}(t_1^n, v_{t_1^n}) = (I + h_n A(t_1^n, v_{t_1^n}))^{-1}.$$

Then the mapping is a strict contraction on S_1. In fact, by $(A1)_1$

$$\begin{aligned}
&\|Q_1x_1 - Q_1x_2\|_{T_1} \\
&\quad\le \Big\| J_{h_n}(t_1^n, v_{t_1^n})\Big(x_0^n + h_nG\Big(t_1^n, (x_1)_{t_1^n}, \int_0^{t_1^n} k(t_1^n, \sigma, (x_1)_\sigma)d\sigma\Big)\Big) \\
&\qquad - J_{h_n}(t_1^n, v_{t_1^n})\Big(x_0^n + h_nG\Big(t_1^n, (x_2)_{t_1^n}, \int_0^{t_1^n} k(t_1^n, \sigma, (x_2)_\sigma)d\sigma\Big)\Big)\Big\| \\
&\quad\le h_n\Big\| G\Big(t_1^n, (x_1)_{t_1^n}, \int_0^{t_1^n} k(t_1^n, \sigma, (x_1)_\sigma)d\sigma\Big) \\
&\qquad - G\Big(t_1^n, (x_2)_{t_1^n}, \int_0^{t_1^n} k(t_1^n, \sigma, (x_2)_\sigma)d\sigma\Big)\Big\| \\
&\quad\le h_nC_2\Big(\|(x_1)_{t_1^n} - (x_2)_{t_1^n}\|_{PC} \\
&\qquad + \Big\| \int_0^{t_1^n} k(t_1^n, \sigma, (x_1)_\sigma)d\sigma - \int_0^{t_1^n} k(t_1^n, \sigma, (x_2)_\sigma)d\sigma \Big\|\Big) \\
&\quad\le h_nC_2(\|(x_1)_{t_1^n} - (x_2)_{t_1^n}\|_{PC} + C_1T_1\|(x_1)_{t_1^n} - (x_2)_{t_1^n}\|_{PC} \\
&\quad\le h_n(1 + C_1T_1)C_2\|x_1 - x_2\|_{T_1}
\end{aligned}$$

for every $x_i \in S_1$, $i = 1, 2$. Here we have used

$$\begin{aligned}
&\|(x_1)_{t_1^n} - (x_2)_{t_1^n}\|_{PC} \\
&\quad= \sup_{\tau\in[-r,0]} \|x_1(t_1^n + \tau) - x_2(t_1^n + \tau)\| \\
&\quad= \sup_{t\in[t_1^n - r, t_1^n]} \|x_1(t) - x_2(t)\| \\
&\quad\le \sup_{t\in[-r,T_1]} \|x_1(t) - x_2(t)\| \\
&\quad= \|x_1 - x_2\|_{T_1}.
\end{aligned}$$

Since $h_n(1 + C_1T_1)C_2 < 1$, Q_1 is a strict contraction on S_1. Using Banach's contraction principle, there exists a unique fixed point $\overline{x_1^n} \in S_1$ with $Q_1\overline{x_1^n} = \overline{x_1^n}$. From the definition of Q_1

$$\overline{x_1^n}(t) = \begin{cases} \phi(t), \quad t \in [-r, 0] \\ J_{h_n}(t_1^n, v_{t_1^n})\Big(x_0^n + h_nG\Big(t_1^n, (\overline{x_1^n})_{t_1^n}, \int_0^{t_1^n} k(t_1^n, \sigma, (\overline{x_1^n})_\sigma)d\sigma\Big)\Big), \\ \qquad\qquad t \in (0, T_1]. \end{cases}$$

If we put

$$x_1^n = J_{h_n}(t_1^n, v_{t_1^n})\Big(x_0^n + h_nG\Big(t_1^n, (\overline{x_1^n})_{t_1^n}, \int_0^{t_1^n} k(t_1^n, \sigma, (\overline{x_1^n})_\sigma)d\sigma\Big)\Big) \tag{4.2.6}$$

then $x_1^n \in D(A(t_1^n, v_{t_1^n}))$,

$$\overline{x_1^n}(t) = \begin{cases} \phi(t), & t \in [-r, 0] \\ x_1^n, & t \in (0, T_1], \end{cases}$$

and

$$\frac{x_1^n - x_0^n}{h_n} + A(t_1^n, v_{t_1^n})x_1^n \ni G\Big(t_1^n, (\overline{x_1^n})_{t_1^n}, \int_0^{t_1^n} k(t_1^n, \sigma, (\overline{x_1^n})_\sigma)d\sigma\Big). \quad (4.2.7)$$

Proceeding in a similar manner we define the spaces S_i, $i = 2, 3, \cdots, n$ by

$$S_i = \{x : [-r, T_1] \to X \mid x(t) = \phi(t),\ t \in [-r, 0] \text{ and } x(t) \text{ is a constant on each interval } (0, t_1^n], \cdots, (t_{i-1}^n, T_1]\}.$$

Then $S_i \neq \emptyset$ and S_i is a real Banach space with the norm (4.2.3). Moreover, the strict contractions $Q_i : S_i \to S_i$ by

$$(Q_i x)(t) = \begin{cases} \phi(t), & t \in [-r, 0] \\ x_j^n, & t \in (t_{j-1}, t_j^n],\ j = 1, 2, \cdots, i-1 \\ J_{h_n}(t_i^n, v_{t_i^n})\Big(x_{i-1}^n + h_n G\Big(t_i^n, x_{t_i^n}, \int_0^{t_i^n} k(t_i^n, \sigma, x_\sigma)d\sigma\Big)\Big), \\ \qquad t \in (t_i^n, T_1] \end{cases} \quad (4.2.8)$$

for every $x \in S_i$. By Banach's contraction principle there exists a unique fixed point $\overline{x_i^n} \in S_i$ with $Q_i\overline{x_i^n} = \overline{x_i^n}$. From (4.2.8)

$$\overline{x_i^n}(t) = \begin{cases} \phi(t), & t \in [-r, 0] \\ x_j^n, & t \in (t_j^n - 1, t_j^n],\ j = 1, 2, \cdots, i-1 \\ J_{h_n}(t_i^n, v_{t_i^n})\Big(x_{i-1}^n + h_n G\Big(t_i^n, (\overline{x_i^n})_{t_i^n}, \int_0^{t_i^n} k(t_i^n, \sigma, (\overline{x_i^n})_\sigma)d\sigma\Big)\Big), \\ \qquad t \in (t_i^n, T_1]. \end{cases}$$

If we put

$$x_i^n = J_{h_n}(t_i^n, v_{t_i^n}) \times \Big(x_{i-1}^n + h_n G\Big(t_i^n, (\overline{x_i^n})_{t_i^n}, \int_0^{t_i^n} k(t_i^n, \sigma, (\overline{x_i^n})_\sigma)d\sigma\Big)\Big) \quad (4.2.9)$$

then $x_i^n \in D(A(t_i^n, v_{t_i^n}))$

$$\overline{x_i^n}(t) = \begin{cases} \phi(t), & t \in [-r, 0] \\ x_j^n, & t \in (t_{j-1}^n, t_j^n],\ j = 1, 2, \cdots, i-1 \\ x_i^n, & t \in (t_i^n, T_1] \end{cases}$$

and

$$\frac{x_i^n - x_{i-1}^n}{h_n} + A(t_i^n, v_{t_i^n})x_i^n \ni G\Big(t_i^n, (\overline{x_i^n})_{t_i^n}, \int_0^{t_i^n} k(t_i^n, \sigma, (\overline{x_i^n})_\sigma)d\sigma\Big) \quad (4.2.10)$$

for $i = 2, 3, \cdots, n$. Using (4.2.6), (4.2.7), (4.2.9), and (4.2.10) we have (4.2.4) and (4.2.5).

(1) Let $v \in S$. From $(A2)_2$ we have

$$\begin{aligned} |A(t, v_t)\phi(0)| \;\; &\leq \;\; |A(0, \phi)\phi(0)| \\ &\quad +L(\|v_t\|_{PC}, \|\phi\|_{PC}, \|\phi(0)\|) \\ &\quad\;\; \times(t(1 + |A(0, \phi)\phi(0)| + \|v_t - \phi\|_{PC}) \quad (4.2.11) \end{aligned}$$

for every $t \in [0, T_1]$.

We estimate $\|v_t - \phi\|_{PC}$. For $t + \tau \geq 0$

$$\begin{aligned} \|v_t - \phi\|_{PC} \;\; &= \;\; \sup_{\tau\in[-r,0]} \|v(t + \tau) - \phi(\tau)\| \\ &\leq \;\; \sup_{\tau\in[-r,0]} \|v(t + \tau) - v(0)\| + \|\phi(0) - \phi(\tau)\| \\ &\leq \;\; \sup_{\tau\in[-r,0]} (M|t + \tau| + c|\tau|) \\ &\leq \;\; \sup_{\tau\in[-r,0]} M(t + \tau - \tau) \\ &= \;\; Mt \leq MT_1. \end{aligned}$$

For $t + \tau < 0$

$$\begin{aligned} \|v_t - \phi\|_{PC} \;\; &= \;\; \sup_{\tau\in[-r,0]} \|v(t + \tau) - \phi(\tau)\| \\ &= \;\; \sup_{\tau\in[-r,0]} \|\phi(t + \tau) - \phi(\tau)\| \\ &\leq \;\; ct \leq MT_1. \end{aligned}$$

Thus

$$\|v_t - \phi\|_{PC} \leq MT_1 \quad (4.2.12)$$

for every $t \in [0, T_1]$, and then

$$\|v_t\|_{PC} \leq \|\phi\|_{PC} + \leq MT_1 \quad (4.2.13)$$

for every $t \in [0, T_1]$.

By (4.2.11)–(4.2.13) we obtain

$$\begin{aligned} |A(t, v_t)\phi(0)| &\leq |A(0, \phi)\phi(0)| \\ &\quad +L(\|\phi\|_{PC}, \|\phi\|_{PC}, \|\phi(0)\|)(1 + |A(0, \phi)\phi(0)| + M)T_1 \\ &\equiv C_3 \end{aligned}$$

for every $t \in [0, T_1]$. Thus

$$\sup_{t\in[0,T_1]} |A(t, v_t)\phi(0)| \leq C_3.$$

If $t \in (t_{i-1}^n, T_1]$ for every $i = 1, 2, \cdots, n$ then using (4.2.5)

$$\begin{aligned}
&\|\overline{x_i^n}(t) - \phi(0)\| \\
&\quad = \|x_i^n - \phi(0)\| \\
&\quad = \Big\| J_{h_n}(t_i^n, v_{t_i^n})\Big(x_{i-1}^n + h_n G\Big(t_i^n, (\overline{x_i^n})_{t_i^n}, \int_0^{t_i^n} k(t_i^n, \sigma, (\overline{x_i^n})_\sigma) d\sigma\Big)\Big) \\
&\qquad\quad - J_{h_n}(t_i^n, v_{t_i^n})\phi(0)\Big\| + \|J_{h_n}(t_i^n, v_{t_i^n})\phi(0) - \phi(0)\| \\
&\quad \leq \|x_{i-1}^n - \phi(0)\| \\
&\qquad + h_n \Big\| G\Big(t_i^n, (\overline{x_i^n})_{t_i^n}, \int_0^{t_i^n} k(t_i^n, \sigma, (\overline{x_i^n})_\sigma) d\sigma\Big) - G(t_i^n, \bar{0}, 0)\Big\| \\
&\qquad + h_n \|G(t_i^n, \bar{0}, 0)\| + h_n |A_{h_n}(t_i^n, , v_{t_i^n})\phi(0)|. \qquad (4.2.14)
\end{aligned}$$

Since

$$\begin{aligned}
\|k(t_1, t_2, \bar{0}) &\leq \|k(0, 0, \bar{0})\| + 2l_1(0)T_1 \equiv C_4, \\
\|G(t, \bar{0}, 0)\| &\leq \|G(0, \bar{0}, 0\| + l_2(0, 0)T_1 \equiv C_5
\end{aligned}$$

and

$$\begin{aligned}
&\Big\| G\Big(t_i^n, (\overline{x_i^n})_{t_i^n}, \int_0^{t_i^n} k(t_i^n, \sigma, (\overline{x_i^n})_\sigma) d\sigma\Big) - G(t_i^n, \bar{0}, 0)\Big\| \\
&\quad \leq a_2(t_i^n)\Big(\|\overline{x_i^n})_{t_i^n}\|_{PC} + \int_0^{t_i^n} \|k(t_i^n, \sigma, (\overline{x_i^n})_\sigma)\| d\sigma\Big) \\
&\quad \leq C_2\Big(\|(\overline{x_i^n})_{t_i^n}\|_{PC} + \int_0^{t_i^n} \|k(t_i^n, \sigma, (\overline{x_i^n})_\sigma) - k(t_i^n, \sigma, \bar{0})\| d\sigma \\
&\qquad\quad + \int_0^{t_i^n} k(t_i^n, \sigma, \bar{0})\| d\sigma\Big) \\
&\quad \leq C_2\Big(\|(\overline{x_i^n})_{t_i^n}\|_{PC} + \int_0^{t_i^n} a_1(t_i^n, \sigma)\|(\overline{x_i^n})_\sigma)\|_{PC} d\sigma + C_4 t_i^n\Big) \\
&\quad \leq C_2\Big(\|\overline{x_i^n}\|_{T_1} + C_1 \int_0^{t_i^n} \|\overline{x_i^n}\|_{T_1} d\sigma + C_4 T_1\Big) \\
&\quad = C_2(1 + C_1 T_1)\|\overline{x_i^n}\|_{T_1} + C_2 C_4 T_1
\end{aligned}$$

from (4.2.14)

$$\begin{aligned}
&\|\overline{x_i^n}(t) - \phi(0)\| \\
&\quad = \|x_i^n - \phi(0)\| \\
&\quad \leq \|x_{i-1}^n - \phi(0)\| + h_n C_2(1 + C_1 T_1)\|\overline{x_i^n}\|_{T_1} \\
&\qquad + h_n C_2 C_4 T_1 + h_n C_5 + h_n C_3,
\end{aligned}$$

which yields

$$\begin{aligned}
&\|\overline{x_i^n}(t) - \phi(0)\| \\
&\quad = \|x_i^n - \phi(0)\| \\
&\quad \leq \|x_{i-1}^n - \phi(0)\| + h_n C_2(1 + C_1 T_1)\|\overline{x_i^n} - \phi(0)\|_{T_1} \\
&\qquad + h_n C_2(1 + C_1 T_1)\|\phi(0)\| + h_n(C_3 + C_5 + C_2 C_4 T_1) \\
&\quad \leq \|\overline{x_{i-1}^n}\|_{T_1} + h_n C_2(1 + C_1 T_1)\|\overline{x_i^n}\|_{T_1} + h_n C_6
\end{aligned}$$

for $i = 1, 2, \cdots, n$, where

$$C_6 = C_3 + C_5 + C_2 C_4 T_1 + C_2(1 + C_1 T_1)\|\phi(0)\|$$

and

$$\overline{x_0^n}(t) = \begin{cases} \phi(t), & t \in [-r, 0] \\ \phi(0), & t \in [0, T_1]. \end{cases}$$

If $t \in [-r, t_{i-1}^n]$ for $i = 1, 2, \cdots, n$ then

$$\|\overline{x_i^n}(t) - \phi(0)\| = \|\overline{x_{i-1}^n}(t) - \phi(0)\| \leq \|\overline{x_{i-1}^n} - \phi(0)\|_{T_1}.$$

Thus

$$\begin{aligned}
&\|\overline{x_i^n}(t) - \phi(0)\| \\
&\quad < \|\overline{x_{i-1}^n} - \phi(0)\|_{T_1} \\
&\qquad + h_n C_2(1 + C_1 T_1)\|\overline{x_i^n}\|_{T_1} + h_n C_6
\end{aligned}$$

for every $t \in [-r, T_1]$ and $i = 1, 2, \cdots, n$, and we have

$$\begin{aligned}
&\|\overline{x_i^n} - \phi(0)\|_{T_1} \\
&\quad \leq \|\overline{x_{i-1}^n} - \phi(0)\|_{T_1} \\
&\qquad + h_n C_2(1 + C_1 T_1)\|\overline{x_i^n}\|_{T_1} + h_n C_6
\end{aligned}$$

for $i = 1, 2, \cdots, n$, which yields

$$\begin{aligned}
&(1 - h_n C_2(1 + C_1 T_1))\|\overline{x_i^n} - \phi(0)\|_{T_1} \\
&\quad \leq \|\overline{x_{i-1}^n} - \phi(0)\|_{T_1} + h_n C_6 \qquad (4.2.15)
\end{aligned}$$

for $i = 1, 2, \cdots, n$. Applying (4.2.15) repeatedly we obtain

$$
\begin{aligned}
&(1 - h_n C_2(1 + C_1 T_1))\|\overline{x_i^n} - \phi(0)\|_{T_1} \\
&\quad \le \frac{\|\overline{x_{i-2}^n} - \phi(0)\|_{T_1}}{1 - h_n C_2(1 + C_1 T_1)} + \frac{h_n C_6}{1 - h_n C_2(1 + C_1 T_1)} + h_n C_6 \\
&\quad \cdots \\
&\quad \le \frac{\|\overline{x_0^n} - \phi(0)\|_{T_1}}{(1 - h_n C_2(1 + C_1 T_1))^{n-1}} + \sum_{j=0}^{i-1} \frac{h_n C_6}{(1 - h_n C_2(1 + C_1 T_1))^j}
\end{aligned}
$$

for $i = 1, 2, \cdots, n$. Thus

$$
\begin{aligned}
&\|\overline{x_i^n} - \phi(0)\|_{T_1} \\
&\quad \le \frac{\|\overline{x_0^n} - \phi(0)\|_{T_1}}{(1 - h_n C_2(1 + C_1 T_1))^n} + \sum_{j=1}^{i} \frac{h_n C_6}{(1 - h_n C_2(1 + C_1 T_1))^j} \\
&\quad \le \frac{\|\phi - \phi(0)\|_{PC}}{(1 - h_n C_2(1 + C_1 T_1))^n} + \sum_{j=1}^{n} \frac{h_n C_6}{(1 - h_n C_2(1 + C_1 T_1))^n}
\end{aligned}
$$

for $i = 1, 2, \cdots, n$, and

$$
\begin{aligned}
\|\overline{x_i^n}\|_{T_1} \le \ & \|\phi(0)\| + \frac{\|\phi\|_{PC}\| + \|\phi(0)\|}{\left(1 - C_2(1 + C_1 T_1)\dfrac{T_1}{n}\right)^n} \\
& + \frac{C_6 T_1}{\left(1 - C_2(1 + C_1 T_1)\dfrac{T_1}{n}\right)^n}.
\end{aligned}
$$

Since

$$
\lim_{n \to \infty} \left(1 - C_2(1 + C_1 T_1)\frac{T_1}{n}\right)^{-n} = e^{C_2(1 + C_1 T_1) T_1}
$$

there exists a constant $C_7 = C_7(\phi, T_1)$ such that

$$
\sup_{n \ge 1} \max_{0 \le i \le n} \|x_i^n\| \le \sup_{n \ge 1} \max_{0 \le i \le n} \|\overline{x_i^n}\|_{T_1} \le C_7.
$$

(2) Let $h_n C_2(1 + C_1 T_1) < 1$, $h_n < 1$ and

$$
\begin{aligned}
&(C_2 + L(\|\phi\|_{PC} + M T_1, \|\phi\|_{PC} + M T_1, \\
&\qquad C_7 + C_2 C_7(1 + C_1 T_1) + C_5 + C_2 C_4 T_1)) h_n < 1
\end{aligned}
$$

for sufficiently large n. Using (4.2.5) for $i = 1$

$$
\|x_1^n - x_0^n\| \le
$$

$$\le h_n\Big\|G\Big(t_1^n,(\overline{x_1^n})_{t_1^n},\int_0^{t_1^n}k(t_1^n,\sigma,(\overline{x_1^n})_\sigma)d\sigma\Big)-G(t_1^n,\bar{0},0)\Big\|$$
$$+h_n\|G(t_1^n,\bar{0},0)\|+h_n|A(t_1^n,v_{t_1^n})\phi(0)|$$
$$\le h_nC_2(1+C_1T_1)\|\overline{x_i^n}\|_{T_1}+h_nC_2C_4T_1+h_nC_5+h_nC_3$$
$$\le h_nC_2(1+C_1T_1)C_7+h_nC_2C_4T_1+h_nC_5+h_nC_3$$
$$\equiv h_nC_8,$$

where

$$C_8=C_2(1+C_1T_1)C_7+C_2C_4T_1+C_5+C_3,$$

and thus

$$\sup_{n\ge1}\frac{\|x_1^n-x_0^n\|}{h_n}\le C_8. \tag{4.2.16}$$

By (4.2.5) for $i=2,3,\cdots,n$ we obtain

$$\|x_i^n-x_{i-1}^n\|$$
$$=\Big\|J_{h_n}(t_i^n,v_{t_i^n})\Big(x_{i-1}^n+h_nG\Big(t_i^n,(\overline{x_i^n})_{t_i^n},\int_0^{t_i^n}k(t_i^n,\sigma,(\overline{x_i^n})_\sigma)d\sigma\Big)\Big)$$
$$-J_{h_n}(t_{i-1}^n,v_{t_{i-1}^n})$$
$$\times\Big(x_{i-2}^n+h_nG\Big(t_{i-1}^n,(\overline{x_{i-1}^n})_{t_{i-1}^n},\int_0^{t_{i-1}^n}k(t_{i-1}^n,\sigma,(\overline{x_{i-1}^n})_\sigma)d\sigma\Big)\Big)\Big\|$$
$$\le\Big\|J_{h_n}(t_i^n,v_{t_i^n})\Big(x_{i-1}^n+h_nG(t_i^n,(\overline{x_i^n})_{t_i^n},\int_0^{t_i^n}k(t_i^n,\sigma,(\overline{x_i^n})_\sigma)d\sigma\Big)\Big)$$
$$-J_{h_n}(t_i^n,v_{t_i^n})$$
$$\times\Big(x_{i-2}^n+h_nG\Big(t_{i-1}^n,(\overline{x_{i-1}^n})_{t_{i-1}^n},\int_0^{t_{i-1}^n}k(t_{i-1}^n,\sigma,(\overline{x_{i-1}^n})_\sigma)d\sigma\Big)\Big)\Big\|$$
$$+\Big\|J_{h_n}(t_i^n,v_{t_i^n})$$
$$\times\Big(x_{i-2}^n+h_nG\Big(t_{i-1}^n,(\overline{x_{i-1}^n})_{t_{i-1}^n},\int_0^{t_{i-1}^n}k(t_{i-1}^n,\sigma,(\overline{x_{i-1}^n})_\sigma)d\sigma\Big)\Big)\Big\|$$
$$-J_{h_n}(t_{i-1}^n,v_{t_{i-1}^n})\times$$
$$\times\Big(x_{i-2}^n+h_nG\Big(t_{i-1}^n,(\overline{x_{i-1}^n})_{t_{i-1}^n},\int_0^{t_{i-1}^n}k(t_{i-1}^n,\sigma,(\overline{x_{i-1}^n})_\sigma)d\sigma\Big)\Big)\Big\|$$
$$\le\|x_{i-1}^n-x_{i-2}^n\|$$
$$+h_n\Big\|G\Big(t_i^n,(\overline{x_i^n})_{t_i^n},\int_0^{t_i^n}k(t_i^n,\sigma,(\overline{x_i^n})_\sigma)d\sigma\Big)$$
$$-G\Big(t_{i-1}^n,(\overline{x_{i-1}^n})_{t_{i-1}^n},\int_0^{t_{i-1}^n}k(t_{i-1}^n,\sigma,(\overline{x_{i-1}^n})_\sigma)d\sigma\Big)\Big\|+$$

$$
\begin{aligned}
&+h_n\Big\|A_{h_n}(t_{i-1}^n, v_{t_{i-1}^n})\\
&\quad\times\Big(x_{i-2}^n + h_n G\Big(t_{i-1}^n, (\overline{x_{i-1}^n})_{t_{i-1}^n}, \int_0^{t_{i-1}^n} k(t_{i-1}^n, \sigma, (\overline{x_{i-1}^n})_\sigma)d\sigma\Big)\Big)\\
&-A_{h_n}(t_i^n, v_{t_i^n})\\
&\quad\times\Big(x_{i-2}^n + h_n G\Big(t_{i-1}^n, (\overline{x_{i-1}^n})_{t_{i-1}^n}, \int_0^{t_{i-1}^n} k(t_{i-1}^n, \sigma, (\overline{x_{i-1}^n})_\sigma)d\sigma\Big)\Big)\Big\|.
\end{aligned}
$$

From $(G1)_2$ and $(G2)_2$

$$
\begin{aligned}
&\Big\|G\Big(t_i^n, (\overline{x_i^n})_{t_i^n}, \int_0^{t_i^n} k(t_i^n, \sigma, (\overline{x_i^n})_\sigma)d\sigma\Big)\\
&\qquad -G\Big(t_{i-1}^n, (\overline{x_{i-1}^n})_{t_{i-1}^n}, \int_0^{t_{i-1}^n} k(t_{i-1}^n, \sigma, (\overline{x_{i-1}^n})_\sigma)d\sigma\Big)\Big\|\\
&\le \Big\|G\Big(t_i^n, (\overline{x_i^n})_{t_i^n}, \int_0^{t_i^n} k(t_i^n, \sigma, (\overline{x_i^n})_\sigma)d\sigma\Big)\\
&\qquad -G\Big(t_i^n, (\overline{x_{i-1}^n})_{t_{i-1}^n}, \int_0^{t_{i-1}^n} k(t_{i-1}^n, \sigma, (\overline{x_{i-1}^n})_\sigma)d\sigma\Big)\Big\|\\
&\quad +\Big\|G\Big(t_i^n, (\overline{x_{i-1}^n})_{t_{i-1}^n}, \int_0^{t_{i-1}^n} k(t_{i-1}^n, \sigma, (\overline{x_{i-1}^n})_\sigma)d\sigma\Big)\\
&\qquad -G\Big(t_{i-1}^n, (\overline{x_{i-1}^n})_{t_{i-1}^n}, \int_0^{t_{i-1}^n} k(t_{i-1}^n, \sigma, (\overline{x_{i-1}^n})_\sigma)d\sigma\Big)\Big\|\\
&\le a_2(t_i^n)\Big(\|(\overline{x_i^n})_{t_i^n} - (\overline{x_{i-1}^n})_{t_{i-1}^n}\|_{PC}\\
&\qquad +\Big\|\int_0^{t_i^n} k(t_i^n, \sigma, (\overline{x_i^n})_\sigma)d\sigma\\
&\qquad\quad -\int_0^{t_{i-1}^n} k(t_{i-1}^n, \sigma, (\overline{x_{i-1}^n})_\sigma)d\sigma\Big\|\Big)\\
&\quad +|t_i^n - t_{i-1}^n| l_2\Big(\|(\overline{x_{i-1}^n})_{t_{i-1}^n}\|_{PC}, \int_0^{t_{i-1}^n} k(t_{i-1}^n, \sigma, (\overline{x_{i-1}^n})_\sigma)d\sigma\Big).
\end{aligned}
$$

Since, by $(k1)_2$, $(k2)_2$ and the fact $\|(\overline{x_i^n})_\sigma - (\overline{x_{i-1}^n})_\sigma\|_{PC} = 0$ for $0 \le \sigma \le t_{i-1}^n$

$$
\begin{aligned}
&\Big\|\int_0^{t_i^n} k(t_i^n, \sigma, (\overline{x_i^n})_\sigma)d\sigma - \int_0^{t_{i-1}^n} k(t_{i-1}^n, \sigma, (\overline{x_{i-1}^n})_\sigma)d\sigma\Big\|\\
&\le \int_0^{t_{i-1}^n} \|k(t_i^n, \sigma, (\overline{x_i^n})_\sigma) - k(t_{i-1}^n, \sigma, (\overline{x_i^n})_\sigma)\|d\sigma\\
&\quad + \int_0^{t_{i-1}^n} \|k(t_{i-1}^n, \sigma, (\overline{x_i^n})_\sigma) - k(t_{i-1}^n, \sigma, (\overline{x_{i-1}^n})_\sigma)\|d\sigma +
\end{aligned}
$$

$$+\int_{t_{i-1}^n}^{t_i^n} \|k(t_i^n,\sigma,(\overline{x_i^n})_\sigma) - k(t_i^n,\sigma,\bar{0})\|d\sigma + \int_{t_{i-1}^n}^{t_i^n} \|k(t_i^n,\sigma,\bar{0})\|d\sigma$$
$$\leq \int_0^{t_{i-1}^n} |t_i^n - t_{i-1}^n| l_1(\|(\overline{x_i^n})_\sigma\|_{PC})d\sigma$$
$$+\int_0^{t_{i-1}^n} a_1(t_{i-1}^n,\sigma)\|(\overline{x_i^n})_\sigma - (\overline{x_{i-1}^n})_\sigma\|_{PC}d\sigma$$
$$+\int_{t_{i-1}^n}^{t_i^n} a_1(t_i^n,\sigma)\|(\overline{x_i^n})_\sigma\|_{PC}d\sigma + C_4|t_i^n - t_{i-1}^n|$$
$$\leq l_1(\|\overline{x_i^n}\|_{T_1})h_nT_1 + C_1C_7h_n + C_4h_n$$
$$\leq l_1(C_7)h_nT_1 + C_1C_7h_n + C_4h_n$$

and

$$\int_0^{t_{i-1}^n} \|k(t_{i-1}^n,\sigma,(\overline{x_{i-1}^n})_\sigma)\|d\sigma$$
$$\leq \int_0^{t_{i-1}^n} \|k(t_{i-1}^n,\sigma,(\overline{x_{i-1}^n})_\sigma) - k(t_{i-1}^n,\sigma,\bar{0})\|d\sigma$$
$$+\int_0^{t_{i-1}^n} \|k(t_{i-1}^n,\sigma,\bar{0})\|d\sigma$$
$$\leq \int_0^{t_{i-1}^n} a_1(t_i^n,\sigma)\|(\overline{x_{i-1}^n})_\sigma\|_{PC}d\sigma + C_4T_1$$
$$\leq C_1T_1\|\overline{x_{i-1}^n}\|_{T_1} + C_4T_1$$
$$\leq C_1C_7T_1 + C_4T_1$$

we have

$$\Big\|G\Big(t_i^n,(\overline{x_i^n})_{t_i^n},\int_0^{t_i^n} k(t_i^n,\sigma,(\overline{x_i^n})_\sigma)d\sigma\Big)$$
$$-G\Big(t_{i-1}^n,(\overline{x_{i-1}^n})_{t_{i-1}^n},\int_0^{t_{i-1}^n} k(t_{i-1}^n,\sigma,(\overline{x_{i-1}^n})_\sigma)d\sigma\Big)\Big\|$$
$$\leq C_2(\|(\overline{x_i^n})_{t_i^n} - (\overline{x_{i-1}^n})_{t_{i-1}^n}\|_{PC}$$
$$+l_1(C_7)T_1h_n + C_1C_7h_n + C_4h_n)$$
$$+L_2(C_7, C_1C_7T_1 + C_4T_1)h_n.$$

Thus by $(A_2)_2$ we have

$$\|x_i^n - x_{i-1}^n\|$$
$$\leq \|x_{i-1}^n - x_{i-2}^n\| +$$

$$
\begin{aligned}
&+h_nC_2(\|(\overline{x_i^n})_{t_i^n}-(\overline{x_{i-1}^n})_{t_{i-1}^n}\|_{PC}\\
&\qquad +l_1(C_7)T_1h_n+C_1C_7h_n+C_4h_n)\\
&+l_2(C_7,C_1C_7+C_4)T_1)h_n^2\\
&+L\Big(\|v_{t_i^n}\|_{PC},\|v_{t_{i-1}^n}\|_{PC},\\
&\qquad \Big\|x_{i-2}^n+h_nG\Big(t_{i-1}^n,(\overline{x_{i-1}^n})_{t_{i-1}^n},\int_0^{t_{i-1}^n}k(t_{i-1}^n,\sigma,(\overline{x_{i-1}^n})_\sigma)d\sigma\Big)\Big\|\Big)\\
&\quad\times\Big(|t_i^n-t_{i-1}^n|\Big(1+\Big\|A_{h_n}(t_{i-1}^n,v_{t_{i-1}^n})\\
&\quad\times\Big(x_{i-2}^n+h_nG\Big(t_{i-1}^n,(\overline{x_{i-1}^n})_{t_{i-1}^n},\int_0^{t_{i-1}^n}k(t_{i-1}^n,\sigma,(\overline{x_{i-1}^n})_\sigma)d\sigma\Big)\Big\|\Big)\\
&+\|v_{t_i^n}-v_{t_{i-1}^n}\|_{PC}\Big)h_n.
\end{aligned}
$$

We note that, by the definition of the Yosida approximant

$$
\begin{aligned}
&A_{h_n}(t_{i-1}^n,v_{t_{i-1}^n})\\
&\times\Big(x_{i-2}^n+h_nG\Big(t_{i-1}^n,(\overline{x_{i-1}^n})_{t_{i-1}^n},\int_0^{t_{i-1}^n}k(t_{i-1}^n,\sigma,(\overline{x_{i-1}^n})_\sigma)d\sigma\Big)\Big)\\
&=G\Big(t_{i-1}^n,(\overline{x_{i-1}^n})_{t_{i-1}^n},\int_0^{t_{i-1}^n}k(t_{i-1}^n,\sigma,(\overline{x_{i-1}^n})_\sigma)d\sigma\Big)\\
&\quad-\frac{x_{i-1}^n-x_{i-2}^n}{h_n}
\end{aligned}
$$

and, using $(k1)_2$, $(k2)_2$, $(G1)_2$, and $(G2)_2$

$$
\begin{aligned}
&\Big\|G\Big(t_{i-1}^n,(\overline{x_{i-1}^n})_{t_{i-1}^n},\int_0^{t_{i-1}^n}k(t_{i-1}^n,\sigma,(\overline{x_{i-1}^n})_\sigma)d\sigma\Big)\Big\|\\
&\qquad\le C_2(1+C_1T_1)C_7+C_5+C_2C_4T_1.
\end{aligned}
$$

Thus

$$
\begin{aligned}
&\|x_i^n-x_{i-1}^n\|\\
&\quad\le\|x_{i-1}^n-x_{i-2}^n\|\\
&\qquad+h_nC_2\|(\overline{x_i^n})_{t_i^n}-(\overline{x_{i-1}^n})_{t_{i-1}^n}\|_{PC}+l_1(C_7)C_2T_1h_n^2\\
&\qquad+C_2(C_1C_7+C_4)h_n^2+l_2(C_7,C_1C_7+C_4)T_1)h_n^2\\
&\qquad+L(\|\phi\|_{PC}+MT_1,\|\phi\|_{PC}+MT_1,\\
&\qquad\qquad C_7+(C_7C_2(1+C_1T_1)+C_5+C_2C_4T_1)h_n\\
&\qquad\quad\times\Big(1+C_2C_7(1+C_1T_1)+C_5+C_2C_4T_1\\
&\qquad\qquad+\Big\|\frac{x_{i-1}^n-x_{i-2}^n}{h_n}\Big\|+M\Big)h_n^2.
\end{aligned}
$$

Here we have used

$$\|v_{t_i^n} - v_{t_{i-1}^n}\|_{PC} \le M|t_i^n - t_{i-1}^n| = Mh_n$$

as in the proof of Theorem 4.1.1. Moreover, from the condition on L we obtain

$$\begin{aligned}
&\|x_i^n - x_{i-1}^n\| \\
&\quad \le \|x_{i-1}^n - x_{i-2}^n\| \\
&\qquad + h_n C_2(\|(\overline{x_i^n})_{t_i^n} - (\overline{x_{i-1}^n})_{t_{i-1}^n}\|_{PC} + l_1(C_7)C_2T_1h_n^2 \\
&\qquad + C_2(C_1C_7 + C_4)h_n^2 + l_2(C_7, C_1C_7 + C_4)T_1)h_n^2 \\
&\qquad + L(\|\phi\|_{PC} + MT_1, \|\phi\|_{PC} + MT_1, \\
&\qquad\qquad C_7 + C_7C_2(1 + C_1T_1) + C_5 + C_2C_4T_1) \\
&\qquad\quad \times\Big(1 + C_2C_7(1 + C_1T_1) + C_5 + C_2C_4T_1 \\
&\qquad\qquad + \Big\|\frac{x_{i-1}^n - x_{i-2}^n}{h_n}\Big\| + M\Big)h_n^2.
\end{aligned}$$

Dividing both sides by h_n we obtain

$$\begin{aligned}
&\frac{\|x_i^n - x_{i-1}^n\|}{h_n} \\
&\quad \le \frac{\|x_{i-1}^n - x_{i-2}^n\|}{h_n} \\
&\qquad + C_2\|(\overline{x_i^n})_{t_i^n} - (\overline{x_{i-1}^n})_{t_{i-1}^n}\|_{PC} \\
&\qquad + h_n(l_1(C_7)C_2T_1 + C_2(C_7C_1 + C_4) + l_2(C_7, (C_7C_1 + C_4)T_1) \\
&\qquad + L(\|\phi\|_{PC} + MT_1, \|\phi\|_{PC} + MT_1, \\
&\qquad\qquad C_7 + C_7C_2 + C_7C_1C_2T_1 + C_5 + C_2C_4T_1) \\
&\qquad\quad \times\Big(1 + C_2C_7(1 + C_1T_1) + C_5 + M + C_2C_4T_1 \\
&\qquad\qquad + \frac{\|x_{i-1}^n - x_{i-2}^n\|}{h_n}\Big).
\end{aligned}$$

From the definition of $\overline{x_i^n}$, $(\overline{x_{i-1}^n})_{t_{i-1}^n}(\tau) = (\overline{x_i^n})_{t_{i-1}^n}(\tau)$ for every $\tau \in [-r, 0]$. We observe that by (3.3.18)

$$\begin{aligned}
&\|(\overline{x_i^n})_{t_i^n} - (\overline{x_{i-1}^n})_{t_{i-1}^n}\|_{PC} \\
&\quad = \sup_{\tau\in[-r,0]} \|\overline{x_i^n}(t_i^n + \tau) - \overline{x_i^n}(t_{i-1}^n + \tau)\| \\
&\quad \le \max_{1\le j\le i} \|x_j^n - x_{j-1}^n\| + ch_n.
\end{aligned}$$

It follows that

$$\begin{aligned}
\max_{1\le j\le i}\frac{\|x_j^n - x_{j-1}^n\|}{h_n} \\
&\le \max_{1\le j\le i-1}\frac{\|x_j^n - x_{j-1}^n\|}{h_n} + C_2\max_{1\le j\le i}\|x_j^n - x_{i-1}^n\| + C_2ch_n \\
&\quad + L(\|\phi\|_{PC}+MT_1, \|\phi\|_{PC}+MT_1, C_{10})\max_{1\le j\le i}\|x_j^n - x_{j-1}^n\| \\
&\quad + h_nL(\|\phi\|_{PC}+MT_1, \|\phi\|_{PC}+MT_1, C_{10})C_{11} + h_nC_9,
\end{aligned}$$

where

$$\begin{aligned}
C_9 &= a_1(C_7)C_2T_1 + C_2(C_7C_1 + C_4) + a_2(C_7,(C_7C_1+C_4)T_1), \\
C_{10} &= C_7 + C_2(1+C_1T_1)C_7 + C_5 + C_2C_4T_1 \\
C_{11} &= 1 + C_2C_7(1+C_1T_1) + C_5 + M + C_2C_4T_1.
\end{aligned}$$

If we put $P_n = 1-(C_2 + L(\|\phi\|_{PC}+MT_1, \|\phi\|_{PC}+MT_1, C_{10}))h_n$ and $C_{12} = L(\|\phi\|_{PC}+MT_1, \|\phi\|_{PC}+MT_1, C_{10})C_{11}+C_9+C_2c$ then $P_n\in(0,1)$ and we obtain

$$P_n\max_{1\le j\le i}\frac{\|x_j^n - x_{j-1}^n\|}{h_n} \le h_nC_{12} + \max_{1\le j\le i-1}\frac{\|x_j^n - x_{j-1}^n\|}{h_n}$$

and thus

$$\begin{aligned}
P_n\max_{1\le j\le i}\frac{\|x_j^n - x_{j-1}^n\|}{h_n} \\
&\le h_nC_{12} + \frac{C_{12}h_n}{P_n} + \frac{1}{P_n}\max_{1\le j\le i-2}\frac{\|x_j^n - x_{j-1}^n\|}{h_n} \\
&\le C_{12}h_n\sum_{k=0}^{i-2}\frac{1}{P_n^k} + \frac{1}{P_n^{i-2}}\frac{\|x_1^n - x_0^n\|}{h_n} \\
&\le C_{12}h_n\sum_{k=0}^{i-1}\frac{1}{P_n^k} + \frac{1}{P_n^{i-1}}\frac{\|x_1^n - x_0^n\|}{h_n},
\end{aligned}$$

which yields, for $i = 2, 3, \cdots, n$

$$\begin{aligned}
\max_{1\le j\le i}\frac{\|x_j^n - x_{j-1}^n\|}{h_n} \\
&\le C_{12}h_n\sum_{k=1}^{i}\frac{1}{P_n^k} + C_8\frac{1}{P_n^n}C_8 \\
&\le C_{12}h_n\sum_{k=1}^{n}\frac{1}{P_n^n} + C_8\frac{1}{P_n^n} =
\end{aligned}$$

$$= \frac{C_{12}T_1 + C_8}{P_n^n}$$
$$= \frac{C_{12}T_1 + C_8}{\left(1 - \left(C_2 + L(\|\phi\|_{PC} + MT_1, \|\phi\|_{PC} + MT_1, C_{10})\frac{T_1}{n}\right)\right)^n}.$$

Since

$$\lim_{n\to\infty}\left(1 - \left(C_2 + L(\|\phi\|_{PC} + MT_1, \|\phi\|_{PC} + MT_1, C_{10})\frac{T_1}{n}\right)\right)^{-n}$$
$$= \exp(C_2 + L(\|\phi\|_{PC} + MT_1, \|\phi\|_{PC} + MT_1, C_{10})T_1),$$

there exists a constant $C_{13} = C_{13}(\phi, T_1) > 0$ such that, for $i = 2, 3, \cdots, n$,

$$\sup_{n\geq 1}\max_{1\leq j\leq i}\frac{\|x_j^n - x_{j-1}^n\|}{h_n} \leq C_{13}. \tag{4.2.17}$$

We put $C_{14} = \max\{C_8, C_{13}\}$. Then we have, using (4.2.16) and (4.2.17)

$$\sup_{n\geq 1}\max_{1\leq i\leq n}\frac{\|x_i^n - x_{i-1}^n\|}{h_n} \leq C_{14}.$$

LEMMA 4.2.2 *Assume that the assumptions are satisfied. Put $x_n(t) = \overline{x_n^n}(t)$ for every $t \in [-r, T]$ and $n = 1, 2, \cdots$. Then there exists a constant $C_{16} > 0$ such that*

$$\|x_n(t_1) - x_n(t_2)\| \leq C_{16}(|t_1 - t_2| + h_n)$$

for every $t_i \in [-r, T_1]$, $i = 1, 2,$ and $n = 1, 2, \cdots$.

Proof. We define Rothe's functions $x_{(n)}(t) : [-r, T_1] \to X$ by

$$x_{(n)}(t) = \begin{cases} \phi(t), & t \in [-r, 0] \\ x_{i-1}^n + (t - t_{i-1}^n)\dfrac{x_i^n - x_{i-1}^n}{h_n}, & t \in (t_{i-1}^n, t_i^n],\ i = 1, 2, \cdots, n. \end{cases}$$

Then $x_{(n)}(t)$ is Lipschitz continuous with Lipschitz constant $C_{15} = \max\{C, C_{14}\}$. Since for every $t \in [0, T_1]$ there exists $1 \leq i \leq n$ such that $t \in (t_{i-1}^n, t_i^n]$ and

$$\|x_n(t) - x_{(n)}(t)\|$$
$$= \left\|x_i^n - x_{i-1}^n - (t - t_{i-1}^n)\frac{x_i^n - x_{i-1}^n}{h_n}\right\|$$
$$\leq C_{14}h_n \leq C_{15}h_n,$$

we obtain

$$\begin{aligned}
&\|x_n(t_1) - x_n(t_2)\| \\
&\quad \leq \|x_n(t_1) - x_{(n)}(t_1)\| + \|x_{(n)}(t_1) - x_{(n)}(t_2)\| \\
&\qquad + \|x_{(n)}(t_2) - x_n(t_2)\| \\
&\quad \leq C_{15}|t_1 - t_2| + 2C_{15}h_n \\
&\quad \leq C_{16}(|t_1 - t_2| + h_n),
\end{aligned}$$

for $n = 1, 2, \cdots$, where $C_{16} = 2C_{15}$. Now we show the uniform convergence of the method of lines as $n \to \infty$. We first estimate the difference of two arbitrary methods of lines

PROPOSITION 4.2.2 *Assume that the assumptions of Theorem 4.2.1 are satisfied. Let* ϕ, $\psi : [-r, 0] \to X$ *be Lipschitz continuous with Lipschitz constants* c, $d > 0$, *respectively. Let* v, $w \in S$. *Let* $\phi(0)$, $\psi(0) \in \widehat{D}$. *If* $\{x_i^n\}$ *with* $t_i^n = ih_n$ *for* $i = 0, 1, 2, \cdots, n$ *and* $\{y_j^m\}$ *with* $s_j^m = jk_m$ *for* $m = 0, 1, 2, \cdots, m$, *are two sequences corresponding to* (v, ϕ) *and* (w, ψ) *in Proposition 4.2.1, respectively, then there exists a positive constant* $C_{18} \equiv C_{18}(\phi, \psi, T_1)$ *such that*

$$\begin{aligned}
&\|x_i^n - y_j^m\| \\
&\quad \leq \frac{h_n}{h_n + k_m}\|x_{i-1}^n - y_j^m\| + \frac{k_m}{h_n + h_k}\|x_i^n - y_{j-1}^m\| \\
&\qquad + \frac{h_n k_m}{h_n + k_m}\Big(C_{18}(|t_i^n - s_j^m| + \|v_{t_i^n} - w_{s_j^m}\|_{PC}) \\
&\qquad\qquad + \Big[G\Big(t_i^n, (\overline{x_i^n})_{t_i^n}, \int_0^{t_i^n} k(t_i^n, \sigma, (\overline{x_i^n})_\sigma)d\sigma\Big) \\
&\qquad\qquad - G\Big(s_j^m, (\overline{y_j^m})_{s_j^m}, \\
&\qquad\qquad\qquad \int_0^{s_j^m} k(s_j^m, \sigma, (\overline{y_j^m})_\sigma)d\sigma\Big), x_i^n - y_j^m\Big]_+\Big).
\end{aligned}$$

Proof. We choose $\lambda \in (0, 1)$ and let $h_{n,m} = \dfrac{h_n k_m}{h_n + k_m}$. Then (4.2.3) implies

$$A(t_i^n, v_{t_i^n})x_i^n \ni G\Big(t_i^n, (\overline{x_i^n})_{t_i^n}, \int_0^{t_i^n} k(t_i^n, \sigma, (\overline{x_i^n})_\sigma)d\sigma\Big) + \frac{x_{i-1}^n - x_i^n}{h_n}$$

and

$$A(s_j^m, w_{s_j^m})y_j^m \ni G\Big(s_j^m, (\overline{y_j^m})_{s_j^m}, \int_0^{s_j^m} k(s_j^m, \sigma, (\overline{y_j^m})_\sigma)d\sigma\Big) + \frac{y_{j-1}^m - y_j^m}{k_m},$$

from which

$$x_i^n = J_{\lambda h_{n,m}}(t_i^n, v_{t_i^n})a_i^n$$

and

$$y_j^m = J_{\lambda h_{n,m}}(s_j^m, w_{s_j^m})b_j^m,$$

where

$$a_i^n = x_i^n + \lambda h_{n,m}\Big(G\Big(t_i^n, (\overline{x_i^n})_{t_i^n}, \int_0^{t_i^n} k(t_i^n, \sigma, (\overline{x_i^n})_\sigma)d\sigma\Big) + \frac{x_{i-1}^n - x_i^n}{h_n}\Big),$$

and

$$b_j^m = y_j^m + \lambda h_{n,m}\Big(G\Big(s_j^m, (\overline{y_j^m})_{s_j^m}, \int_0^{s_j^m} k(s_j^m, \sigma, (\overline{y_j^m})_\sigma)d\sigma\Big) + \frac{y_{j-1}^m - y_j^m}{h_m}\Big).$$

Then by $(A2)_2$

$$\begin{aligned}
&\|x_i^n - y_j^m\| \\
&\quad = \|J_{\lambda h_{n,m}}(t_i^n, v_{t_i^n})a_i^n - J_{\lambda h_{n,m}}(s_j^m, w_{s_j^m})b_j^m\| \\
&\quad \le \|J_{\lambda h_{n,m}}(t_i^n, v_{t_i^n})a_i^n - J_{\lambda h_{n,m}}(t_i^n, v_{t_i^n})b_j^m\| \\
&\qquad + \|J_{\lambda h_{n,m}}(t_i^n, v_{t_i^n})b_j^m - J_{\lambda h_{n,m}}(s_j^m, w_{s_j^m})b_j^m\| \\
&\quad \le \|a_i^n - b_j^m\| + \lambda h_{n,m}\|A_{\lambda h_{n,m}}(t_i^n, v_{t_i^n})b_j^m - A_{\lambda h_{n,m}}(s_j^m, w_{s_j^m})b_j^m\| \\
&\quad \le \|a_i^n - b_j^m\| + \lambda h_{n,m} L(\|v_{t_i^n}\|_{PC}, \|w_{s_j^m}\|_{PC}, \|b_j^m\|) \\
&\qquad \times (|t_i^n - s_j^m|(1 + \|A_{\lambda h_{n,m}}(s_j^m, w_{s_j^m})b_j^m\|) + \|v_{t_i^n} - w_{s_j^m}\|_{PC}).
\end{aligned}$$

Letting

$$G_i^n = G\Big(t_i^n, (\overline{x_i^n})_{t_i^n}, \int_0^{t_i^n} k(t_i^n, \sigma, (\overline{x_i^n})_\sigma)d\sigma\Big)$$

and

$$G_j^m = G\Big(s_j^m, (\overline{y_j^m})_{s_j^m}, \int_0^{s_j^m} k(s_j^m, \sigma, (\overline{y_j^m})_\sigma)d\sigma\Big),$$

we find

$$\begin{aligned}
&\|a_i^n - b_j^m\| \\
&\quad = \Big\|x_i^n - x_j^m + \lambda h_{n,m}\Big(G_i^n - G_j^m + \frac{x_{i-1}^n - x_i^n}{h_n} - \frac{y_{j-1}^m - y_j^m}{h_m}\Big)\Big\| \\
&\quad \le \Big\|x_i^n - y_j^m + \lambda h_{n,m}(G_i^n - G_j^m) \\
&\qquad + \frac{\lambda k_m}{h_n + k_m}(x_{i-1}^n - x_i^n) - \frac{\lambda h_n}{h_n + k_m}(y_{j-1}^m - y_j^m)\Big\| \\
&\quad = \Big\|(1-\lambda)(x_i^n - y_j^m) + \lambda h_{n,m}(G_i^n - G_j^m) +
\end{aligned}$$

$$
\begin{aligned}
&+\frac{\lambda k_m}{h_n+k_m}(x_{i-1}^n - y_j^m) + \frac{\lambda h_n}{h_n+k_m}(x_i^n - y_{j-1}^m)\Big\| \\
&\leq \|(1-\lambda)(x_i^n - y_j^m) + \lambda h_{n,m}(G_i^n - G_j^m)\| \\
&+\frac{\lambda h_m}{h_n+h_m}\|x_{i-1}^n - y_j^m\| + \frac{\lambda h_n}{h_n+k_m}\|x_i^n - y_{j-1}^m\|
\end{aligned}
$$

and thus

$$
\begin{aligned}
&\lambda\|x_i^n - y_j^m\| + (1-\lambda)\|x_i^n - y_j^m\| \\
&= \|x_i^n - y_j^m\| \\
&\leq \|(1-\lambda)(x_i^n - y_j^m) + \lambda h_{n,m}(G_i^n - G_j^m)\| \\
&+\frac{\lambda k_m}{h_n+k_m}\|x_{i-1}^n - y_j^m\| + \frac{\lambda h_n}{h_n+k_m}\|x_i^n - y_{j-1}^m\| + U,
\end{aligned}
$$

where

$$
\begin{aligned}
U = {} & \lambda h_{n,m} L(\|v_{t_i^n}\|_{PC}, \|v_{s_j^m}\|_{PC}, \|b_j^m\|) \\
&\times(|t_i^n - s_j^m|(1 + \lambda h_{n,m}\|A_{\lambda h_{n,m}}(s_j^m, w_{s_j^m})b_j^m\|) \\
&+\|v_{t_i^n} - w_{s_j^m}\|_{PC}).
\end{aligned}
$$

We now let $\mu = \dfrac{\lambda}{1-\lambda}$, rearrange and then divide by λ to arrive at

$$
\begin{aligned}
&\|x_i^n - y_j^m\| \\
&\leq \frac{k_m}{h_n+k_m}\|x_{i-1}^n - y_j^m\| + \frac{h_n}{h_n+k_m}\|x_i^n - y_{j-1}^m\| \\
&+\frac{\|(x_i^n - y_j^m) + \mu h_{n,m}(G_i^n - G_j^m)\| - \|x_i^n - y_j^m\|}{\mu} + \frac{U}{\lambda} \\
&= \frac{k_m}{h_n+k_m}\|x_{i-1}^n - y_j^m\| + \frac{h_n}{h_n+k_m}\|x_i^n - y_{j-1}^m\| \\
&+h_{n,m}[G_i^n - G_j^m, x_i^n - y_j^m]_\mu + \frac{U}{\lambda}. \qquad (4.2.18)
\end{aligned}
$$

Now we estimate U. We observe that

$$
\begin{aligned}
&\|A_{\lambda h_{n,m}}(s_j^m, w_{s_j^m})b_j^m\| \\
&= \Big\|A_{\lambda h_{n,m}}(s_j^m, w_{s_j^m})\Big(y_j^m + \lambda h_{n,m}\Big(G_j^m + \frac{y_{j-1}^m - y_j^m}{k_m}\Big)\Big)\Big\| \\
&= \Big\|G_j^m + \frac{y_{j-1}^m - y_j^m}{k_m}\Big\| \\
&\leq \|G_j^m\| + \Big\|\frac{y_{j-1}^m - y_j^m}{k_m}\Big\|.
\end{aligned}
$$

But then, from $(k1)_2$ and $(G1)_2$

$$\begin{aligned}
\|G_j^m\| &= \left\|G\Big(s_j^m, (\overline{y_j^m})_{s_j^m}, \int_0^{s_j^m} k(s_j^m, \sigma, (\overline{y_j^m})_\sigma)d\sigma\Big)\right\| \\
&\le \left\|G\Big(s_j^m, (\overline{y_j^m})_{s_j^m}, \int_0^{s_j^m} k(s_j^m, \sigma, (\overline{y_j^m})_\sigma)d\sigma\Big) - G(t_j^m, \bar{0}, 0)\right\| \\
&\quad +\|G(s_j^m, \bar{0}, 0)\| \\
&\le a_2(s_j^m)\Big(\|\overline{y_j^m})_{s_j^m}\|_{PC} + \int_0^{s_j^m} \|k(s_j^m, \sigma, (\overline{y_j^m})_\sigma) - k(s_j^m, \sigma, \bar{0})\|d\sigma \\
&\qquad + \int_0^{t_j^m} \|k(s_j^m, \sigma, \bar{0})\|d\sigma\Big) + \|G(s_j^m, \bar{0}, 0)\| \\
&\le C_2(C_7 + C_1C_7T_1 + C_4T_1) + C_5 \\
&= C_{17}.
\end{aligned}$$

Hence

$$\begin{aligned}
\frac{U}{\lambda} \le h_{n,m}L(\|\phi\|_{PC} + MT_1, \|\psi\|_{PC} + MT_1,& \\
C_7 + \lambda h_{n,m}(C_{17} + C_{14}))& \\
\times(|t_i^n - t_j^m|(1 + C_{17} + C_{14}) + \|v_{t_i^n} - w_{s_j^m}\|_{PC}).&
\end{aligned}$$

Since L is continuous as $\lambda \to 0+$, from (4.2.18) we obtain

$$\begin{aligned}
&\|x_i^n - y_n^m\| \\
&\quad = \frac{k_m}{h_n + k_m}\|x_{i-1}^n - y_j^m\| + \frac{h_n}{h_n + k_m}\|x_i^n - y_{j-1}^m\| \\
&\quad +h_{n,m}[G_i^n - G_j^m, x_i^n - y_j^m]_+ \\
&\quad +h_{n,m}L(\|\phi\|_{PC} + MT_1, \|\psi\|_{PC} + MT_1, C_7) \\
&\quad \times(|t_i^n - s_j^m|(1 + C_{17} + C_{14}) + \|v_{t_i^n} - w_{s_j^m}\|_{PC}).
\end{aligned}$$

We put

$$C_{18} = L(\|\phi\|_{PC} + MT_1, \|\psi\|_{PC} + MT_1, C_7)(1 + C_{17} + C_{14}).$$

Then the result holds.

PROPOSITION 4.2.3 *Assume that the assumptions of Proposition 4.2.2 are satisfied. Then there exist positive constants $C_{21} \equiv C_{21}(\phi, \psi, T_1)$ and $\epsilon_{n,m} > 0$ such that*

$$\|x_i^n - y_j^m\| \le C_{21}D_{i,j} + E_i + ih_n(C_{21}D_{i,j} + \epsilon_{n,m}) \qquad (4.2.19)$$

for $i = 0, 1, 2, \cdots, n$ *and* $j = 0, 1, 2, \cdots, m$. *Here,*

$$D_{i,k} = \{(t_i^n - s_j^m)^2 + h_n t_i^n + k_m s_j^m\}^{\frac{1}{2}}$$

and

$$E_i = C_2(1 + C_1 T_1) \sum_{k=1}^{i} \sup_{t \in [-r, t_k^n]} \|x_n(t) - y_m(t)\| h_n.$$

Proof. We observe from $(G1)_2$ and $(G2)_2$ that

$$\begin{aligned}
&\Big\| G\Big(t_i^n, (\overline{x_i^n})_{t_i^n}, \int_0^{t_i^n} k(t_i^n, \sigma, (\overline{x_i^n})_\sigma) d\sigma\Big) \\
&\qquad - G\Big(s_j^m, (\overline{y_j^m})_{s_j^m}, \int_0^{s_j^m} k(s_j^m, \sigma, (\overline{y_j^m})_\sigma) d\sigma\Big) \Big\| \\
&\le \Big\| G\Big(t_i^n, (\overline{x_i^n})_{t_i^n}, \int_0^{t_i^n} k(t_i^n, \sigma, (\overline{x_i^n})_\sigma) d\sigma\Big) \\
&\qquad - G\Big(t_i^n, (\overline{y_j^m})_{s_j^m}, \int_0^{s_j^m} k(s_j^m, \sigma, (\overline{y_j^m})_\sigma) d\sigma\Big) \Big\| \\
&\quad + \Big\| G\Big(t_i^n, (\overline{y_j^m})_{s_j^m}, \int_0^{s_j^m} k(s_j^m, \sigma, (\overline{y_j^m})_\sigma) d\sigma\Big) \\
&\qquad - G\Big(s_j^m, (\overline{y_j^m})_{s_j^m}, \int_0^{s_j^m} k(s_j^m, \sigma, (\overline{y_j^m})_\sigma) d\sigma\Big) \Big\| \\
&\le C_2\Big(\|(\overline{x_i^n})_{t_i^n} - (\overline{y_j^m})_{s_j^m}\|_{PC} \\
&\qquad + \Big\| \int_0^{t_i^n} k(t_i^n, \sigma, (\overline{x_i^n})_\sigma) d\sigma - \int_0^{s_j^m} k(s_j^m, \sigma, (\overline{y_j^m})_\sigma) d\sigma \Big\| \Big) \\
&\qquad + |t_i^n - s_j^m| l_2\Big(\|(\overline{y_j^m})_{s_j^m}\|_{PC}, \int_0^{s_j^m} \|k(s_j^m, \sigma, (\overline{x_j^m})_\sigma)\| d\sigma \Big).
\end{aligned}$$

Since

$$\begin{aligned}
|t_i^n - s_j^m| &\le |(s_i^n - s_j^m) - h_n| + h_n \le |t_{i-1}^n - s_j^m| + h_n \\
&\le D_{i-1,j} + h_n
\end{aligned}$$

we have

$$\begin{aligned}
&\|(\overline{x_i^n})_{t_i^n} - (\overline{y_j^m})_{s_j^m}\|_{PC} \\
&= \|(\overline{x_n^n})_{t_i^n} - (\overline{y_m^m})_{s_j^m}\|_{PC} \\
&\le \|(\overline{x_n^n})_{t_i^n} - (\overline{y_m^m})_{t_i^n}\|_{PC} + \|(\overline{y_m^m})_{t_i^n} - (\overline{y_m^m})_{s_j^m}\|_{PC} =
\end{aligned}$$

$$= \sup_{t \in [t_i^n - r, t_i^n]} \|x_n(t) - y_m(t)\| + \sup_{\tau \in [-r,0]} \|y_m(t_i^n + \tau) - y_m(s_j^m + \tau)\|$$
$$\leq \sup_{t \in [-r, t_i^n]} \|x_n(t) - y_m(t)\| + C_{16}(|t_i^n - s_j^m| + k_m)$$
$$\leq \sup_{t \in [-r, t_i^n]} \|x_n(t) - y_m(t)\| + C_{16}(D_{i-1,j} + h_n + k_m).$$

Without loss of generality we may assume that $t_i^n \geq s_j^m$. Then by $(k1)_2$ and $(k2)_2$

$$\Big\| \int_0^{t_i^n} k(t_i^n, \sigma, (\overline{x_i^n})_\sigma) d\sigma - \int_0^{s_j^m} k(s_j^m, \sigma, (\overline{y_j^m})_\sigma) d\sigma \Big\|$$
$$\leq \int_0^{s_j^m} \|k(t_i^n, \sigma, (\overline{x_i^n})_\sigma) - k(s_j^m, \sigma, (\overline{y_j^m})_\sigma)\| d\sigma$$
$$+ \int_{s_j^m}^{t_i^n} \|k(t_i^n, \sigma, (\overline{x_i^n})_\sigma)\| d\sigma$$
$$\leq \int_0^{s_j^m} \|k(s_j^m, \sigma, (\overline{y_j^m})_\sigma) - k(s_j^m, \sigma, (\overline{x_i^n})_\sigma)\| d\sigma$$
$$+ \int_0^{s_j^m} \|k(s_j^m, \sigma, (\overline{x_i^n})_\sigma) - k(t_i^n, \sigma, (\overline{x_i^n})_\sigma)\| d\sigma$$
$$+ \int_{s_j^m}^{t_i^n} \|k(t_i^n, \sigma, (\overline{x_i^n})_\sigma) - k(t_i^n, \sigma, \bar{0})\| d\sigma + \int_{s_j^m}^{t_i^n} \|k(t_i^n, \sigma, \bar{0})\| d\sigma$$
$$\leq \int_0^{s_j^m} a_1(s_j^m, \sigma) \|(\overline{x_i^n})_\sigma - (\overline{y_j^m})_\sigma\|_{PC} d\sigma$$
$$+ \int_0^{s_j^m} l_1(\|(\overline{x_i^n})_\sigma\|_{PC}) |t_i^n - s_j^m| d\sigma$$
$$+ \int_{s_j^m}^{t_i^n} \|(\overline{x_i^n})_\sigma\|_{PC} d\sigma + C_4 |t_i^n - s_j^m|$$
$$\leq C_1 \int_0^{s_j^m} \|(\overline{x_i^n})_\sigma - (\overline{y_j^m})_\sigma\|_{PC} d\sigma$$
$$+ l_1(C_7) |t_i^n - s_j^m| T_1 + (C_7 C_1 + C_4) |t_i^n - s_j^m| T_1$$
$$\leq C_1 \int_0^{s_j^m} \sup_{t \in [-r, s_j^m]} \|x_n(t) - y_m(t)\| d\sigma$$
$$+ (l_1(C_7) T_1 + C_7 C_1 + C_4)(D_{i-1,j} + h_n)$$
$$\leq C_1 T_1 \sup_{t \in [-r, s_j^m]} \|x_n(t) - y_m(t)\|$$
$$+ (l_1(C_7) T_1 + C_7 C_1 + C_4)(D_{i-1,j} + h_n).$$

Here we have used the fact that for $0 \leq \sigma \leq s_j^m \leq t_i^n$

$$\begin{aligned}
&\|(\overline{x_i^n})_\sigma - (\overline{y_j^m})_\sigma\|_{PC} \\
&= \sup_{\tau\in[-r,0]} \|\overline{x_i^n}(\sigma+\tau) - \overline{y_j^m}(\sigma+\tau)\| \\
&= \sup_{t\in[\sigma-r,\sigma]} \|\overline{x_n^n}(t) - \overline{y_m^m}(t)\| \\
&= \sup_{t\in[\sigma-r,\sigma]} \|x_n(t) - y_m(t)\| \leq \sup_{t\in[-r,s_j^m]} \|x_n(t) - y_m(t)\| \\
&\leq \sup_{t\in[-r,t_i^n]} \|x_n(t) - y_m(t)\|.
\end{aligned}$$

Thus

$$\begin{aligned}
&\Big\|G\Big(t_i^n, (\overline{x_i^n})_{t_i^n}, \int_0^{t_i^n} k(t_i^n,\sigma,(\overline{x_i^n})_\sigma)d\sigma\Big) \\
&\quad -G\Big(s_j^m, (\overline{y_j^m})_{t_j^m}, \int_0^{s_j^m} k(s_j^m,\sigma,(\overline{y_j^m})_\sigma)d\sigma\Big)\Big\| \\
&\leq C_2\Big(\sup_{t\in[-r,t_i^n]} \|x_n(t)-y_m(t)\| + C_{16}(D_{i-1,j}+h_n+k_m) \\
&\quad +C_1T_1 \sup_{t\in[-r,t_i^n]} \|x_n(t)-y_m(t)\| \\
&\quad +(l_1(C_7)T_1 + C_1C_7 + C_4)(D_{i-1,j}+h_n)\Big) \\
&\quad +l_2(C_7,(C_1C_7+C_4)T_1)(D_{i-1,j}+h_n).
\end{aligned}$$

We set $A_{i,j} = \|x_i^n - y_j^m\|$. Then, using Proposition 4.2.2

$$\begin{aligned}
A_{i,j} \;\leq\; & \frac{h_n}{h_n+h_m}A_{i,j-1} + \frac{k_m}{h_n+k_m}A_{i-1,j} \\
& +\frac{h_nk_m}{h_n+k_m}\Big(C_{19}D_{i-1,j} \\
& \qquad +C_2(1+C_1T_1)\sup_{t\in[-r,t_i^n]}\|x_n(t)-y_m(t)\| + \epsilon_{n,m}\Big),
\end{aligned} \tag{4.2.20}$$

where

$$C_{19} = C_2C_{16} + C_2l_1(C_7)T_1 + C_1C_2C_7 + C_2C_4 + l_2(C_7,(C_1C_7+C_4)T_1)$$

and

$$\begin{aligned}
\epsilon_{n,m} \;=\; & (C_2C_{16} + C_2(l_1(C_7)T_1 + C_1C_7 + C_4) \\
& +l_2(C_7,(C_1C_7+C_4)T_1) + C_{18})h_n + C_2C_{16}k_m.
\end{aligned}$$

For $i = 1, 2, \cdots, n$, we obtain

$$\begin{aligned}
&\|x_i^n - \phi(0)\| \\
&\leq \Big\| J_{h_n}(t_i^n, v_{t_i^n})\Big(x_{i-1}^n + h_n G\Big(t_i^n, (\overline{x_i^n})_{t_i^n}, \int_0^{t_i^n} k(t_i^n, \sigma, (\overline{x_i^n})_\sigma) d\sigma\Big)\Big) \\
&\qquad - J_{h_n}(t_i^n, v_{t_i^n})\phi(0)\Big\| + \|J_{h_n}(t_i^n, v_{t_i^n})\phi(0) - \phi(0)\| \\
&\leq \|x_{i-1}^n - \phi(0)\| + h_n \Big\| G\Big(t_i^n, (\overline{x_i^n})_{t_i^n}, \int_0^{t_i^n} k(t_i^n, \sigma, (\overline{x_i^n})_\sigma) d\sigma\Big)\Big\| \\
&\quad + h_n |A(t_i^n, v_{t_i^n})\phi(0)| \\
&\leq \|x_{i-1}^n - \phi(0)\| + h_n(C_2 C_7(1 + C_1 T_1) + C_5 + C_3) \\
&= \|x_{i-1}^n - \phi(0)\| + h_n C_{20} \\
&\leq \|x_{i-1}^n - \phi(0)\| + h_n C_8,
\end{aligned}$$

where

$$C_{20} = C_5 + C_2 C_7(1 + C_1 T_1) + C_3 \leq C_8.$$

Thus

$$\|x_i^n - x_0^n\| \leq \|x_{i-1}^n - x_0^n\| + h_n C_8$$

for $i = 1, 2, \cdots, n$. Hence $A_{i,0} \leq C_{21} D_{i,0}$ for $i = 0, 1, 2, \cdots, n$, where $C_{21} = \max\{C_8, C_{19}\}$. Similarly $A_{0,j} \leq C_{21} D_{0,j}$ for $j = 0, 1, 2, \cdots, m$. Thus (4.2.19) holds for $(i, 0)$ when $i = 0, 1, 2, \cdots, n$, and for $(0, j)$ when $j = 0, 1, 2, \cdots, m$. Let $1 \leq i \leq n$ and $1 \leq k \leq m$. We assume that (4.2.19) holds for $(i-1, j)$ and $(i, j-1)$. From showing that (4.2.19) holds for (i, j), we may conclude that (4.2.19) holds for every $0 \leq i \leq n$ and $0 \leq j \leq m$. From (4.2.19) and (4.2.20)

$$\begin{aligned}
A_{i,j} \leq\ & \frac{h_n}{h_n + k_m}(C_{21} D_{i,j-1} + E_i + i h_n(C_{21} D_{i,j-1} + \epsilon_{n,m})) \\
& + \frac{k_m}{h_n + k_m}(C_{21} D_{i-1,j} + E_{i-1} + (i-1) h_n(C_{21} D_{i-1,j} + \epsilon_{n,m})) \\
& + \frac{h_n k_m}{h_n + k_m}(C_{19} D_{i-1,j} \\
& \qquad + C_2(1 + C_1 T_1) \sup_{t \in [-r, t_i^n]} \|x_n(t) - y_m(t)\| + \epsilon_{n,m}) \\
\leq\ & C_{21} D_{i,j} + E_i + i h_n(C_{21} D_{i,j} + \epsilon_{n,m}).
\end{aligned}$$

Here we used

$$\frac{h_n}{h_n + k_m} D_{i,j-1} + \frac{k_m}{h_n + k_m} D_{i,j-1} \leq D_{i,j}$$

for $i = 0, 1, 2, \cdots, n$ and $j = 0, 1, 2, \cdots, m$.

We now prove the convergence theorem for discrete scheme approximant solutions of (4.2.2).

THEOREM 4.2.1 *Assume that* $(A1)_2$, $(A2)_2$, $(k1)_2$, $(k2)_2$, $(G1)_2$ *and* $(G2)_2$ *are satisfied. Let* $\phi : [-r, 0] \to X$ *be a Lipschitz continuous function with Lipschitz constant* $c > 0$ *and* $\phi(0) \in \widehat{D}$. *Then there exists a local limit solution* $x(t)$ *of* (4.2.2) *on* $[-r, T]$. *Moreover,* $x(t)$ *is Lipschitz continuous on* $[-r, T_1]$ *for some* $T_1 \in (0, T]$.

Proof. We suppose that n and m are sufficiently large. Let $x_n(t)$ and $x_m(t)$ be two discrete scheme approximant solutions of (4.2.2) in Proposition 4.2.1 for partitions $\{t_i^n\}$ with $t_i^n = ih_n = i\frac{T_1}{n}$, $i = 0, 1, 2, \cdots, n$, and $\{t_j^m\}$ with $t_j^m = jh_m = j\frac{T_1}{m}$, $j = 0, 1, 2, \cdots, m$, of $[0, T_1]$, respectively. Let $t \in (t_{i-1}^n, t_i^n] \cap (t_{j-1}^m, t_j^m]$.

We show that the limit $x(t) = \lim_{n\to\infty} x_n(t)$ exists uniformly on $[-r, T_1]$. Since

$$((t_i^n - t_j^m)^2 + t_i^n h_n + t_j^m h_m)^{\frac{1}{2}} \le ((h_n + h_m)^2 + (h_n + h_m)T_1)^{\frac{1}{2}}$$

by the virtue of Proposition 4.2.3 with $v = w$ and $\phi = \psi$ we have

$$\begin{aligned}
&\|x_n(t) - x_m(t)\| \\
&\quad = \|x_i^n - x_j^m\| \\
&\quad \le C_{21}\Big((h_n + h_m)^2 + (h_n + h_m)T_1\Big)^{\frac{1}{2}} \\
&\qquad + C_2(1 + C_1T_1)\sum_{k=1}^{i} \sup_{t\in[-r,t_k^n]} \|x_n(t) - x_m(t)\| h_n \\
&\qquad + T_1\Big(C_{21}\Big((h_n + h_m)^2 + (h_n + h_m)T_1\Big)^{\frac{1}{2}} + \epsilon_{n,m}\Big). \qquad (4.2.21)
\end{aligned}$$

We define

$$F_{n,m}(\sigma) = \begin{cases} 0, \quad \sigma = 0 \\ \sup_{s\in[-r,t_k^n]} \|x_n(s) - x_m(s)\|, \quad \sigma \in (t_{k-1}^n, t_k^n], \\ \qquad\qquad\qquad\qquad\qquad k = 1, 2, \cdots, n. \end{cases}$$

Then we obtain that

$$F_{n,m}(\sigma) \le \sup_{s\in[-r,\sigma]} \|x_n(s) - x_m(s)\| + 3C_{16}h_n + C_{16}h_m \qquad (4.2.22)$$

for every $\sigma \in [0, T_1]$.

Indeed, letting $\sigma \in (t_{k-1}^n, t_k^n]$ we note that

$$F_{n,m}(\sigma) = \max\Big\{ \sup_{s\in[-r,\sigma]} \|x_n(s) - x_m(s)\|, \sup_{s\in[\sigma,t_k^n]} \|x_n(s) - x_m(s)\| \Big\}.$$

Let $s \in [\sigma, t_k^n]$. From Lemma 4.2.2 we obtain

$$\begin{aligned}
&\|x_n(s) - x_m(s)\| \\
&\quad \le \|x_n(\sigma) - x_m(\sigma)\| + \|x_m(\sigma) - x_m(s)\| + \|x_n(\sigma) - x_(s)\| \\
&\quad \le \sup_{s\in[-r,\sigma]} \|x_n(\sigma) - x_m(\sigma)\| + 2C_{16}|s-\sigma| + C_{16}(h_n + h_m) \\
&\quad \le \sup_{s\in[-r,\sigma]} \|x_n(\sigma) - x_m(\sigma)\| + 2C_{16}h_n + C_{16}(h_n + h_m).
\end{aligned}$$

Thus

$$\begin{aligned}
&\sup_{s\in[\sigma,t_k^n]} \|x_n(s) - x_m(s)\| \\
&\quad \le \sup_{s\in[-r,\sigma]} \|x_n(\sigma) - x_m(\sigma)\| + 2C_{16}h_n + C_{16}(h_n + h_m).
\end{aligned}$$

Hence we have (4.2.22). We also have

$$\begin{aligned}
&\sum_{k=1}^{i} \sup_{s\in[-r,t_k^n]} \|x_n(s) - x_m(s)\| h_n \\
&\quad = \sum_{k=1}^{i} \sup_{s\in[-r,t_k^n]} \|x_n(s) - x_m(s)\| (t_k^n - t_{k-1}^n) \\
&\quad = \sum_{k=1}^{i} \int_{t_{k-1}^n}^{t_k^n} F_{n,m}(\sigma) d\sigma \\
&\quad = \int_0^t F_{n,m}(\sigma) d\sigma + \int_t^{t_k^n} F_{n,m}(\sigma) d\sigma \\
&\quad < \int_0^t F_{n,m}(\sigma) d\sigma + 2C_7 h_n \\
&\quad \le \int_0^t \sup_{s\in[-r,\sigma]} \|x_n(s) - x_m(s)\| d\sigma + (3C_{16}h_n + C_{16}h_m)T_1 + 2C_7 h_n.
\end{aligned}$$

It follows from (4.2.21) that

$$\begin{aligned}
&\|x_n(t) - x_m(t)\| \\
&\quad \le \delta_{n,m} + C_2(1 + C_1 T_1) \int_0^t \sup_{s\in[-r,\sigma]} \|x_n(s) - x_m(s)\| d\sigma
\end{aligned}$$

for every $t \in [0, T_1]$, where

$$\delta_{n,m} = C_{21}((h_n + h_m)^2 + (h_n + h_m)T_1)^{\frac{1}{2}} \\ +T_1(C_{21}((h_n + h_m)^2 + (h_n + h_m)T_1)^{\frac{1}{2}} + \epsilon_{n,m}) \\ +C_{16}(3h_n + h_m)C_2(1 + C_1T_1)T_1 + 2C_2C_7(1 + C_1T_1)h_n.$$

In other words we obtain that

$$\sup_{s\in[-r,t]} \|x_n(s) - x_m(s)\| \\ \leq \delta_{n,m} + C_2(1 + C_1T_1) \int_0^t \sup_{s\in[-r,\sigma]} \|x_n(s) - x_m(s)\| d\sigma$$

for every $t \in [0, T_1]$. Consequently from Gronwall's inequality we have

$$\sup_{s\in[-r,t]} \|x_n(s) - x_m(s)\| \\ \leq \delta_{n,m}\exp\Big(\int_0^t C_2(1 + C_1T_1)d\sigma\Big) \\ \leq \delta_{n,m}\exp(C_2(1 + C_1T_1)T_1),$$

which implies that $x_n(t)$ converges to a function $x(t)$ uniformly on $[-r, T_1]$. Finally, since by Lemma 4.2.2

$$\|x_n(t_1) - x_n(t_2)\| \leq C_{16}(|t_1 - t_2| + h_n)$$

for every $t_i \in [-r, T_1]$, $i = 1, 2$, letting $n \to \infty$ we obtain

$$\|x(t_1) - x(t_2)\| \leq C_{16}|t_1 - t_2|$$

for every $t_i \in [-r, T_1]$, $i = 1, 2$. This shows that $x(t)$ is Lipschitz continuous on $[-r, T_1]$. Hence $x(t)$ is a limit solution of (4.2.2) on $[-r, T_1]$.

PROPOSITION 4.2.4 *Assume that the assumptions of Proposition 4.2.2 are satisfied. Then there exists a positive constant* $C_{23} = C_{23}(\phi, \psi, T_1)$ *such that*

$$\|x_i^n - y_j^m\| \\ \leq \|x_p^n - y_q^m\| \\ +C_{23}D_{i,j} + \sum_{l=p}^{i} \delta_k^n h_n + \sum_{l=q}^{j} \epsilon_k^m k_m \\ +ih_n\Big(\Big(\frac{1}{\delta}\rho(T_1) + C_{23}\Big)(D_{i,j} + |t_p^n - s_q^m|) \\ +\rho(2\delta) + C_{23}h_n + C_{23}\|v - w\|_{T_1}\Big) \qquad (4.2.23)$$

for $i = p, p+1, \cdots, n$ *and* $j = q, q+1, \cdots, m$. *Here, the symbols used above are defined by*

$$\begin{aligned}
\delta_i^n &= \Big[G\Big(t_i^n, (\overline{x_i^n})_{t_i^n}, \int_0^{t_i^n} k(t_i^n, \sigma, (\overline{x_i^n})_\sigma)d\sigma\Big) \\
&\quad -G\Big(t_i^n, y_{t_i^n}, \int_0^{t_i^n} k(t_i^n, \sigma, y_\sigma)d\sigma\Big), x_i^n - y(t_i^n)\Big]_\lambda, \\
\epsilon_j^m &= \Big\|G\Big(s_j^m, (\overline{y_j^m})_{s_j^m}, \int_0^{s_j^m} k(s_j^m, \sigma, (\overline{y_j^m})_\sigma)d\sigma\Big) \\
&\quad -G\Big(s_j^m, y_{s_j^m}, \int_0^{s_j^m} k(s_j^m, \sigma, y_\sigma)d\sigma\Big)\Big\| \\
&\quad +\frac{2}{\lambda}\|y_j^m - y(s_j^m)\|, \\
\rho(t) &= \sup\Big\{\frac{2}{\lambda}\|y(t_1) - y(t_2)\| + \Big\|G\Big(t_1, y_{t_1}, \int_0^{t_1} k(t_1, \sigma, y_\sigma)d\sigma\Big) \\
&\quad -G\Big(t_2, y_{t_2}, \int_0^{t_2} k(t_2, \sigma, y_\sigma)d\sigma\Big)\Big\| \;\Big|\; |t_1 - t_2| \le t\Big\}, \\
D_{i,j} &= \Big(((t_i^n - t_p^n) - (s_j^m - s_q^m))^2 + (t_i^n - t_p^n)h_n + (s_j^m - s_q^m)k_m\Big)^{\frac{1}{2}} \\
&\quad +\Big(((t_i^n - t_p^n) - (s_j^m - s_q^m))^2 + (t_i^n - t_p^n)h_n + (s_j^m - s_q^m)k_m\Big)
\end{aligned}$$

and $\delta \in (0, \frac{T_1}{2})$.

Proof. Using Proposition 4.2.2

$$\begin{aligned}
&\|x_i^n - y_j^m\| \\
&\le \frac{h_n}{h_n + k_m}\|x_{i-1}^n - y_j^m\| + \frac{k_m}{h_n + h_k}\|x_i^n - y_{j-1}^m\| \\
&\quad +\frac{h_n k_m}{h_n + k_m}\Big(C_{18}(|t_i^n - s_j^m| + \|v_{t_i^n} - w_{s_j^m}\|_{PC}) \\
&\qquad +\Big[G\Big(t_i^n, (x_i^n)_{t_i^n}, \int_0^{t_i^n} k(t_i^n, \sigma, (\overline{x_i^n})_\sigma)d\sigma\Big) \\
&\qquad -G\Big(s_j^m, (\overline{y_j^m})_{s_j^m}, \int_0^{s_j^m} k(s_j^m, \sigma, (\overline{y_j^m})_\sigma)d\sigma\Big), x_i^n - y_j^m\Big]_+\Big)
\end{aligned}$$

for n and m sufficiently large such that $\max\{h_n, k_m\} < \delta$. Noting that

$$[b_1, b_2]_+ \le [b_1, b_2]_\lambda \le [b_3, b_4]_\lambda + \|b_2 - b_4\| + \frac{2}{\lambda}\|b_1 - b_3\|$$

for every $b_l \in X$, $l = 1, 2, 3, 4$ we obtain

$$\begin{aligned}
&\Big[G\Big(t_i^n, (\overline{x_i^n})_{t_i^n}, \int_0^{t_i^n} k(t_i^n, \sigma, (\overline{x_i^n})_\sigma) d\sigma\Big) \\
&\qquad -G\Big(s_j^m, (\overline{y_j^m})_{s_j^m}, \int_0^{s_j^m} k(s_j^m, \sigma, (\overline{y_j^m})_\sigma) d\sigma\Big), x_i^n - y_j^m\Big]_+ \\
&\leq \Big[G\Big(t_i^n, (\overline{x_i^n})_{t_i^n}, \int_0^{t_i^n} k(t_i^n, \sigma, (\overline{x_i^n})_\sigma) d\sigma\Big) \\
&\quad -G\Big(s_j^m, (\overline{y_j^m})_{s_j^m}, \int_0^{s_j^m} k(s_j^m, \sigma, (\overline{y_j^m})_\sigma) d\sigma\Big), x_i^n - y(t_i^n)\Big]_\lambda \\
&\quad +\Big\|G\Big(s_j^m, (\overline{y_j^m})_{s_j^m}, \int_0^{s_j^m} k(s_j^m, \sigma, (\overline{y_j^m})_\sigma) d\sigma\Big) \\
&\quad -G\Big(s_j^m, y_{s_j^m}, \int_0^{s_j^m} k(s_j^m, \sigma, y_\sigma) d\sigma\Big)\Big\| \\
&\quad +\Big\|G\Big(s_j^m, y_{s_j^m}, \int_0^{s_j^m} k(s_j^m, \sigma, y_\sigma) d\sigma\Big) \\
&\quad -G\Big(t_i^n, (\overline{x_i^n})_{t_i^n}, \int_0^{t_i^n} k(t_i^n, \sigma, (\overline{x_i^n})_\sigma) d\sigma\Big)\Big\| \\
&\quad +\frac{2}{\lambda}\|y_j^m - y(s_j^m)\| + \frac{2}{\lambda}\|y(s_j^m) - y(t_i^n)\|.
\end{aligned}$$

Let $A_{i,j} = \|x_i^n - y_j^m\|$. Then

$$\begin{aligned}
A_{i,j} \;\leq\; & \frac{h_n}{h_n + k_m} A_{i,j-1} + \frac{h_n}{h_n + k_m} A_{i-1,j} \\
& + \frac{h_n k_m}{h_n + k_m} (C_{18}(|t_i^n - s_j^m| + \|v_{t_i^n} - w_{s_j^m}\|_{PC}) \\
& \qquad + \delta_i^n + \epsilon_j^m + \rho(|t_i^n - s_j^m|)).
\end{aligned}$$

Since

$$\begin{aligned}
|t_i^n - s_j^m| &\leq |(t_i^n - s_j^m) - h_n| + h_n \\
&\leq |(t_i^n - t_p^n) - (s_j^m - s_q^m) - h_n| + |t_p^n - s_q^m| + h_n \\
&\leq D_{i-1,j} + |t_p^n - s_q^m| + h_n, \\
\rho(|t_i^n - s_j^m|) &\leq \frac{1}{\delta}\rho(T_1)(|t_i^n - s_j^m| - h_n) + \rho(2\delta) \\
&\leq \frac{1}{\delta}\rho(T_1)(D_{i-1,j} + |t_p^n - s_q^m|) + \rho(2\delta)
\end{aligned}$$

and

$$\|v_{t_i^n} - w_{s_j^m}\|_{PC} \leq$$

$$\begin{aligned}
&\leq \|v_{t_i^n} - v_{s_j^m}\|_{PC} + \|v_{s_j^m} - w_{s_j^m}\|_{PC} \\
&\leq M|t_i^n - s_j^m| + \|v_{s_j^m} - w_{s_j^m}\|_{PC} \\
&\leq MD_{i-1,j} + M|t_p^n - s_j^m| + Mj_n + \|v - w\|_{T_1}
\end{aligned}$$

it follows that

$$\begin{aligned}
A_{i,j} \quad \leq \quad & \frac{h_n}{h_n + k_m} A_{i,j-1} + \frac{h_n}{h_n + k_m} A_{i-1,j} \\
& + \frac{h_n k_m}{h_n + k_m} \Big(\Big(C_{18} + \frac{1}{\delta}\rho(T_1) \Big) (D_{i-1,j} + |t_p^n - s_q^m|) \\
& \qquad + C_{18} h_n \delta_i^n + \epsilon_j^m + \rho(2\delta) \\
& \qquad + C_{18}\|v - w\|_{T_1} \Big). \qquad (4.2.24)
\end{aligned}$$

Since

$$A(t_i^n, v_{t_i^n})x_i^n \ni G\Big(t_i^n, (\overline{x_i^n})_{t_i^n}, \int_0^{t_i^n} k(t_i^n, \sigma, (\overline{x_i^n})_\sigma) d\sigma\Big) - \frac{x_i^n - x_{i-1}^n}{h_n}$$

we obtain that

$$\begin{aligned}
|A(t_i^n, v_{t_i^n})x_i^n| \quad \leq \quad & \Big\| G\Big(t_i^n, (\overline{x_i^n})_{t_i^n}, \int_0^{t_i^n} k(t_i^n, \sigma, (\overline{x_i^n})_\sigma) d\sigma\Big) \Big\| \\
& + \Big\| \frac{x_i^n - x_{i-1}^n}{h_n} \Big\| \\
\leq \quad & C_{17}
\end{aligned}$$

for $i = 1, 2, \cdots, n$. Thus it follows from $(A2)_2$ that

$$\begin{aligned}
|A(t_i^n, v_{t_i^n})x_i^n| &\leq |A(t_p^n, v_{t_p^n})x_p^n| \\
&< L(\|v_{t_i^n}\|_{PC}, \|v_{t_p^n}\|_{PC}, \|x_p^n\|) \\
&\quad \times (|t_i^n - t_p^n|(1 + |A(t_p^n, v_{t_p^n})x_p^n|) + \|v_{t_i^n} - v_{t_p^n}\|_{PC}) \\
&\leq C_{17} + L(\|\phi\|_{PC} + MT_1, \|\phi\|_{PC} + MT_1, \|x_p^n\|) \\
&\qquad \times |t_i^n - t_p^n|(1 + C_{17} + M).
\end{aligned}$$

From (4.2.5)

$$\begin{aligned}
&\|x_l^n - x_p^n\| \\
&\quad \leq \Big\| J_{h_n}(t_l^n, v_{t_l^n}) \Big(x_{l-1}^n + h_n G\Big(t_l^n, (\overline{x_l^n})_{t_l^n}, \int_0^{t_l^n} k(t_l^n, \sigma, (\overline{x_l^n})_\sigma) d\sigma\Big)\Big) \\
&\qquad - J_{h_n}(t_l^n, v_{t_l^n}) x_p^n \Big\| + \|J_{h_n}(t_l^n, v_{t_l^n}) x_p^n - x_p^n\| \leq
\end{aligned}$$

$$\leq \|x_{l-1}^n - x_p^n\| + h_n \Big\| G\Big(t_l^n, (\overline{x_l^n})_{t_l^n}, \int_0^{t_l^n} k(t_l^n, \sigma, (\overline{x_l^n})_\sigma) d\sigma\Big) \Big\|$$
$$+ h_n |A(t_p^n, v_{t_p^n}) x_p^n|$$
$$\leq \|x_{l-1}^n - x_p^n\|$$
$$+ 2h_n C_{17} + h_n L(\|\phi\|_{PC} + MT_1, \|\phi\|_{PC} + MT_1, \|x_p^n\|)$$
$$\times |t_i^n - t_p^n|(1 + C_{17} + M)$$
$$\leq \|x_{l-1}^n - x_p^n\| + h_n C_{23} |t_l^n - t_p^n| + C_{23} h_n$$

for $l = p+1, \cdots, n$, where $C_{23} = \max\{2C_{17}, C_{18}\}$. Adding this inequality for $l = p+1, \cdots, i$ we obtain

$$\begin{aligned} \|x_i^n - x_p^n\| &\leq C_{23} h_n (i-p) + C_{23} h_n \textstyle\sum_{l=p+1}^{i} |t_l^n - t_p^n| \\ &\leq C_{23}(i-p) + C_{23}(i-p)^2 h_n^2 \\ &= C_{23}|t_i^n - t_p^n| + C_{23}|t_i^n - t_p^n|^2 \\ &\leq C_{23} D_{i,q}. \end{aligned}$$

Thus

$$\|x_i^n - y_q^m\| \leq \|x_p^n - y_q^m\| + C_{23} D_{i,q}$$

for $i = p, \cdots, n$. This implies that (4.2.23) holds for $p \leq i \leq\leq n$ and $j = q$. In the same way, (4.2.23) also holds for $i = p$ and $q \leq j \leq m$. Next, let $p+1 \leq i \leq n$ and $q+1 \leq j \leq m$. We assume that (4.2.23) holds for the pairs $(i-1, j)$ and $(i, j-1)$. Since $C_{18} \leq C_{23}$, from (4.2.24)

$$\begin{aligned} A_{i,j} \leq \;& \frac{h_n}{h_n + k_m}\Big(\|x_p^n - y_q^m\| + C_{23} D_{i,j-1} + \sum_{l=p}^{i} \delta_n^l h_n + \sum_{l=q}^{j-1} \epsilon_l^m k_m \\ & + i h_n \Big(\Big(\frac{1}{\delta}\rho(T_1) + C_{23}\Big)(D_{i,j-1} + |t_p^n - s_j^m|) \\ & + C_{23} h_n + \rho(2\delta) + C_{23}\|v - w\|_{T_1} \Big)\Big) \\ & + \frac{k_m}{h_n + k_m}\Big(\|x_p^n - y_q^m\| + C_{23} D_{i-1,j} + \sum_{l=p}^{i} \delta_n^l h_n + \sum_{l=q}^{j-1} \epsilon_l^m k_m \\ & + (i-1) h_n \Big(\Big(\frac{1}{\delta}\rho(T_1) + C_{23}\Big)(D_{i-1,j} + |t_p^n - s_j^m|) \\ & + C_{23} h_n + \rho(2\delta) + C_{23}\|v - w\|_{T_1} \Big)\Big) \\ & + \frac{h_n k_m}{h_n + k_m}\Big(\Big(\frac{1}{\delta}\rho(T_1) + C_{23}\Big)(D_{i-1,j} + |t_p^n - s_j^m|) + C_{23} h_n \\ & + \rho(2\delta) + \delta_i^n + \epsilon_j^m + C_{23}\|v - w\|_{T_1} \Big) = \end{aligned}$$

$$= \|x_i^n - y_j^m\| C_{23}\Big(\frac{h_n}{h_n+k_m} D_{i,j-1} + \frac{k_m}{h_n+k_m} D_{i-1,j}\Big) + \sum_{l=p}^{i} \delta_n^l h_n$$

$$+ \sum_{l=q}^{j-1} \epsilon_l^m k_m i h_n \Big(\Big(\frac{1}{\delta}\rho(T_1) + C_{23}\Big)(D_{i,j} + |t_p^n - s_q^m|)$$

$$+ C_{23} h_n + \rho(2\delta) + C_{23}\|v - w\|_{T_1}\Big)$$

$$\leq \|x_p^n - y_q^m\| + C_{23} D_{i,j} + \sum_{l=p}^{i} \delta_l^n h_n + \sum_{l=q}^{j-1} \epsilon_l^m k_m$$

$$+ i h_n \Big((\frac{1}{\delta}\rho(T_1) + C_{23})(D_{i,j} + |t_p^n - s_q^m|)$$

$$+ C_{23} h_n + \rho(2\delta) + C_{23}\|v - w\|_{T_1}\Big).$$

Hence, (4.2.23) holds for the pair (i, j). Therefore, we conclude that (4.2.23) holds for every $i = p, \cdots, n$ and $j = q, \cdots, m$.

PROPOSITION 4.2.5 *Assume that the assumptions of Theorem 4.2.1 are satisfied. Let ϕ, $\psi : [-r, 0] \to X$ be Lipschitz continuous with Lipschitz constant c, $d > 0$, respectively, and v, $w \in S$. Let $\phi(0)$, $\psi(0) \in \widehat{D}$. If $x(t)$ and $y(t)$ are two local limit solutions of (4.2.2) on $[-r, T]$ corresponding to (v, ϕ) and (w, ψ), respectively, then there exists a positive constant C_{23} such that*

$$\|x(t) - y(t)\| \tag{4.2.25}$$

$$\leq \|x(s) - y(s)\| + C_{23} T_1 \|v - w\|_{T_1}$$

$$+ \int_s^t \Big[G\Big(\tau, x_\tau, \int_0^\tau k(\tau, \sigma, x_\sigma) d\sigma\Big)$$

$$- G\Big(\tau, y_\tau, \int_0^\tau k(\tau, \sigma, y_\sigma) d\sigma\Big), x(\tau) - y(\tau)\Big]_+ d\tau$$

for $0 \leq s \leq t \leq T_1$ for some $T_1 \in (0, T]$.

Proof. Using notations in the proof of Proposition 4.2.4, let $s \in (t_{p-1}^n, t_p^n] \cap (s_{q-1}^m, s_q^m]$ and $t \in (t_{i-1}^n, t_i^n] \cap (s_{j-1}^m, s_j^m]$. Letting n, $m \to \infty$ in (4.2.23) we obtain

$$\|x(t) - y(t)\|$$

$$\leq \|x(s) - y(s)\| + \limsup_{n\to\infty} \sum_{l=p}^{i} \delta_l^n h_n + \limsup_{m\to\infty} \sum_{l=q}^{j} \epsilon_l^m k_m$$

$$+ \rho(2\delta) T_1 + C_{23} T_1 \|v - w\|_{T_1}. \tag{4.2.26}$$

Since

$$\lim_{n\to\infty}\sum_{l=p}^{i}\delta_i^n h_n = \int_s^t \Big[G\Big(\tau, x_\tau, + \int_0^\tau k(\tau,\sigma,x_\sigma)d\sigma\Big) - G\Big(\tau, y_\tau, \int_0^\tau k(\tau,\sigma,y_\sigma)d\sigma\Big), x(\tau)-y(\tau)\Big]_\lambda d\tau$$

and $\lim_{m\to\infty}\sum_{l=q}^{j}\epsilon_j^m k_m = 0$ let $\delta \to 0+$ in (4.2.26). Then we have

$$\begin{aligned}\|x(t)-y(t)\| &\le \|x(s)-y(s)\| + C_{23}T_1\|v-w\|_{T_1} \\ &+ \int_s^t \Big[G\Big(\tau, x_\tau, \int_0^\tau k(\tau,\sigma,x_\sigma)d\sigma\Big) \\ &- G\Big(\tau, y_\tau, \int_0^\tau k(\tau,\sigma,y_\sigma)d\sigma\Big), x(\tau)-y(\tau)\Big]_\lambda d\tau.\end{aligned}$$

Letting $\lambda \to 0+$, we obtain the desired result (4.2.25).

We now consider uniqueness of a limit solution for (4.2.2).

THEOREM 4.2.2 *Assume that the assumptions of Theorem 4.2.1 are satisfied. Then there exists a unique limit solution of* (4.2.2).

Proof. Let $x(t)$ and $y(t)$ be two local limit solutions of (4.2.2) on $[-r, T]$. From (4.2.25) we obtain

$$\begin{aligned}&\|x(t)-y(t)\| \\ &\le \|x(s)-y(s)\| + C_{23}T_1\|v-w\|_{T_1} \\ &\quad + \int_s^t \Big[G\Big(\tau, x_\tau, \int_0^\tau k(\tau,\sigma,x_\sigma)d\sigma\Big) \\ &\quad - G\Big(\tau, y_\tau, \int_0^\tau k(\tau,\sigma,y_\sigma)d\sigma\Big), x(\tau)-y(\tau)\Big]_+ d\tau \\ &\le \|x(s)-y(s)\| + C_{23}T_1\|v-w\|_{T_1} \\ &\quad + \int_s^t \Big\|G\Big(\tau, x_\tau, \int_0^\tau k(\tau,\sigma,x_\sigma)d\sigma\Big) \\ &\quad - G\Big(\tau, y_\tau, \int_0^\tau k(\tau,\sigma,y_\sigma)d\sigma\Big)\Big\| d\tau \\ &\le \|x(s)-y(s)\| + C_{23}T_1\|v-w\|_{T_1} \\ &\quad + \int_s^t a_2(\tau)\Big(\|x_\tau - y_\tau\|_{PC} + \int_0^\tau \|k(\tau,\sigma,x_\sigma)d\sigma - k(\tau,\sigma,y_\sigma)\|d\sigma\Big)d\tau \\ &\le \|x(s)-y(s)\| + C_{23}T_1\|v-w\|_{T_1} +\end{aligned}$$

$$
\begin{aligned}
&+C_2\int_s^t \Big(\|x_\tau - y_\tau\|_{PC} + \int_0^\tau a_1(\tau,\sigma)\|x_\sigma - y_\sigma\|_{PC}d\sigma\Big)d\tau \\
&\le \|x(s)-y(s)\| + C_{23}T_1\|v-w\|_{T_1} \\
&+C_2\int_s^t \|x_\tau - y_\tau\|_{PC}d\tau + C_2C_1\int_s^t\int_0^\tau \|x_\sigma - y_\sigma\|_{PC}d\sigma d\tau. \quad (4.2.27)
\end{aligned}
$$

Suppose $t \ge r$. Then $t+\theta \ge 0$ for every $\theta \in [-r,0]$. By (4.2.27)

$$
\begin{aligned}
&\|x(t+\theta)-y(t+\theta)\| \\
&\quad\le \|x(s)-y(s)\| + C_{23}T_1\|v-w\|_{T_1} \\
&\quad +C_2\int_s^{t+\theta} \|x_\tau - y_\tau\|_{PC}d\tau + C_2C_1\int_s^{t+\theta}\int_0^\tau \|x_\sigma - y_\sigma\|_{PC}d\sigma d\tau,
\end{aligned}
$$

which yields

$$
\begin{aligned}
&\|x_t - y_t\| \qquad\qquad (4.2.28)\\
&\quad\le \|x(s)-y(s)\| + C_{23}T_1\|v-w\|_{T_1} \\
&\quad +C_2\int_s^{t+\theta} \|x_\tau - y_\tau\|_{PC}d\tau + C_2C_1\int_s^{t+\theta}\int_0^\tau \|x_\sigma - y_\sigma\|_{PC}d\sigma d\tau.
\end{aligned}
$$

Suppose $0 \le t < r$. Let $\theta \in (-r,-t]$. Then $t+\theta < 0$ and thus

$$\|x(t+\theta)-y(t+\theta)\| = \|\phi(t+\theta)-\phi(t+\theta)\| = 0.$$

Let $\theta \in [-t,0]$. Then $t+\theta \ge 0$ and (4.2.28) holds. Hence, for every $t \in [0,T_1]$, we have

$$
\begin{aligned}
&\|x_t - y_t\| \le \\
&\quad\le \|x(s)-y(s)\| + C_{23}T_1\|v-w\|_{T_1} \\
&\quad +C_2\int_s^{t+\theta} \|x_\tau - y_\tau\|_{PC}d\tau + C_2C_1\int_s^{t+\theta}\int_0^\tau \|x_\sigma - y_\sigma\|_{PC}d\sigma d\tau \\
&\quad\le \|x(s)-y(s)\| + C_{23}T_1\|v-w\|_{T_1} \\
&\quad +C_{24}\int_0^{t-s} \|x_\xi - y_\xi\|_{PC}d\xi + \int_0^{t-s} C_{24}\int_0^{\xi+s} \|x_\sigma - y_\sigma\|_{PC}d\sigma d\xi
\end{aligned}
$$

by changing variables, where $C_{24} = \max\{C_2, C_1C_2\}$. From the Gronwall–Bellmann inequality in Pachpatte [112],

$$
\begin{aligned}
&\|x_t - y_t\|_{PC} \\
&\quad\le (\|x(s)-y(s)\| + C_{23}T_1\|v-w\|_{T_1}) \\
&\quad\times\Big(1+\int_0^{t-s} C_{24}\exp\Big(\int_0^{\xi+s}(C_{24}+1)d\sigma\Big)d\xi\Big) =
\end{aligned}
$$

$$= (\|x(s) - y(s)\| + C_{23}T_1\|v - w\|_{T_1})$$
$$\times\Big(1 + \int_0^{t-s} C_{24}\exp(C_{24}+1)(\xi + s)d\xi\Big)$$
$$= (\|x(s) - y(s)\| + C_{23}T_1\|v - w\|_{T_1})\Big(1 + \int_s^t C_{24}\exp(C_{24}+1)\eta d\eta\Big).$$

Hence

$$\|x_t - y_t\|_{PC} \le (\|x(s) - y(s)\| + C_{23}T_1\|v - w\|_{T_1}) \times\Big(1 + \frac{C_{24}}{C_{24}+1}\exp(C_{24}+1)T_1\Big) \tag{4.2.29}$$

for every $t \in [0, T_1]$. In particular, if $s = 0$, $\phi = \psi$ and $v = w$ then $x(t) = y(t)$ for every $t \in [-r, T_1]$.

PROPOSITION 4.2.6 *Assume that the assumptions of Theorem 4.2.1 are satisfied. Then there exist $T_1 \in (0, T]$ and a local limit solution $x(t)$ of (4.2.2) on $[-r, T]$ with $v = x$ and $x(t) \in \widehat{D}$ for every $t \in [0, T_1]$.*

Proof. Let $T_1 \in (0, T]$ be sufficiently small so that

$$C_{25} = C_{23}T_1\Big(1 + \frac{C_{24}}{C_{24}+1}\exp(C_{24}+1)T_1\Big) < 1$$

and $C_{16} \le M$. Let $x_v(t)$ be a limit solution of (4.2.2) for $v \in S$. Then from Lemma 4.2.2

$$\|x_v(t_1) - x_v(t_2)\| \le C_{16}|t_1 - t_2| \le M|t_1 - t_2|$$

for every $t_i \in [-r, T_1]$, $i = 1, 2$. Thus $x_v \in S$.

Let $x_v(t)$ and $y_w(t)$ be limit solutions of (4.2.2) for v and w in S. By (4.2.29) with $s = 0$ and $\phi = \psi$ we obtain

$$\|x_v(t) - y_w(t)\| \le \|x_t - y_t\|_{PC} \le C_{25}\|v - w\|_{T_1}$$

for every $t \in [-r, T_1]$, and thus

$$\|x_v - y_w\|_{T_1} \le C_{25}\|v - w\|_{T_1}. \tag{4.2.30}$$

We define an operator $Q : S \to S$ by $Qv = x_v$ for every $v \in S$, where x_v is the limit solution of (4.2.2) for $v \in S$. Then (4.2.30) means that Q is a strict contraction on a complete metric space S. Using Banach's contraction principle there exists a unique fixed point of Q in S. We put $x = x_v$. Then $v = x$ and $x(t)$ is the desired limit solution on $[-r, T_1]$.

In order to show that the limit solution in Theorem 4.2.1 is contained in $\widehat{D}$, we need the following result.

PROPOSITION 4.2.7 *Assume that the assumptions of Theorem 4.2.1 are satisfied. Let* $v \in S$. *Then* $|A(t_i^n, v_{t_i^n})x_{i-1}^n|$ *is bounded independently of* n *and* $i = 1, 2, \cdots, n$.

Proof. We fix n and define $B(t, \psi) : X \supset D(A(t, \psi)) \to 2^X$ by

$$B(t,\psi)u = A(t,\psi)u - G\Big(t, (\overline{x_i^n})_t, \int_0^t k(t,\sigma,(\overline{x_i^n})_\sigma)d\sigma\Big)$$

for every $t \in [0, T_1]$, $\psi \in PC([-r, 0]; X)$ and $u \in X$. Then $B(t, \psi)$ is m-accretive in X for every $t \in [0, T_1]$ and $\psi \in PC([-r, 0]; X)$. Since

$$\begin{aligned} u &= (I + \lambda B(t,\psi))J_\lambda^B(t,\psi)u \\ &= \Big(I + \lambda A(t,\psi) - \lambda G\Big(t, (\overline{x_i^n})_t, \int_0^t k(t,\sigma,(\overline{x_i^n})_\sigma)d\sigma\Big)\Big)J_\lambda^B(t,\psi)u \\ &= (I + \lambda A(t,\psi))J_\lambda^B(t,\psi)u - \lambda G\Big(t, (\overline{x_i^n})_t, \int_0^t k(t,\sigma,(\overline{x_i^n})_\sigma)d\sigma\Big) \end{aligned}$$

and

$$u + \lambda G\Big(t, (\overline{x_i^n})_t, \int_0^t k(t,\sigma,(\overline{x_i^n})_\sigma)d\sigma\Big) \in (I + \lambda A(t,\psi))J_\lambda^B(t,\psi)u$$

for every $\lambda > 0$, $t \in [0, T_1]$, $\psi \in PC([-r, 0]; X)$ and $u \in X$, where $J_\lambda(t,\psi) = (I + \lambda A(t,\psi))^{-1}$ and $J_\lambda^B(t,\psi) = (I + \lambda B(t,\psi))^{-1}$. We have

$$J_\lambda^B(t,v_t)u = J_\lambda(t,v_t)\Big(u + \lambda G\Big(t, (\overline{x_i^n})_t, \int_0^t k(t,\sigma,(\overline{x_i^n})_\sigma)d\sigma\Big)\Big)$$

for every $\lambda > 0$, $t \in [0, T_1]$, $\psi = v_t \in PC([-r, 0]; X)$ and $u \in X$. From $(A1)_2$ and $(A2)_2$

$$\begin{aligned} &\|B_\lambda(t,v_t)u - B_\lambda(s,v_s)u\| \\ &\quad\le \Big\|A_\lambda(t,v_t)\Big(u + \lambda G(t, (\overline{x_i^n})_t, \int_0^t k(t,\sigma,(\overline{x_i^n})_\sigma)d\sigma)\Big) \\ &\qquad -A_\lambda(t,v_t)\Big(u + \lambda G(s, (\overline{x_i^n})_s, \int_0^s k(s,\sigma,(\overline{x_i^n})_\sigma)d\sigma)\Big)\Big\| \\ &\quad\le \Big\|A_\lambda(t,v_t)\Big(u + \lambda G\Big(s, (\overline{x_i^n})_s, \int_0^s k(s,\sigma,(\overline{x_i^n})_\sigma)d\sigma\Big)\Big) \\ &\qquad -A_\lambda(s,v_s)\Big(u + \lambda G\Big(s, (\overline{x_i^n})_s, \int_0^s k(s,\sigma,(\overline{x_i^n})_\sigma)d\sigma\Big)\Big)\Big\| \le \end{aligned}$$

$$
\begin{aligned}
&\leq 2\Big\|G\Big(t, (\overline{x_i^n})_t, \int_0^t k(t,\sigma,(\overline{x_i^n})_\sigma)d\sigma\Big) \\
&\quad -G\Big(s, (\overline{x_i^n})_s, \int_0^s k(s,\sigma,(\overline{x_i^n})_\sigma)d\sigma\Big)\Big\| \\
&\quad +L\Big(\|v_t\|_{PC}, \|v_s\|_{PC}, \|u\| + \lambda\Big\|G\Big(s, (\overline{x_i^n})_s, \int_0^s k(s,\sigma,(\overline{x_i^n})_\sigma)d\sigma\Big)\Big\|\Big) \\
&\qquad \times(|t-s|(1+\|B_\lambda(s,v_s)u\|) + \|v_t - v_s\|_{PC} \\
&\leq 2\Big\|G\Big(t, (\overline{x_i^n})_t, \int_0^t k(t,\sigma,(\overline{x_i^n})_\sigma)d\sigma\Big) \\
&\quad -G\Big(s, (\overline{x_i^n})_s, \int_0^s k(s,\sigma,(\overline{x_i^n})_\sigma)d\sigma\Big)\Big\| \\
&\quad +L(\|\phi\|_{PC} + MT_1, \|\phi\|_{PC} + MT_1, \\
&\qquad \|u\| + \lambda(C_7C_1(1+C_1T_1) + C_2C_4T_1 + C_5)) \\
&\qquad \times|t-s|(1+\|B_\lambda(s,v_s)u)\| + M)),
\end{aligned}
$$

where $B_\lambda(t, v_t)$ is the Yosida approximant of $B(t, v_t)$ for every $t \in [0, T_1]$, $v \in S$ and $\lambda > 0$. Since L is continuous, as $\lambda \to 0+$ in the above inequality, we obtain

$$
\begin{aligned}
|B(t,v_t)u| \leq\ & |B(s,v_s)u| \\
& +|t-s|L(\|\phi\|_{PC} + MT_1, \|\phi\|_{PC} + MT_1, \|u\|) \\
& \quad \times(1 + M + |B(s,v_s)u|) \\
& +2\Big\|G\Big(t, (\overline{x_i^n})_t, \int_0^t k(t,\sigma,(\overline{x_i^n})_\sigma)d\sigma\Big) \\
& \quad -G\Big(s, (\overline{x_i^n})_s, \int_0^s k(s,\sigma,(\overline{x_i^n})_\sigma)d\sigma\Big)\Big\|. \qquad (4.2.31)
\end{aligned}
$$

By (4.2.4) and the definition of $B(t, v_t)$ we obtain

$$
\frac{x_i^n - x_{i-1}^n}{h_n} + B(t_i^n, v_{t_i^n})x_i^n \ni 0 \qquad (4.2.32)
$$

for $i = 1, 2, \cdots, n$. From (4.2.31)

$$
\begin{aligned}
|B(t_i^n, v_{t_i^n})x_{i-1}^n| \leq\ & |B(t_{i-1}^n, v_{t_{i-1}^n})x_{i-1}^n| \\
& +|t-s|L(\|\phi\|_{PC} + MT_1, \|\phi\|_{PC} + MT_1, \|x_{i-1}^n\|) \\
& \qquad \times(1 + M + |B(t_{i-1}^n, v_{t_{i-1}^n})x_{i-1}^n|) \\
& +2\Big\|G\Big(t_i^n, (\overline{x_i^n})_{t_i^n}, \int_0^{t_i^n} k(t_i^n,\sigma,(\overline{x_i^n})_\sigma)d\sigma\Big) \\
& \quad -G\Big(t_{i-1}^n, (\overline{x_i^n})_{t_{i-1}^n}, \int_0^{t_{i-1}^n} k(t_{i-1}^n,\sigma,(\overline{x_i^n})_\sigma)d\sigma\Big)\Big\|.
\end{aligned}
$$

Since from (4.2.32)

$$B(t_{i-1}^n, v_{t_{i-1}^n})x_{i-1}^n \ni B_{h_n}(t_{i-1}^n, v_{t_{i-1}^n})x_{i-2}^n$$

we obtain

$$|B(t_{i-1}^n, v_{t_{i-1}^n})x_{i-1}^n)| \leq \|B_{h_n}(t_{i-1}^n, v_{t_{i-1}^n})x_{i-2}^n\| \leq |B(t_{i-1}^n, v_{t_{i-1}^n})x_{i-2}^n|.$$

We put

$$a_i = |B(t_i^n, v_{t_i^n})x_{i-1}^n|,$$

$$\begin{aligned} b_i \ = \ & |t_i^n - t_{i-1}^n| L(\|\phi\|_{PC} + MT_1, \|\phi\|_{PC} + MT_1, \|x_{i-1}^n\|)(1+M) \\ & +2\Big\| G\Big(t_i^n, (\overline{x_i^n})_{t_i^n}, \int_0^{t_i^n} k(t_i^n, \sigma, (\overline{x_i^n})_\sigma) d\sigma\Big) \\ & \quad -G\Big(t_{i-1}^n, (\overline{x_i^n})_{t_{i-1}^n}, \int_0^{t_{i-1}^n} k(t_{i-1}^n, \sigma, (\overline{x_i^n})_\sigma) d\sigma\Big)\Big\| \end{aligned}$$

and

$$c_i = 1 + b_i.$$

Then it follows from (4.2.33) that

$$a_i \leq a_{i-1} + b_i(1 + a_{i-1}) = c_i a_{i-1} + b_i.$$

for $i = 1, 2, \cdots, n$. Hence we have

$$a_i \leq \Big(\prod_{l=2} c_l\Big) a_1 + \sum_{l=2} \Big(\prod_{k=l+1} c_k\Big) b_l$$

for $i = 1, 2, \cdots, n$. Since, for $k \geq 2$

$$\prod_{i=k}^{l} c_i = \prod_{i=k}^{l}(1 + b_i) \leq \exp\Big(\sum_{i=l}^{l} b_i\Big) \leq \exp\Big(\sum_{i=2}^{k} b_i\Big)$$

we obtain

$$a_i \leq \Big(a_1 + \sum_{l=2}^{i} b_l\Big) \exp\Big(\sum_{l-2}^{i} b_l\Big). \tag{4.2.33}$$

Since, as in the proof of Propositon 4.2.1,

$$\begin{aligned} & \Big\| G\Big(t_i^n, (\overline{x_i^n})_{t_i^n}, \int_0^{t_i^n} k(t_i^n, \sigma, (\overline{x_i^n})_\sigma) d\sigma\Big) \\ & \quad -G\Big(t_{i-1}^n, (\overline{x_i^n})_{t_{i-1}^n}, \int_0^{t_{i-1}^n} k(t_{i-1}^n, \sigma, (\overline{x_i^n})_\sigma) d\sigma\Big)\Big\| \\ & \leq C_2(C_{14}h_n + ch_n + l_1(C_7)T_1 h_n + C_7 C_1 h_n + C_4 h_n) \\ & \quad + l_2(C_7, C_1 C_7 T_1 + C_4 T_1) h_n, \end{aligned}$$

we obtain

$$\sum_{l=2}^{i} b_l \leq \sum_{l=2}^{i} C_{26} h_n \leq C_{26} T_1,$$

where

$$\begin{aligned} C_{26} &= C_{18} + 2(C_2 C_{14} + C_2 c + C_2 l(C_7) T_1 + C_1 C_2 C_7 \\ &\quad + C_4 + l_2(C_7, (C_1 C_7 + C_4) T_1)). \end{aligned}$$

Since $\phi(0) \in \widehat{D}$ and

$$|B(t, v_t)u| \leq |A(t, v_t)u| + \left\| G\Big(t, (\overline{x_i^n})_t, \int_0^t k(t, \sigma, (\overline{x_i^n})_\sigma) d\sigma\Big) \right\|$$

for every $t \in [0, T_1]$ and $u \in X$, we have $a_1 = |B(t_1^n, v_{t_1^n})\phi(0)| \leq C_8$. Thus (4.2.35) implies that $a_i = |B(t_i^n, v_{t_i^n})\phi(0)| \leq C_8$ is bounded independently of i and n. By

$$\begin{aligned} &|A(t_i^n, v_{t_i^n}) x_{i-1}^n| \\ &\quad \leq |B(t_i^n, v_{t_i^n}) x_{i-1}^n| + \left\| G\Big(t_i^n, (\overline{x_i^n})_{t_i^n}, \int_0^{t_i^n} k(t_i^n, \sigma, (\overline{x_i^n})_\sigma) d\sigma\Big) \right\| \end{aligned}$$

we finally obtain that $|A(t_i^n, v_{t_i^n}) x_{i-1}^n|$ is uniformly bounded for i and n.

THEOREM 4.2.3 *Assume that the assumptions of Theorem 4.2.1 are satisfied. Let $v \in S$. Let $x(t)$ be a local limit solution of* (4.2.2) *on* $[-r, T]$. *Then $x(t) \in \widehat{D}$ for every $t \in [0, T_1]$ with some $T_1 \in (0, T]$.*

Proof. We note that $x_{i-1}^n \in D(A(t_{i-1}^n, v_{t_{i-1}^n})) \subset \widehat{D}$. Fix $t \in [0, T_1]$. Using $(A2)_2$ and Proposition 4.2.7

$$\begin{aligned} &|A(t, v_t) x_{i-1}^n| \\ &\quad \leq |A(t_i^n, v_{t_i^n}) x_{i-1}^n| \\ &\qquad + L(\|v_t\|_{PC}, \|v_{t_i^n}\|_{PC}, \|x_{i-1}^n\|) \\ &\qquad\quad \times (|t - t_i^n|(1 + |A(t_i^n, v_{t_i^n}) x_{i-1}^n|) + \|v_t - v_{t_i^n}\|_{PC}) \\ &\quad \leq |A(t_i^n, v_{t_i^n}) x_{i-1}^n| \\ &\qquad + L(\|\phi\|_{PC} + M T_1, \|\phi\|_{PC} + M T_1, C_7) \\ &\qquad\quad \times (|t - t_i^n|(1 + |A(t_i^n, v_{t_i^n}) x_{i-1}^n|) + M) \end{aligned}$$

is bounded independently of i and n. Choose $i = i(n)$ for every n such that $t_i^n \to t$ as $n \to \infty$. Then $x_{i-1}^n \to x(t)$. From (4) of Proposition 1.4.1, $x(t) \in \widehat{D}$.

The next theorem states that a local limit solution $x(t) = x_x(t)$ of (4.2.2) on $[-r, T]$ with $v = x$ is a local integral solutuon of (4.2.1) on $[-r, T]$.

THEOREM 4.2.4 *Assume that the assumptions of Theorem 4.2.1 are satisfied. Let $x(t) = x_x(t)$ the local limit solution of* (4.2.2.) *on* $[-r, T]$ *with* $x = v$. *Then* $x(t)$ *is a local integral solution of* (4.2.1) *on* $[-r, T]$.

Proof. Let $x_n(t)$ be the discrete scheme approximate solution of (4.2.2) with $\lim_{n\to\infty} x_n(t) = x(t)$ unifomly on $[-r, T_1]$ for some $T_1 \in (0, T]$. Let $[x, y] \in A(\tau, x_\tau)$, $\tau \in [0, T_1]$. From (4.2.4) there exists $[x_i^n, y_i^n] \in A(t_i^n, x_{t_i^n})$ such that

$$(x_i^n - x) - (x_{i-1}^n - x) = h_n\Big(G(t_i^n, (\overline{x_i^n})_{t_i^n}, \int_0^{t_i^n} k(t_i^n, \sigma, (\overline{x_i^n})_\sigma)d\sigma) - y_i^n\Big).$$

Then

$$\begin{aligned}
&h_n[y - y_i^n, x_i^n - x]_- \\
&\quad + h_n\Big[G\Big(t_i^n, (\overline{x_i^n})_{t_i^n}, \int_0^{t_i^n} k(t_i^n, \sigma, (\overline{x_i^n})_\sigma)d\sigma\Big) - y, x_i^n - x\Big]_+ \\
&\geq h_n\Big[G\Big(t_i^n, (\overline{x_i^n})_{t_i^n}, \int_0^{t_i^n} k(t_i^n, \sigma, (\overline{x_i^n})_\sigma)d\sigma\Big) - y_i^n, x_i^n - x\Big]_- \\
&= [(x_i^n - x) - (x_{i-1}^n), x_i^n - x]_- \\
&\geq [x_i^n - x, x_i^n - x]_- - [x_{i-1}^n - x, x_i^n - x]_+ \\
&\geq \|x_i^n - x\| - \|x_{i-1}^n - x\|. \qquad (4.2.34)
\end{aligned}$$

We observe from (4.2.5) that

$$\begin{aligned}
\|x_i^n - x\| &\leq \|J_\lambda(t_i^n, x_{t_i^n})(x_i^n + \lambda y_i^n) - J_\lambda(t_i^n, x_{t_i^n})(x + \lambda y)\| \\
&\quad + \|J_\lambda(t_i^n, x_{t_i^n})(x + \lambda y) - J_\lambda(\tau, x_\tau)(x + \lambda y)\| \\
&\leq \|(x_i^n + \lambda y_i^n) - (x + \lambda y)\| \\
&\quad + \lambda\|A_\lambda(t_i^n, x_{t_i^n})(x + \lambda y) - A_\lambda(\tau, x_\tau)(x + \lambda y)\|
\end{aligned}$$

for every $\lambda > 0$. Thus, from $(A2)_2$,

$$\begin{aligned}
&\frac{\|x_i^n - x\| - \|(x_i^n - x) + \lambda(y_i^n - y)\|}{\lambda} \\
&\leq L(\|\phi\|_{PC} + MT_1, \|\phi\|_{PC} + MT_1, \|x + \lambda y\|) \\
&\qquad \times\Big(|t_i^n - \tau|(1 + \|A_\lambda(\tau, x_\tau)(x + \lambda y)\|) + \|x_{t_i^n} - x_\tau\|_{PC}\Big) \\
&\leq |t_i^n - \tau| L(\|\phi\|_{PC} + MT_1, \|\phi\|_{PC} + MT_1, \|x + \lambda y\|) \\
&\qquad \times(1 + \|y\| + M).
\end{aligned}$$

is an unknown function for $T > 0$ and $r > 0$, $\bar{r} > 0$, $G : [0,T] \times PC([-r,0];\overline{B_{\bar{r}}(0)}) \times X \to X$, $L_t : PC([-r,t];X) \to X$ for every $t \in [0,T]$ and $\phi \in PC([-r,0];\overline{B_{\bar{r}}(0)})$. Here, $PC([-r,0];\overline{B_{\bar{r}}(0)})$ is the space of all piecewise continuous functions from $[-r,0]$ into $\overline{B_{\bar{r}}(0)}$ with norm $\|\cdot\|_{PC}$ defined by

$$\|\psi\|_{PC} = \sup_{\tau\in[-r,0]} \|\psi(\tau)\|$$

for every $\psi \in PC([-r,0];\overline{B_{\bar{r}}(0)})$ and $\overline{B_{\bar{r}}(0)})$ is the ball of X with radius $\bar{r}$ and center 0, and $PC([-r,t];X)$ for every $t \in [0,T]$ is the space of all piecewise continuous function from $[-r,t]$ into X with norm $\|\cdot\|_t$ defined by

$$\|\psi\|_t = \sup_{\tau\in[-r,t]} \|\psi(\tau)\|$$

for every $\psi \in PC([-r,t];X)$.

We are now in a position to give definitions for solutions of (4.3.1).

DEFINITION 4.3.1 *1. A function $x(t) : [-r,T] \to X$ is called a strong solution of* (4.3.1) *on $[-r,T]$ if:*

(1) $x(t) = \phi(t)$ for every $t \in [-r,0]$;

(2) $x(t)$ is continuous on $[0,T]$;

(3) $x(t)$ is absolutely continuous on every compact interval of $[0,T]$;

(4) $x(t)$ is differentiable a.e. $t \in (0,T)$;

(5) $x(t)$ satisfies

$$\frac{dx}{dt}(t) + A(t,x_t)x(t) \ni G(t,x_t,L_t(x))$$

a.e. $t \in (0,T)$.

2. A function $x(t) : [-r,T] \to X$ is called an integral solution of (4.3.1) *on $[-r,T]$ if:*

(1) $x(t) = \phi(t)$ for every $t \in [-r,0]$;

(2) $x(t)$ is continuous on $[0,T]$;

(3) $x(t)$ satisfies the inequality

$$\|x(t) - u\| - \|x(s) - u\| \\ \leq \int_\tau^t ([G(\sigma,x_\sigma,L_t(x)) - v, x(\sigma) - u]_+ + K|\sigma - \tau|)d\sigma$$

for every $[u, v] \in A(\tau, x_\tau)$, $\tau \in [0, T]$, t, $s \in [0, T]$ *and a constant* $K > 0$.

3. A function $x(t) : [-r, T] \to X$ *is called a limit solution of* (4.3.1) *on* $[-r, T]$ *if:*

(1) $x(t) = \phi(t)$ *for every* $t \in [-r, 0]$;

(2) $x(t)$ *is continuous on* $[0, T]$;

(3) $x(t) = \lim_{n\to\infty} x_n(t)$ *uniformly on* $[0, T]$,

where $x_n(t) = \overline{x_n^n}(t)$ *for every* $t \in [-r, T]$ *with*

$$\overline{x_i^n}(t) = \begin{cases} \phi(t), & t \in [-r, 0] \\ x_j^n, & t \in (t_{j-1}^n, t_j^n],\ j = 1, 2, \cdots, i-1 \\ x_i^n, & t \in (t_{i-1}^n, T] \end{cases}$$

satisfying

$$\frac{x_i^n - x_{i-1}^n}{t_i^n - t_{i-1}^n} + A(t_i^n, (\overline{x_{i-1}^n})_{t_{i-1}^n})x_i^n \ni G(t_i^n, (\overline{x_i^n})_{t_i^n}, L_{t_i^n}(\overline{x_i^n}))$$

for $x_i^n \in D(A(t_i^n, (\overline{x_{i-1}^n})_{t_{i-1}^n}))$ *and a partition* $\{t_i^n\}$ *of* $[0, T]$ *such that*

$$0 = t_0^n < t_1^n < \cdots < t_n^n = T$$

and

$$\lim_{n\to\infty} \max_{1\le i\le n} (t_i^n - t_{i-1}^n) = 0,$$

for $i = 1, 2, \cdots, n$; $n = 1, 2, \cdots$.

To show the existence for limit solutions, integral solutions and strong solutions we impose the following conditions.

$(A1)_3$ $A(t, \psi) : X \supset D(A(t, \psi)) \to 2^X$ is m-accretive in X for every $t \in [0, T]$ and $\psi \in PC([-r, 0]; \overline{B_{\bar{r}}(0)})$;

$(A2)_3$ There exists an increasing continuous function $L : [0, \infty) \times [0, \infty) \times [0, \infty) \to [0, \infty)$ such that

$$\begin{aligned} &\|A_\lambda(t_1, \psi_1)u - A_\lambda(t_2, \psi_2)u\| \\ &\quad \le L(\|\psi_1\|_{PC}, \|\psi_2\|_{PC}, \|u\|)(|t_1 - t_2|(1 + \|A_\lambda(t_2, \psi_2)u\|) \\ &\qquad + \|\psi_1 - \psi_2\|_{PC}) \end{aligned}$$

for every $u \in X$, $t_i \in [0, T]$, $\psi_i \in PC([-r, 0]; \overline{B_{\bar{r}}(0)})$, $i = 1, 2$ and $\lambda > 0$, where $A_\lambda(t, \psi)$ is the Yosida approximant of $A(t, \psi)$ for every $t \in [0, T]$ and $\psi \in PC([-r, 0]; \overline{B_{\bar{r}}(0)})$;

$(L1)_3$ There exists a continuous function $a_1 : [0,T] \times [0,T] \to [0,\infty)$ such that

$$\|L_t(\phi_1) - L_t(\phi_2)\| \leq a_1(t)\|\psi_1 - \psi_2\|_t$$

for every $t \in [0,T]$ and $\psi_i \in PC([-r,T];X)$, $i = 1,2$;

$(L2)_3$ There exists a non-decreasing function $l_1; [0,\infty) \to [0,\infty)$ such that

$$\|L_{t_1}(\psi) - L_{t_2}(\psi)\| \leq |t_1 - t_2| l_1(\|\psi\|_T)$$

for every $t_i \in [0,T]$, $i = 1,2$ and $\psi \in PC([-r,T];X)$;

$(G1)_3$ There exists a continuous function $a_2 : [0,T] \to [0,\infty)$ such that

$$\|G(t,\psi_1,u_1) - G(t,\psi_2,u_2)\| \leq a_2(t)(\|\psi_1 - \psi_2\|_{PC} + \|u_1 - u_2\|)$$

for every $t \in [0,T]$, $\psi_i \in PC([-r,0];\overline{B_{\bar{r}}(0)})$ and $u_i \in X$, $i = 1,2$;

$(G2)_3$ There exists a non-decreasing function $l_2 : [0,\infty) \times [0,\infty) \to [0,\infty)$ such that

$$\|G(t_1,\psi,u) - G(t_2,\psi,u)\| \leq |t_1 - t_2| l_2(\|\psi\|_{PC}, \|u\|)$$

for every $t_i \in [0,T]$, $i = 1,2$, $\psi \in PC([-r,0];\overline{B_{\bar{r}}(0)})$ and $u \in X$.

In this section we do not make use of the intermediate step involving the analogue of (4.2.2) for solutions of (4.2.1) as in Section 4.2 and directly constitute solutions for (4.3.1).

$\widehat{D}(A)$ now denotes the generalized domain of $D(A(t,\psi))$, which is independent of $t \in [0,T]$ and $\psi \in PC([-r,T];X)$ from Lemma 4.2.1. Let $K_1 = \max_{t\in[0,T]} a_1(t)$ and $K_2 = \max_{t\in[0,T]} a_2(t)$.

We consider the existence of a method of lines for (4.3.1) with some properties.

PROPOSITION 4.3.1 *Assume that* $(A1)_3$, $(A2)_3$, $(L1)_3$, $(L2)_3$, $(G1)_3$ *and* $(G2)_3$ *are satisfied. Let* $\{t_i^n\}$ *be a partition of* $[0,T]$ *for* $i = 1,2,\cdots,n$, $n = 1,2,\cdots$ *with* $0 = t_0^n < t_1^n < \cdots < t_n^n = T$ *and* $h_n = \dfrac{T}{n}$. *Let* $\phi \in PC([-r,0];\overline{B_{\bar{r}}(0)})$ *be a Lipschitz continuous function with Lipschitz constant* $K = K_\phi > 0$ *with* $\phi(0) \in \widehat{D}$. *Assume furthermore that for every* $\psi \in PC([-r,0];\overline{B_{\bar{r}}(0)})$, $\eta \in PC([-r,T];\overline{B_{\bar{r}}(0)})$, $u \in D(A(t,\psi))$ *with* $\|u\| > \bar{r}$, $t \in [0,T]$ *and* $v \in A(t,\psi)u$ *there exists* $u^* \in F(u)$ *such that*

$$(v - G(t,\eta_t,L_t(\eta)),u^*) \geq 0. \tag{4.3.2}$$

Then there exists $\{x_i^n\} \in \overline{B_{\bar{r}}} \cap D(A(t_i^n, (\overline{x_{i-1}^n})_{t_{i-1}^n}))$ *such that*

$$\frac{x_i^n - x_{i-1}^n}{t_i^n - t_{i-1}^n} + A(t_i^n, (\overline{x_{i-1}^n})_{t_{i-1}^n})x_i^n \ni G(t_i^n, (\overline{x_i^n})_{t_i^n}, L_{t_i^n}(\overline{x_i^n}))$$

and thus

$$x_i^n = J_{h_n}(t_i^n, (\overline{x_{i-1}^n})_{t_{i-1}^n})(x_{i-1}^n + h_n G(t_i^n, (\overline{x_i^n})_{t_i^n}), L_{t_i^n}(\overline{x_i^n})))$$

with

$$\overline{x_i^n}(t) = \begin{cases} \phi(t), & t \in [-r, 0] \\ x_j^n, & t \in (t_{j-1}^n, t_j^n],\ j = 1, 2, \cdots, i-1 \\ x_i^n, & t \in (t_{i-1}^n, T] \end{cases}$$

for $i = 1, 2, \cdots, n$, *satisfying the following properties*

(1) $\{x_i^n\}$ *is uniformly bounded;*

(2) $\left\{\dfrac{x_i^n - x_{i-1}^n}{h_n}\right\}$ *is uniformly bounded.*

Proof. Assume that $x_0^n = \phi(0) \in \widehat{D}$. Then $x_0^n \in \overline{B_{\bar{r}}(0)} \cap \widehat{D}$. We put

$$\overline{x_0^n}(t) = \begin{cases} \phi(t), & t \in [-r, 0] \\ \phi(0), & t \in (0, T]. \end{cases}$$

We set

$$(g_1(u))(t) = \chi_{[-r,0]}(t)\overline{x_0^n}(t) + \chi_{(0,T]}(t)u$$

for every $t \in [-r, T]$ and $u \in X$ and

$$G_1(t) = G(t, (g_1(u))_t, L_t(g_1(u)))$$

for every $t \in [0, T]$ and $u \in \overline{B_{\bar{r}}(0)}$. Then

$$\begin{aligned} &\|G_1(t)u_1 - G_1(t)u_2\| \\ &\quad = \|G(t, (g_1(u_1))_t, L_t(g_1(u_1))) - G(t, (g_1(u_2))_t, L_t(g_1(u_2)))\| \\ &\quad \le K_2(\|(g_1(u_1))_t - (g_1(u_2))_t\|_{PC} + \|L_t(g_1(u_1)) - L_t(g_1(u_2))\| \\ &\quad \le K_2(\|u_1 - u_2\| + K_1\|u_1 - u_2\|) \\ &\quad = K_2(1 + K_1)\|u_1 - u_2\| \end{aligned}$$

for every $t \in [0,T]$ and $u_i \in \overline{B_{\bar{r}}(0)}$, $i = 1,2$. Thus the function $u \to G_1(t)u$ is Lipschitz continuous on $\overline{B_{\bar{r}}(0)}$ with Lipschitz constant $K_2(1+K_1)$. We define $Q_1 : \overline{B_{\bar{r}}(0)} \to X$ by

$$Q_1 u = \Big(A(t_1^n, (\overline{x_0^n})_{t_0^n}) + \frac{1}{h_n} I\Big)^{-1} \Big(G_1(t_0^n)u + \frac{1}{h_n} x_0^n\Big)$$

for every $u \in \overline{B_{\bar{r}}(0)}$. Then $G_1(t_1^n)u$ is Lipschitz continuous with Lipschitz constant $K_2(1+K_1)$ on $\overline{B_{\bar{r}}(0)}$. We also observe that the mapping $u \to \Big(A(t_1^n, (\overline{x_0^n})_{t_0^n}) + \frac{1}{h_n} I\Big)^{-1} u$ is Lipschitz continuous with Lipschitz constant h_n. Thus $Q_1 : \overline{B_{\bar{r}}(0)} \to X$ is Lipschitz continuous with Lipschitz constant $K_2(1+K_1)h_n$. We choose n so large that $K_2(1+K_1)h_n < 1$, and we show that $Q_1 : \overline{B_{\bar{r}}(0)} \to \overline{B_{\bar{r}}(0)}$. In fact, given $u \in \overline{B_{\bar{r}}(0)}$ let $v = Q_1 u$. Then for some $w \in A(t_1^n, (\overline{x_0^n})_{t_0^n})v$ we have

$$w + \frac{1}{h_n} v - G_1(t_1^n)u - \frac{1}{h_n} x_0^n = 0.$$

Assuming that $\|v\| > \bar{r}$ and picking an appropriate $v^* \in F(v)$ we apply (4.3.2) to obtain

$$\begin{aligned} 0 &= (w - G_1(t_1^n)u, v^*) + \frac{1}{h_n}(v - x_0^n, v^*) \\ &\geq \frac{1}{h_n}(\|v\|^2 - \|x_0^n\|\|v\|) \\ &\geq \frac{1}{h_n}(\|v\| - \bar{r})\|v\| > 0, \end{aligned}$$

which is a contradiction. Here we have used the fact $\|(g_1(u))_{t_1^n}\|_{PC} \leq \bar{r}$. From Banach's contraction principle Q_1 has a unique fixed point $x_1^n \in \overline{B_{\bar{r}}(0)}$ such that $Q_1 x_1^n = x_1^n$. Thus $x_1^n \in \overline{B_{\bar{r}}(0)} \cap D(A(t_1^n, (\overline{x_0^n})_{t_0^n})) \subset \overline{B_{\bar{r}}(0)} \cap \widehat{D}$ and it is a solution of the inclusion

$$\Big(A(t_1^n, (\overline{x_0^n})_{t_0^n}) - G_1(t_1^n) + \frac{1}{h_n} I\Big) u \ni \frac{1}{h_n} x_0^n.$$

We set

$$\overline{x_1^n}(t) = \begin{cases} \phi(t), & t \in [-r, 0] \\ x_1^n & t \in (0, T]. \end{cases}$$

Then $\overline{x_1^n}(t) \in \overline{B_{\bar{r}}(0)}$ for every $t \in [-r, T]$. Similarly we define for each $i = 2, 3, \cdots, n$ the functions g_i and G_i as follows

$$(g_i(u))(t) = \chi_{[-r, t_{i-1}^n]}(t)\overline{x_{i-1}^n}(t) + \chi_{(t_{i-1}^n, T]}(t)u$$

for every $t \in [-r, T]$ and $u \in X$ and

$$G_i(t) = G(t, (g_i(u))_t, L_t(g_i(u)))$$

for every $t \in [0, T]$ and $u \in \overline{B_{\bar{r}}(0)}$. The functions $G_i(t)u$ then are Lipschitz continuous with respect to u on $\overline{B_{\bar{r}}(0)}$ with Lipschitz constant $K_2(1 + K_1)$. We define $Q_i : \overline{B_{\bar{r}}(0)} \to X$ by

$$Q_i u = \left(A(t_1^n, (\overline{x_{i-1}^n})_{t_{i-1}^n}) + \frac{1}{h_n} I\right)^{-1} \left(G_1(t_{i-1}^n)u + \frac{1}{h_n} x_{i-1}^n\right)$$

for every $u \in \overline{B_{\bar{r}}(0)}$. Then $Q_i : \overline{B_{\bar{r}}(0)} \to X$ is Lipschitz continuous with Lipschitz constant $K_2(1+K_1)h_n$. We also choose n so large that $K_2(1+K_1)h_n < 1$, and thus $Q_i : \overline{B_{\bar{r}}(0)} \to \overline{B_{\bar{r}}(0)}$. From Banach's contraction principle Q_i has a unique fixed point $x_i^n \in \overline{B_{\bar{r}}(0)} \cap D(A(t_i^n, (\overline{x_{i-1}^n})_{t_{i-1}^n}) \subset \overline{B_{\bar{r}}(0)} \cap \widehat{D}$ such that $Q_i x_i^n = x_i^n$ and it is a solution of the inclusion

$$\left(A(t_i^n, (\overline{x_{i-1}^n})_{t_{i-1}^n}) - G_i(t_i^n) + \frac{1}{h_n} I\right) u \ni \frac{1}{h_n} x_{i-1}^n.$$

We set

$$\overline{x_i^n}(t) = \begin{cases} \phi(t), & t \in [-r, 0] \\ x_j^n & t \in (t_{j-1}^n, t_j^n], \ j = 1, 2, \cdots, i-1 \\ x_i^n, & t \in (t_{i-1}^n, T] \end{cases}$$

for $i = 2, 3, \cdots, n$. Then $\overline{x_i^n}(t) \in \overline{B_{\bar{r}}(0)}$ for every $t \in [-r, T]$. Moreover,

$$\frac{x_i^n - x_{i-1}^n}{t_i^n - t_{i-1}^n} + A(t_i^n, (\overline{x_{i-1}^n})_{t_{i-1}^n}) x_i^n \ni G(t_i^n, (\overline{x_i^n})_{t_i^n}, L_{t_i^n}(\overline{x_i^n}))$$

and thus

$$x_i^n = J_{h_n}(t_i^n, (\overline{x_{i-1}^n})_{t_{i-1}^n})(x_{i-1}^n + h_n G(t_i^n, (\overline{x_i^n})_{t_i^n}, L_{t_i^n}(\overline{x_i^n}))) \qquad (4.3.3)$$

for $i = 1, 2, \cdots, n$.

(1) Since $x_i^n \in \overline{B_{\bar{r}}(0)} \cap \widehat{D}$ $\{x_i^n\}$ is uniformly bounded.

(2) From (4.3.3) for $i = 1$ we have

$$\begin{aligned} &\|x_1^n - x_0^n\| \\ &= \|J_{h_n}(t_1^n, (\overline{x_0^n})_{t_0^n})(x_0^n + h_n G(t_1^n, (\overline{x_1^n})_{t_1^n}, L_{t_1^n}(\overline{x_1^n}))) - x_0^n\| \\ &\le \|J_{h_n}(t_1^n, (\overline{x_0^n})_{t_0^n})(x_0^n + h_n G(t_1^n, (\overline{x_1^n})_{t_1^n}, L_{t_1^n}(\overline{x_1^n}))) \\ &\quad - J_{h_n}(t_1^n, (\overline{x_0^n})_{t_0^n}) x_0^n\| + \|J_{h_n}(t_1^n, (\overline{x_0^n})_{t_0^n}) x_0^n - x_0^n\| \\ &\le h_n \|G(t_1^n, (\overline{x_1^n})_{t_1^n}, L_{t_1^n}(\overline{x_1^n}))\| + h_n |A(t_1^n, (\overline{x_0^n})_{t_0^n}) \phi(0)|. \end{aligned}$$

Since $\phi \in PC([-r,0];\overline{B_{\bar{r}}(0)})$ from $(A2)_3$ we see that

$$\begin{aligned}|A(t,\psi)\phi(0)| &\le |A(0,\bar{0})\phi(0)| \\ &\quad +L(\|\psi\|_{PC},\|\bar{0}\|_{PC},\|\phi(0)\|) \\ &\qquad \times((1+|A(0,\bar{0})\phi(0)|)t+\|\phi\|_{PC}) \\ &\le |A(0,\bar{0})\phi(0)|+L(\bar{r},0,\bar{r})((1+|A(0,\bar{0})\phi(0)|)T+\bar{r}) \\ &= K_3\end{aligned}$$

for every $t \in [0,T]$ and $\psi \in PC([-r,0];\overline{B_{\bar{r}}(0)})$, where $\bar{0}$ denotes the zero function in $PC([-r,0];X)$. Thus

$$\sup_{t\in[0,T]} |A(t,\psi)\phi(0)| \le K_3$$

for every ϕ, $\psi \in PC([-r,0];\overline{B_{\bar{r}}(0)})$. Thus from $(G1)_3$

$$\begin{aligned}\|x_1^n - x_0^n\| &\le h_n\|G(t_1^n,(\overline{x_1^n})_{t_1^n},L_{t_1^n}(\overline{x_1^n}))-G(t_1^n,\bar{0},0)\| \\ &\quad +h_n\|G(t_1^n,\bar{0},0)\|+K_3h_n \\ &\le K_2h_n(\|(\overline{x_1^n})_{t_1^n}\|+\|L_{t_1^n}(\overline{x_1^n})\|)+K_4h_n+K_3h_n,\end{aligned}$$

where $K_4 = \max_{t\in[0,T]}\|G(t,\bar{0},0)\|$. Since, from $(L1)_3$ and $(L2)_3$

$$\begin{aligned}\|L_t\psi\| &\le \|L_t(\psi)-L_t(0)\|+\|L_t(0)\| \\ &\le K_1\|\psi\|_T+\|L_t(0)-L_0(0)\|+\|L_0(0)\| \\ &\le K_1\bar{r}+l_1(0)t+\|L_0(0)\| \\ &\le K_1\bar{r}+l_1(0)T+\|L_0(0)\| = K_5,\end{aligned}$$

$$\|x_1^n - x_0^n\| \le K_2h_n(\bar{r}+K_5)+(K_3+K_4)h_n = K_6h_n,$$

where $K_6 = K_2(\bar{r}+K_5)+K_3+K_4$. From (4.3.3) for $i = 2,3,\cdots,n$ we obtain

$$\begin{aligned}&\|x_i^n - x_{i-1}^n\| \\ &= \|J_{h_n}(t_i^n,(\overline{x_{i-1}^n})_{t_{i-1}^n})(x_{i-1}^n+h_nG(t_i^n,(\overline{x_i^n})_{t_i^n},L_{t_i^n}(\overline{x_i^n}))) \\ &\quad -J_{h_n}(t_{i-1}^n,(\overline{x_{i-2}^n})_{t_{i-2}^n})(x_{i-2}^n+h_nG(t_{i-1}^n,(\overline{x_{i-1}^n})_{t_{i-1}^n},L_{t_{i-1}^n}(\overline{x_{i-1}^n})))\| \\ &\le \|J_{h_n}(t_i^n,(\overline{x_{i-1}^n})_{t_{i-1}^n})(x_{i-1}^n+h_nG(t_i^n,(\overline{x_i^n})_{t_i^n},L_{t_i^n}(\overline{x_i^n}))) \\ &\quad -J_{h_n}(t_i^n,(\overline{x_{i-1}^n})_{t_{i-1}^n})(x_{i-2}^n+h_nG(t_{i-1}^n,(\overline{x_{i-1}^n})_{t_{i-1}^n},L_{t_{i-1}^n}(\overline{x_{i-1}^n})))\| \\ &\quad +\|J_{h_n}(t_i^n,(\overline{x_{i-1}^n})_{t_{i-1}^n})(x_{i-2}^n+h_nG(t_{i-1}^n,(\overline{x_{i-1}^n})_{t_{i-1}^n},L_{t_{i-1}^n}(\overline{x_{i-1}^n})))\| \\ &\quad -J_{h_n}(t_{i-1}^n,(\overline{x_{i-2}^n})_{t_{i-2}^n})(x_{i-2}^n+h_nG(t_{i-1}^n,(\overline{x_{i-1}^n})_{t_{i-1}^n},L_{t_{i-1}^n}(\overline{x_{i-1}^n})))\| \le\end{aligned}$$

$$
\begin{aligned}
&\le \|x_{i-1}^n - x_{i-2}^n\| \\
&\quad + h_n\|G(t_i^n, (\overline{x_i^n})_{t_i^n}, L_{t_i^n}(\overline{x_i^n})) - G(t_{i-1}^n, (\overline{x_{i-1}^n})_{t_{i-1}^n}, L_{t_{i-1}^n}(\overline{x_{i-1}^n}))\| \\
&\quad + h_n\|A_{h_n}(t_i^n, (\overline{x_{i-1}^n})_{t_{i-1}^n})(x_{i-2}^n + h_n G(t_{i-1}^n, (\overline{x_{i-1}^n})_{t_{i-1}^n}, L_{t_{i-1}^n}(\overline{x_{i-1}^n})))\| \\
&\qquad - A_{h_n}(t_{i-1}^n, (\overline{x_{i-2}^n})_{t_{i-2}^n})(x_{i-2}^n + h_n G(t_{i-1}^n, (\overline{x_{i-1}^n})_{t_{i-1}^n}, L_{t_{i-1}^n}(\overline{x_{i-1}^n})))\| \\
&\le \|x_{i-1}^n - x_{i-2}^n\| \\
&\quad + h_n\|G(t_i^n, (\overline{x_i^n})_{t_i^n}, L_{t_i^n}(\overline{x_i^n})) - G(t_i^n, (\overline{x_{i-1}^n})_{t_{i-1}^n}, L_{t_{i-1}^n}(\overline{x_{i-1}^n}))\| \\
&\quad + h_n\|G(t_i^n, (\overline{x_{i-1}^n})_{t_{i-1}^n}, L_{t_{i-1}^n}(\overline{x_{i-1}^n})) \\
&\qquad - G(t_{i-1}^n, (\overline{x_{i-1}^n})_{t_{i-1}^n}, L_{t_{i-1}^n}(\overline{x_{i-1}^n}))\| \\
&\quad + h_n L(\|(\overline{x_{i-1}^n})_{t_{i-1}^n}\|_{PC}, \|(\overline{x_{i-2}^n})_{t_{i-2}^n}\|_{PC}, \\
&\qquad\qquad \|x_{i-2}^n + h_n G(t_{i-1}^n, (\overline{x_{i-1}^n})_{t_{i-1}^n}, L_{t_{i-1}^n}(\overline{x_{i-1}^n}))\|) \\
&\qquad\qquad \times |t_i^n - t_{i-1}^n|(1 + A_{h_n}(t_{i-1}^n, (\overline{x_{i-2}^n})_{t_{i-2}^n}) \\
&\qquad\qquad\qquad \times (x_{i-2}^n + h_n G(t_{i-1}^n, (\overline{x_{i-1}^n})_{t_{i-1}^n}, L_{t_{i-1}^n}(\overline{x_{i-1}^n})))\|) \\
&\quad + \|(\overline{x_{i-1}^n})_{t_{i-1}^n} - (\overline{x_{i-2}^n})_{t_{i-2}^n}\|_{PC} \\
&\le \|x_{i-1}^n - x_{i-2}^n\| \\
&\quad + K_2 h_n(\|(\overline{x_i^n})_{t_i^n} - (\overline{x_{i-1}^n})_{t_{i-1}^n}\|_{PC} + \|L_{t_i^n}(\overline{x_i^n}) - L_{t_{i-1}^n}(\overline{x_{i-1}^n})\|) \\
&\quad + h_n l_2(\|(\overline{x_{i-1}^n})_{t_{i-1}^n}\|_{PC}, L_{t_{i-1}^n}(\overline{x_{i-1}^n})\|)|t_i^n - t_{i-1}^n| \\
&\quad + h_n L(\bar{r}, \bar{r}, \|x_{i-2}^n + h_n G(t_{i-1}^n, (\overline{x_{i-1}^n})_{t_{i-1}^n}, L_{t_{i-1}^n}(\overline{x_{i-1}^n})\|) \\
&\qquad \times h_n(1 + \|A_{h_n}(t_{i-1}^n, (\overline{x_{i-2}^n})_{t_{i-2}^n}) \\
&\qquad\qquad \times (x_{i-2}^n + h_n G(t_{i-1}^n, (\overline{x_{i-1}^n})_{t_{i-1}^n}, L_{t_{i-1}^n}(\overline{x_{i-1}^n}))\|) \\
&\quad + \|(\overline{x_{i-1}^n})_{t_{i-1}^n} - (\overline{x_{i-2}^n})_{t_{i-2}^n}\|_{PC}).
\end{aligned}
$$

Since

$$
\begin{aligned}
&\|L_{t_i^n}(\overline{x_i^n}) - L_{t_{i-1}^n}(\overline{x_{i-1}^n})\| \\
&\qquad \le \|L_{t_i^n}(\overline{x_i^n}) - L_{t_i^n}(\overline{x_{i-1}^n})\| + \|L_{t_i^n}(\overline{x_{i-1}^n}) - L_{t_{i-1}^n}(\overline{x_{i-1}^n})\| \\
&\qquad \le K_1\|\overline{x_i^n} - \overline{x_{i-1}^n}\|_{t_i^n} + l_1(\|\overline{x_{i-1}^n}\|_T)|t_i^n - t_{i-1}^n| \\
&\qquad \le K_1\|\overline{x_i^n} - \overline{x_{i-1}^n}\|_{t_i^n} + l_1(\bar{r})h_n,
\end{aligned}
$$

$$
\begin{aligned}
&\|x_{i-2}^n + h_n G(t_{i-1}^n, (\overline{x_{i-1}^n})_{t_{i-1}^n}, L_{t_{i-1}^n}(\overline{x_{i-1}^n}))\| \\
&\qquad \le \bar{r} + h_n\|G(t_{i-1}^n, (\overline{x_{i-1}^n})_{t_{i-1}^n}, L_{t_{i-1}^n}(\overline{x_{i-1}^n})) - G(t_{i-1}^n, \bar{0}, 0)\| \\
&\qquad\quad + \|G(t_{i-1}^n, \bar{0}, 0)\| \\
&\qquad \le \bar{r} + K_2 h_n(\|\overline{x_{i-1}^n})_{t_{i-1}^n}\|_{PC} + \|L_{t_{i-1}^n}(\overline{x_{i-1}^n})\|) + K_4 h_n \\
&\qquad \le \bar{r} + K_2 h_n(\bar{r} + K_5) + K_4 h_n \le
\end{aligned}
$$

$$\leq \bar{r} + K_2\bar{r}T + K_2K_5T + K_4T = K_7,$$

and

$$\begin{aligned}
&\|A_{h_n}(t_{i-1}^n, (\overline{x_{i-2}^n})_{t_{i-2}^n})(x_{i-2}^n + h_nG(t_{i-1}^n, (\overline{x_{i-1}^n})_{t_{i-1}^n}, L_{t_{i-1}^n}(\overline{x_{i-1}^n})))\| \\
&= \frac{1}{h_n}\|x_{i-2}^n + h_nG(t_{i-1}^n, (\overline{x_{i-1}^n})_{t_{i-1}^n}, L_{t_{i-1}^n}(\overline{x_{i-1}^n})) \\
&\qquad - J_{h_n}(t_{i-1}^n, (\overline{x_{i-2}^n})_{t_{i-2}^n}) \\
&\qquad\qquad \times (x_{i-2}^n + h_nG(t_{i-1}^n, (\overline{x_{i-1}^n})_{t_{i-1}^n}, L_{t_{i-1}^n}(\overline{x_{i-1}^n})))\| \\
&= \frac{1}{h_n}\|x_{i-2}^n + h_nG(t_{i-1}^n, (\overline{x_{i-1}^n})_{t_{i-1}^n}, L_{t_{i-1}^n}(\overline{x_{i-1}^n})) - x_{i-1}^n\| \\
&\leq \frac{1}{h_n}\|x_{i-1}^n - x_{i-2}^n\| + \|G(t_{i-1}^n, (\overline{x_{i-1}^n})_{t_{i-1}^n}, L_{t_{i-1}^n}(\overline{x_{i-1}^n}))\| \\
&\leq \frac{1}{h_n}\|x_{i-1}^n - x_{i-2}^n\| + K_2(\bar{r} + K_5) + K_4
\end{aligned}$$

we obtain

$$\begin{aligned}
&\|x_i^n - x_{i-1}^n\| \leq \\
&\quad \leq \|x_{i-1}^n - x_{i-2}^n\| + K_2h_n(\|(\overline{x_i^n})_{t_i^n} - (\overline{x_{i-1}^n})_{t_{i-1}^n}\|_{PC} \\
&\qquad + K_1\|\overline{x_i^n} - \overline{x_{i-1}^n}\|_{t_{i-1}^n} + l_1(\bar{r})h_n) + h_n^2 l_2(\bar{r}, K_5) \\
&\qquad + h_nL(\bar{r}, \bar{r}, K_7)(h_n + \|x_{i-1}^n - x_{i-2}^n\| \\
&\qquad\qquad + (K_2\bar{r} + K_2K_4 + K_4)h_n \\
&\qquad\qquad + \|(\overline{x_{i-1}^n})_{t_{i-1}^n} - (\overline{x_{i-2}^n})_{t_{i-2}^n}\|_{PC}).
\end{aligned}$$

We observe that

$$\begin{aligned}
\|(\overline{x_i^n})_{t_i^n} - (\overline{x_{i-1}^n})_{t_{i-1}^n}\|_{PC} &\leq \max_{1\leq j\leq i}\|x_j^n - x_{j-1}^n\| + Kh_n, \\
\|(\overline{x_{i-1}^n})_{t_{i-1}^n} - (\overline{x_{i-2}^n})_{t_{i-2}^n}\|_{PC} &\leq \max_{1\leq j\leq i-1}\|x_j^n - x_{j-1}^n\| + Kh_n
\end{aligned}$$

and

$$\|\overline{x_i^n} - \overline{x_{i-1}^n}\|_{t_i^n} = \|x_i^n - x_{i-1}^n\|.$$

Thus

$$\begin{aligned}
&\frac{\|x_i^n - x_{i-1}^n\|}{h_n} \\
&\quad \leq \frac{\|x_{i-1}^n - x_{i-2}^n\|}{h_n} \\
&\qquad + K_2h_n\Big(\max_{1\leq j\leq i}\frac{\|x_j^n - x_{j-1}^n\|}{h_n} + K + K_1\frac{\|x_i^n - x_{i-1}^n\|}{h_n} + l_1(\bar{r})\Big) +
\end{aligned}$$

$$+h_n l_2(\bar{r}, K_5)$$
$$+h_n L(\bar{r}, \bar{r}, K_7)\Big(1 + \frac{\|x_{i-1}^n - x_{i-2}^n\|}{h_n} + K_2\bar{r}$$
$$+K_2K_5 + K_4 + K + \max_{1\le j\le i-1} \frac{\|x_j^n - x_{j-1}^n\|}{h_n}\Big).$$

Letting

$$K_8 = K_2(K + l_1(\bar{r})) + l_2(\bar{r}, K_5)$$
$$+L(\bar{r}, \bar{r}, K_7)(1 + K_2\bar{r} + K_2K_5 + K_4 + K)$$

and

$$p_n = 1 - h_n(K_2 + K_1K_2 + 2L(\bar{r}, \bar{r}, K_7)).$$

We assume that n is large enough with $p_n \in (0, 1)$. Then we have

$$p_n \max_{1\le j\le i} \frac{\|x_j^n - x_{j-1}^n\|}{h_n} \le \max_{1\le j\le i-1} \frac{\|x_j^n - x_{j-1}^n\|}{h_n} + K_8 h_n.$$

This implies

$$p_n \max_{1\le j\le i} \frac{\|x_j^n - x_{j-1}^n\|}{h_n}$$
$$\le L_8 h_n + \frac{K_8 h_n}{p_n} + \frac{1}{p_n} \max_{1\le j\le i-2} \frac{\|x_j^n - x_{j-1}^n\|}{h_n}$$
$$\le \cdots$$
$$\le K_8 h_n \sum_{k=0}^{i-2} \frac{1}{p_n^k} + \frac{1}{p_n^{i-2}} \frac{\|x_1^n - x_0^n\|}{h_n}$$
$$\le K_8 h_n \sum_{k=0}^{i-1} \frac{1}{p_n^k} + \frac{1}{p_n^{i-1}} \frac{\|x_1^n - x_0^n\|}{h_n},$$

which yields, for $i = 2, 3, \cdots, n$ we obtain

$$\max_{1\le j\le i} \frac{\|x_j^n - x_{j-1}^n\|}{h_n}$$
$$\le K_8 h_n \sum_{k=1}^{i} \frac{1}{p_n^k} + \frac{1}{p_n^i} K_6 \le K_8 h_n \sum_{k=1}^{n} \frac{1}{p_n^k} + \frac{1}{p_n^n} K_6$$
$$\le K_8 h_n \sum_{k=1}^{n} \frac{1}{p_n^n} + \frac{K_6}{p_n^n} = \frac{K_8 T}{p_n^n} + \frac{K_6}{p_n^n} = \frac{K_8 T + K_6}{p_n^n}.$$

Since

$$\lim_{n\to\infty} p_n^n = \lim_{n\to\infty}\left(1 - \frac{K_2 + K_1K_2 + 2L(\bar{r},\bar{r},K_7)}{n}T\right)^n = \exp(-(K_2 + K_1K_2 + 2L(\bar{r},\bar{r},K_7)))T,$$

there exists a constant $K_9 > 0$ such that

$$\frac{\|x_i^n - x_{i-1}^n\|}{h_n} \le K_9$$

for $i = 0, 1, 2, \cdots, n$ and $n = 1, 2, \cdots$. Hence $\left\{\frac{x_i^n - x_{i-1}^n}{h_n}\right\}$ is uniformly bounded.

LEMMA 4.3.1 *Assume that the assumptions of Proposition 4.3.1 are satisfied. Let* $x_n(t) = \overline{x_n^n}(t)$ *for every* $t \in [-r, T]$ *and* $n = 1, 2, \cdots, n$. *Then there exists a constant* $K_{11} > 0$ *such that*

$$\|x_n(t_1) - x_n(t_2)\| \le K_{11}(|t_1 - t_2| + h_n)$$

for every $t_l \in [-r, T]$, $l = 1, 2$, *and* $n = 1, 2, \cdots, n$.

Proof. We define Rothe's functions $x_{(n)}(t) : [-r, T] \to X$ by

$$x_{(n)}(t) = \begin{cases} \phi(t), & t \in [-r, 0] \\ x_{i-1}^n + (t - t_{i-1}^n)\dfrac{x_i^n - x_{i-1}^n}{h_n}, & t \in [t_{i-1}^n, t_i^n] \end{cases}$$

for $i = 1, 2, \cdots, n$, $n = 1, 2, \cdots$, and then $x_{(n)}(t)$ is Lipschitz continuous on $[-r, T]$ with Lipschitz constant $K_{10} = \max\{K_9, K\}$.

For every $t \in (0, T]$, $t \in (t_{i-1}^n, t_i^n]$ for some $i = 1, 2, \cdots, n$ and we obtain

$$\begin{aligned} \|x_n(t) - x_{(n)}(t)\| &= \left\|x_i^n - x_{i-1}^n - (t - t_{i-1}^n)\frac{x_i^n - x_{i-1}^n}{h_n}\right\| \\ &= (t_i^n - t)\left\|\frac{x_i^n - x_{i-1}^n}{h_n}\right\| \le K_9 h_n \le K_{10} h_n. \end{aligned}$$

Since for every $t \in [-r, 0]$, $x_n(t) = x_{(n)}(t) = \phi(t)$ we have

$$\|x_n(t) - x_{(n)}(t)\| \le K_{10} h_n$$

for every $t \in [-r, T]$. From the Lipschitz continuity of $x_{(n)}(t)$

$$\begin{aligned} &\|x_n(t_1) - x_n(t_2)\| \\ &\quad \le \|x_n(t_1) - x_{(n)}(t_1)\| + \|x_{(n)}(t_1) - x_{(n)}(t_2)\| + \|x_{(n)}(t_2) - x_n(t_2)\| \\ &\quad \le K_{10}h_n + K_{10}|t_1 - t_2| + K_{10}h_n \\ &\quad \le 2K_{10}(|t_1 - t_2| + h_n) \\ &\quad = K_{11}(|t_1 - t_2| + h_n) \end{aligned}$$

for every $t_l \in [-r, T]$, $l = 1, 2$, where $K_{11} = 2K_{10}$.

PROPOSITION 4.3.2 *Assume that the assumptions of Proposition 4.3.1 are satisfied. Let ϕ, $\hat{\phi} \in PC([-r, 0]; \overline{B_{\bar{r}}(0)})$ be Lipschitz continuous with Lipschitz constants $K = K_\phi > 0$, $\widehat{K} = \widehat{K}_{\hat{\phi}} > 0$, respectively, and $\phi(0)$, $\hat{\phi}(0) \in \widehat{D}$. Let x_i^n for $i = 1, 2, \cdots, n$, and y_j^m for $j = 1, 2, \cdots, m$, be as in Proposition 4.3.1 satisfying*

$$\frac{x_i^n - x_{i-1}^n}{h_n} + A(t_i^n, (\overline{x_i^n})_{t_{i-1}^n}) x_i^n \ni G(t_i^n, (\overline{x_i^n})_{t_i^n}, L_{t_i^n}(\overline{x_i^n})), \qquad (4.3.4)$$

where $h_n = t_i^n - t_{i-1}^n = \dfrac{T}{n}$ and $x_0^n = \phi(0)$, and

$$\frac{y_j^m - y_{j-1}^m}{k_m} + A(s_j^m, (\overline{y_j^m})_{s_{j-1}^m}) y_j^m \ni G(s_j^m, (\overline{y_j^m})_{s_j^m}, L_{s_j^m}(\overline{y_j^m})), \quad (4.3.5)$$

where $k_m = s_j^m - s_{j-1}^m = \dfrac{T}{m}$ and $y_0^m = \hat{\phi}(0)$. Then

$$\begin{aligned}
&\|x_i^n - y_j^m\| \\
&\quad \le \frac{h_n}{h_n + k_m}\|x_i^n - y_{j-1}^m\| + \frac{k_m}{h_n + k_m}\|x_{i-1}^n - y_j^m\| \\
&\quad\quad + \frac{h_n k_m}{h_n + k_m}(K_{12}|t_i^n - s_j^m| + L(\bar{r}, \bar{r}, \bar{r})\|(\overline{x_i^n})_{t_{i-1}^n} - (\overline{y_j^m})_{s_{j-1}^m}\|_{PC} \\
&\quad\quad\quad + [G(t_i^n, (\overline{x_i^n})_{t_{i-1}^n}, L_{t_i^n}(\overline{x_i^n})) \\
&\quad\quad\quad - G(s_j^m, (\overline{y_j^m})_{s_{j-1}^m}, L_{s_j^m}(\overline{y_j^m})), x_i^n - y_j^m]_+),
\end{aligned}$$

where $K_{12} = L(\bar{r}, \bar{r}, \bar{r})(1 + K_2(\bar{r} + K_5)K_4 + K_9)$.

Proof. We choose $\lambda \in (0, 1)$ and put $h_{n,m} = \dfrac{h_n k_m}{h_n + k_m}$. Then (4.3.4) and (4.3.5) imply

$$\begin{aligned}
x_i^n &= J_{\lambda h_{n,m}}(t_i^n, (\overline{x_i^n})_{t_{i-1}^n}) \\
&\quad \times \Big(x_i^n + \lambda h_{n,m}\Big(G(t_i^n, (\overline{x_i^n})_{t_i^n}, L_{t_i^n}(\overline{x_i^n})) + \frac{x_{i-1}^n - x_i^n}{h_n}\Big)\Big)
\end{aligned}$$

and

$$\begin{aligned}
y_j^m &= J_{\lambda h_{n,m}}(s_j^m, (\overline{y_j^m})_{s_{j-1}^m}) \\
&\quad \times \Big(y_j^m + \lambda h_{n,m}\Big(G(s_j^m, (\overline{y_j^m})_{s_j^m}, L_{s_j^m}(\overline{y_j^m})) + \frac{y_{j-1}^m - y_j^m}{k_m}\Big)\Big).
\end{aligned}$$

We put

$$a_{n,i} = x_i^n + \lambda h_{n,m}\Big(G(t_i^n, (\overline{x_i^n})_{t_i^n}, L_{t_i^n}(\overline{x_i^n})) + \frac{x_{i-1}^n - x_i^n}{h_n}\Big)$$

and

$$b_{m,j} = y_j^m + \lambda h_{n,m}\Big(G(s_j^m, (\overline{y_j^m})_{s_j^m}, L_{s_j^m}(\overline{y_j^m})) + \frac{y_{j-1}^m - y_j^m}{k_m}\Big).$$

Then from $(A2)_3$

$$\begin{aligned}
&\|x_i^n - y_j^m\| \\
&\quad = \|J_{\lambda h_{n,m}}(t_i^n, (\overline{x_i^n})_{t_{i-1}^n})a_i^n - J_{\lambda h_{n,m}}(s_j^m, (\overline{y_j^m})_{s_{j-1}^m})b_j^m\| \\
&\quad \leq \|J_{\lambda h_{n,m}}(t_i^n, (\overline{x_i^n})_{t_{i-1}^n})a_i^n - J_{\lambda h_{n,m}}(s_j^m, (\overline{y_j^m})_{s_{j-1}^m})a_i^n\| \\
&\qquad + \|J_{\lambda h_{n,m}}(s_j^m, (\overline{y_j^m})_{s_{j-1}^m})a_i^n - J_{\lambda h_{n,m}}(s_j^m, (\overline{y_j^m})_{s_{j-1}^m})b_j^m\| \\
&\quad \leq \|a_i^n - b_j^m\| \\
&\qquad + \lambda h_{n,m}\|A_{\lambda h_{n,m}}(t_i^n, (\overline{x_i^n})_{t_{i-1}^n})a_i^n - A_{\lambda h_{n,m}}(s_j^m, (\overline{y_j^m})_{s_{j-1}^m})a_i^n\| \\
&\quad \leq \|a_i^n - b_j^m\| \\
&\qquad + \lambda h_{n,m}L(\|(\overline{x_i^n})_{t_{i-1}^n}\|_{PC}, \|(\overline{y_j^m})_{s_{j-1}^m}\|_{PC}, \|a_i^n\|) \\
&\qquad\qquad \times(|t_i^n - s_j^m|(1 + \|A_{\lambda h_{n,m}}(t_i^n, (\overline{x_i^n})_{t_{i-1}^n})a_i^n\|) \\
&\qquad + \|(\overline{x_i^n})_{t_{i-1}^n} - (\overline{y_j^m})_{s_{j-1}^m}\|_{PC}).
\end{aligned}$$

Since

$$\begin{aligned}
&\|a_i^n - b_j^m\| \\
&\quad = \Big\|(x_i^n - y_j^m) \\
&\qquad + \lambda h_{n,m}\Big(G(t_i^n, (\overline{x_i^n})_{t_i^n}, L_{t_i^n}(\overline{x_i^n})) - G(s_j^m, (\overline{y_j^m})_{s_j^m}, L_{s_j^m}(\overline{y_j^m})) \\
&\qquad\qquad + \frac{x_{i-1}^n - x_i^n}{h_n} - \frac{y_{j-1}^m - y_j^m}{k_m}\Big)\Big\|
\end{aligned}$$

letting

$$G_i^n = G(t_i^n, (\overline{x_i^n})_{t_i^n}, L_{t_i^n}(\overline{x_i^n}))$$

and

$$\overline{G_j^m} = G(s_j^m, (\overline{y_j^m})_{s_j^m}, L_{s_j^m}(\overline{y_j^m}))$$

we obtain

$$\|a_i^n - b_j^m\| =$$

$$= \Big\| x_i^n - y_j^m + \lambda h_{n,m}(G_i^n - \overline{G_j^m}) + \frac{\lambda k_m}{h_n + k_m}(x_{i-1}^n - x_i^n) - \frac{h_n}{h_n + k_m}(y_{j-1}^m - y_j^m) \Big\|$$
$$= \Big\| (1-\lambda)(x_i^n - y_j^m) + \lambda h_{n,m}(G_i^n - \overline{G_j^m}) + \frac{h_n}{h_n + k_m}(x_i^n - y_{j-1}^m) + \frac{\lambda k_m}{h_n + k_m}(x_{i-1}^n - y_j^m) \Big\|$$
$$\leq \|(1-\lambda)(x_i^n - y_j^m) + \lambda h_{n,m}(G_i^n - \overline{G_j^m})\| + \frac{h_n}{h_n + k_m}\|x_{j-1}^n - y_j^m\| + \frac{\lambda k_m}{h_n + k_m}\|x_{i-1}^n - y_j^m\|.$$

Thus

$$\begin{aligned} &\lambda\|x_i^n - y_j^m\| + (1-\lambda)\|x_i^n - y_j^m\| \\ &= \|x_i^n - y_j^m\| \\ &\leq \|(1-\lambda)(x_i^n - y_j^m) + \lambda h_{n,m}(G_i^n - \overline{G_j^m})\| \\ &\quad + \frac{h_n}{h_n + k_m}\|x_i^n - y_{j-1}^m\| + \frac{\lambda k_m}{h_n + k_m}\|x_{i-1}^n - y_j^m\| + U, \end{aligned}$$

where

$$\begin{aligned} U &= \lambda h_{n,m} L(\|(\overline{x_i^n})_{t_{i-1}^n}\|_{PC}, \|(\overline{y_j^m})_{s_{j-1}^m}\|_{PC}, \|a_i^n\|) \\ &\quad \times (|t_i^n - s_j^m|(1 + \|A_{\lambda h_{n,m}}(t_i^n, (\overline{x_i^n})_{t_{i-1}^n})a_i^n\|) \\ &\quad + \|(\overline{x_i^n})_{t_{i-1}^n} - (\overline{y_j^m})_{s_{j-1}^m}\|_{PC}). \end{aligned}$$

We now let $\eta = \dfrac{\lambda}{1-\lambda}$, rearrange the above and then divide by λ to arrive at

$$\begin{aligned} \|x_i^n - y_j^m\| &\leq \frac{h_n}{h_n + k_m}\|x_i^n - y_{j-1}^m\| + \frac{\lambda k_m}{h_n + k_m}\|x_{i-1}^n - y_j^m\| \\ &\quad + \frac{\|(x_i^n - y_j^m) + \lambda h_{n,m}(G_i^n - \overline{G_j^m})\| - \|x_i^n - y_j^m\|}{\eta} + \frac{U}{\lambda} \\ &= \frac{h_n}{h_n + k_m}\|x_i^n - y_{j-1}^m\| + \frac{\lambda k_m}{h_n + k_m}\|x_{i-1}^n - y_j^m\| \\ &\quad + [h_{n,m}(G_i^n - \overline{G_j^m}), x_i^n - y_j^m]_\eta + \frac{U}{\lambda}. \end{aligned}$$

We now estimate U. We observe that

$$\|A_{\lambda h_{n,m}}(t_i^n, (\overline{x_i^n})_{t_{i-1}^n})a_i^n\| =$$

$$= \Big\| A_{\lambda h_{n,m}}(t_i^n, (\overline{x_i^n})_{t_{i-1}^n})\Big(x_i^n + \lambda h_{n,m}\Big(G_i^n + \frac{x_{i-1}^n - x_i^n}{h_n}\Big)\Big)\Big\|$$
$$= \Big\| G_i^n + \frac{x_{i-1}^n - x_i^n}{h_n}\Big\|$$
$$\leq K_2(\bar{r} + K_5) + K_4 + K_9.$$

Hence

$$\frac{U}{\lambda} \leq h_{n,m} L\Big(\bar{r}, \bar{r}, \bar{r} + \lambda h_{n,m}\Big\| G_i^n + \frac{x_{i-1}^n - x_i^n}{h_n}\Big\|\Big)$$
$$\times(|t_i^n - s_j^m|(1 + K_2(\bar{r} + K_5) + K_4 + K_9)$$
$$+\|(\overline{x_i^n})_{t_{i-1}^n} - (\overline{y_j^m})_{s_{j-1}^m}\|_{PC}).$$

Letting $\lambda \to 0+$ we have

$$\|x_i^n - y_j^m\|$$
$$\leq \frac{h_n}{h_n + k_m}\|x_i^n - y_{j-1}^m\| + \frac{\lambda k_m}{h_n + k_m}\|x_{i-1}^n - y_j^m)\|$$
$$+[h_{n,m}(G_i^n - \overline{G_j^m}), x_i^n - y_j^n]_+$$
$$+h_{n,m}L(\bar{r}, \bar{r}, \bar{r})(|t_i^n - s_j^m|(1 + K_2(\bar{r} + K_5) + K_4 + K_9)$$
$$+\|(\overline{x_i^n})_{t_{i-1}^n} - (\overline{y_j^m})_{s_{j-1}^m}\|_{PC})$$
$$= \frac{h_n}{h_n + k_m}\|x_i^n - y_{j-1}^m\| + \frac{\lambda k_m}{h_n + k_m}\|x_{i-1}^n - y_j^m)\|$$
$$+\frac{h_n k_m}{h_n + k_m}(K_{12}|t_i^n - s_j^m| + L(\bar{r}, \bar{r}, \bar{r})\|(\overline{x_i^n})_{t_{i-1}^n} - (\overline{y_j^m})_{s_{j-1}^m}\|_{PC}$$
$$+[G(t_i^n, (\overline{x_i^n})_{t_i^n}, L_{t_i^n}(\overline{x_i^n}))$$
$$-G(s_j^m, (\overline{y_j^m})_{s_j^m}, L_{s_j^m}(\overline{y_j^m})), x_i^n - y_j^m]_+),$$

with $K_{12} = L(\bar{r}, \bar{r}, \bar{r})(1 + K_2(\bar{r} + K_5) + K_4 + K_9)$.

PROPOSITION 4.3.3 *Assume that the assumptions of Proposition 4.3.1 are satisfied. Let n, m be positive integers. Let x_i^n, $i = 0, 1, 2, \cdots, n$ with $t_0^n = 0$, $t_n^n = T$ and $h_n = \frac{T}{n}$, x_j^m, $j = 0, 1, 2, \cdots, m$ with $t_0^m = 0$, $t_m^m = T$ and $h_m = \frac{T}{m}$ in the way described in Proposition 4.3.1 Let $x_n(t) = \overline{x_n^n}(t)$ and $x_m(t) = \overline{x_m^m}(t)$. Then there exist constants K_{13}, $K_{16} > 0$ and $\epsilon_{n,m} > 0$ such that $\lim_{n\to\infty} \epsilon_{n,m} = 0$ and*

$$\|x_i^n - x_j^m\| = K_{16}D_{i,j} + E_i + ih_n(K_{16}D_{i,j} + \epsilon_{n,m}) \qquad (4.3.6)$$

for $i = 0, 1, 2, \cdots, n$ *and* $j = 0, 1, 2, \cdots, m$, *where*

$$D_{i,j} = \{(t_i^n - t_j^m)^2 + h_n t_i^n + h_m t_j^m\}^{\frac{1}{2}}$$

and

$$E_i = K_{13} \sum_{l=1}^{i} \sup_{t \in [-r, t_l^n]} \|x_n(t) - x_m(t)\| h_n.$$

Proof. Let $x_0^n = x_0^m = \phi(0)$. From Proposition 4.3.2,

$$\begin{aligned}
&\|x_i^n - x_j^m\| \\
&\le \frac{h_n}{h_n + h_m}\|x_i^n - x_{j-1}^m\| + \frac{h_m}{h_n + h_m}\|x_{i-1}^n - x_j^m\| \\
&\quad + \frac{h_n h_m}{h_n + h_m}(K_{12}|t_i^n - t_j^m| + L(\bar r, \bar r, \bar r)\|(\overline{x_i^n})_{t_{i-1}^n} - (\overline{x_j^m})_{t_{j-1}^m}\|_{PC} \\
&\quad + [G(t_i^n, (\overline{x_i^n})_{t_i^n}, L_{t_i^n}(\overline{x_i^n})) - G(t_j^m, (\overline{x_j^m})_{t_j^m}, L_{t_j^m}(\overline{x_j^m})), x_i^n - x_j^m]_+)
\end{aligned}$$

for $i = 1, 2, \cdots, n$ and $j = 1, 2, \cdots, m$. From $(G1)_3$, $(G2)_3$, and Lemma 4.3.1

$$\begin{aligned}
&\|G(t_i^n, (\overline{x_i^n})_{t_i^n}, L_{t_i^n}(\overline{x_i^n})) - G(t_j^m, (\overline{x_j^m})_{t_j^m}, L_{t_j^m}(\overline{x_j^m}))\| \\
&\quad \le \|G(t_i^n, (\overline{x_i^n})_{t_i^n}, L_{t_i^n}(\overline{x_i^n})) - G(t_i^n, (\overline{x_j^m})_{t_j^m}, L_{t_j^m}(\overline{x_j^m}))\| \\
&\qquad + \|G(t_i^n, (\overline{x_j^m})_{t_j^m}, L_{t_j^m}(\overline{x_j^m})) - G(t_j^m, (\overline{x_j^m})_{t_j^m}, L_{t_j^m}(\overline{x_j^m}))\| \\
&\quad \le K_2(\|(\overline{x_i^n})_{t_i^n} - (\overline{x_j^m})_{t_j^m}\|_{PC} + \|L_{t_i^n}(\overline{x_i^n}) - L_{t_j^m}(\overline{x_j^m})\|) \\
&\qquad + l_2(\|(\overline{x_j^m})_{t_j^m}\|_{PC}, \|L_{t_j^m}(\overline{x_j^m})\|)|t_i^n - t_j^m|.
\end{aligned}$$

Now,

$$\|(\overline{x_j^m})_{t_j^m}\|_{PC} \le \bar r, \ \|L_{t_j^m}(\overline{x_j^m})\| \le K_5$$

and

$$\begin{aligned}
\|(\overline{x_i^n})_{t_i^n} - (\overline{x_j^m})_{t_j^m}\|_{PC} &= \|(x_n)_{t_i^n} - (x_m)_{t_j^m}\|_{PC} \\
&< \|(x_n)_{t_i^n} - (x_m)_{t_i^n}\|_{PC} + \\
&\quad + \|(x_m)_{t_i^n} - (x_m)_{t_j^m}\|_{PC} \\
&= \sup_{t \in [t_i^n - r, t_i^n]} \|x_n(t) - x_m(t)\| \\
&\quad + \sup_{\tau \in [-r, 0]} \|x_m(t_i^n + \tau) - x_m(t_j^m + \tau)\| \\
&\le \sup_{t \in [-r, t_i^n]} \|x_n(t) - x_m(t)\| \\
&\quad + K_{11}(|t_i^n - t_j^m| + h_m).
\end{aligned}$$

Thus we arrive at

$$\begin{aligned}
&\|(\overline{x_i^n})_{t_{i-1}^n} - (\overline{x_j^m})_{t_{j-1}^m}\|_{PC} \\
&\quad = \|(\overline{x_{i-1}^n})_{t_{i-1}^n} - (\overline{x_{j-1}^m})_{t_{j-1}^m}\|_{PC} \\
&\quad \le \sup_{t\in[-r,t_{i-1}^n]} \|x_n(t) - x_m(t)\| + K_{11}(|t_{i-1}^n - t_{j-1}^m| + h_m) \\
&\quad \le \sup_{t\in[-r,t_i^n]} \|x_n(t) - x_m(t)\| + K_{11}(|t_i^n - t_j^m| + h_n + 2h_m).
\end{aligned}$$

If $0 \le t_j^m \le t_i^n$ then, from $(L1)_3$ and $(L2)_3$ we obtain

$$\begin{aligned}
&\|L_{t_i^n}(\overline{x_i^n}) - L_{t_j^m}(\overline{x_j^m})\| \\
&\quad \le \|L_{t_i^n}(\overline{x_i^n}) - L_{t_j^m}(\overline{x_i^n})\| + \|L_{t_j^m}(\overline{x_i^n}) - L_{t_j^m}(\overline{x_j^m})\| \\
&\quad \le l_1(\|\overline{x_i^n}\|_T)|t_i^n - t_j^m| + K_1\|\overline{x_i^n} - \overline{x_j^m}\|_{t_j^m} \\
&\quad \le l_1(\bar{r})|t_i^n - t_j^m| + K_1 \sup_{t\in[-r,t_j^m]} \|\overline{x_i^n}(t) - \overline{x_j^m}(t)\| \\
&\quad \le l_1(\bar{r})|t_i^n - t_j^m| + K_1 \sup_{t\in[-r,t_j^m]} \|x_n(t) - x_m(t)\| \\
&\quad \le l_1(\bar{r})|t_i^n - t_j^m| + K_1 \sup_{t\in[-r,t_i^n]} \|x_n(t) - x_m(t)\|.
\end{aligned}$$

Similarly if $0 \le t_i^n \le t_j^m$ then we obtain

$$\begin{aligned}
&\|L_{t_i^n}(\overline{x_i^n}) - L_{t_j^m}(\overline{x_j^m})\| \\
&\quad \le l_1(\bar{r})|t_i^n - t_j^m| + K_1 \sup_{t\in[-r,t_i^n]} \|x_n(t) - x_m(t)\|.
\end{aligned}$$

We observe that

$$\begin{aligned}
|t_i^n - t_j^m| &\le |(t_i^n - t_j^m) - h_n| + h_n \\
&= |t_{i-1}^n - t_j^m| + h_n \le D_{i-1,j} + h_n.
\end{aligned}$$

Hence letting $A_{i,j} = \|x_i^n - x_j^m\|$ we obtain

$$\begin{aligned}
A_{i,j} \;\le\; & \frac{h_n}{h_n + h_m} A_{i,j-1} + \frac{h_m}{h_n + h_m} A_{i-1,j} \\
& + \frac{h_n h_m}{h_n + h_m}\Big(K_{13} \sup_{t\in[-r,t_i^n]} \|x_n(t) - x_m(t)\| \\
& \qquad + K_{14} D_{i-1,j} + \epsilon_{n,m}\Big), \qquad (4.3.7)
\end{aligned}$$

where

$$\begin{aligned}
K_{13} &= L(\bar{r}, \bar{r}, \bar{r}) + K_2(1 + K_1), \\
K_{14} &= K_{12} + K_2 K_{11} + K_2 l_(\bar{r}) + l_2(\bar{r}, K_5) + K_{11} L(\bar{r}, \bar{r}, \bar{r}),
\end{aligned}$$

and

$$\epsilon_{n,m} = (K_{14} + K_{11}L(\bar{r}, \bar{r}, \bar{r}))h_n + K_{11}(K_2 + 2L(\bar{r}, \bar{r}, \bar{r}))h_m.$$

We now employ (4.3.3) in order to obtain

$$\begin{aligned}
&\|x_k^n - \phi(0)\| \\
&= \|J_{\lambda h_{n,m}}(t_k^n, (\overline{x_k^n})_{t_{k-1}^n})(x_k^n + \lambda h_{n,m} G(t_k^n, (\overline{x_k^n})_{t_k^n}, L_{t_k^n}(\overline{x_k^n}))) - \phi(0)\| \\
&\leq \|J_{\lambda h_{n,m}}(t_k^n, (\overline{x_k^n})_{t_{k-1}^n})(x_k^n + \lambda h_{n,m} G(t_k^n, (\overline{x_k^n})_{t_k^n}, L_{t_k^n}(\overline{x_k^n}))) \\
&\quad - J_{\lambda h_{n,m}}(t_k^n, (\overline{x_k^n})_{t_{k-1}^n})\phi(0)\| \\
&\quad + \|J_{\lambda h_{n,m}}(t_k^n, (\overline{x_k^n})_{t_{k-1}^n})\phi(0) - \phi(0)\| \\
&\leq \|x_{k-1}^n - \phi(0)\| + h_n\|G(t_k^n, (\overline{x_k^n})_{t_k^n}, L_{t_k^n}(\overline{x_i^n}))\| \\
&\quad + h_n|A(t_k^n, (\overline{x_k^n})_{t_{k-1}^n})\phi(0)|.
\end{aligned}$$

Since

$$|A(t_k^n, (\overline{x_k^n})_{t_{k-1}^n})\phi(0)| \leq K_3$$

and

$$\|G(t_k^n, (\overline{x_k^n})_{t_k^n}, L_{t_k^n}(\overline{x_k^n}))\| \leq K_2(\bar{r} + K_5) + K_4$$

we conclude that

$$\|x_k^n - \phi(0)\| \leq \|x_{k-1}^n - \phi(0)\| + K_{15}h_n,$$

where $K_{15} = K_2(\bar{r} + K_5) + K_3 + K_4$. Adding these inequalities for $k = 1, 2, \cdots, i$ we obtain

$$\|x_i^n - x_0^n\| = \|x_i^n - \phi(0)\| \leq \Sigma_{k=1}^i K_{15}h_n = t_i^n K_{15} \leq K_{15}D_{i,0}.$$

Thus $A_{i,0} \leq K_{16}D_{i,0}$ for $i = 0, 1, 2, \cdots, n$, where $K_{16} = \max\{K_{14}, K_{15}\}$. In the same way we see that $A_{0,j} \leq K_{16}D_{0,j}$ for $j = 0, 1, 2, \cdots, m$.

This means that (4.3.6) holds for pairs $(i, 0)$ and $(0, j)$. Assume that (4.3.6) holds for the pairs $(i-1, j)$ and $(i, j-1)$. We want to show that (4.3.6) holds for the pair (i, j) as well. From (4.3.7) we have

$$\begin{aligned}
A_{i,j} &\leq \frac{h_n}{h_n + h_m}(K_{16}D_{i,j-1} + E_i + ih_n(K_{16}D_{i,j-1} + \epsilon_{n,m})) \\
&\quad + \frac{h_m}{h_n + h_m}(K_{16}D_{i-1,j} + E_{i-1} + (i-1)h_n(K_{16}D_{i-1,j} + \epsilon_{n,m})) \\
&\quad + \frac{h_n h_m}{h_n + h_m}\Big(K_{13} \sup_{t \in [-r, t_i^n]} \|x_n(t) - x_m(t)\| + K_{16}D_{i-1,j} + \epsilon_{n,m}\Big) \\
&\leq K_{16}D_{i,j} + E_i + ih_n(K_{16}D_{i,j} + \epsilon_{n,m}).
\end{aligned}$$

Here we have used

$$\frac{h_n}{h_n + h_m} D_{i,j-1} + \frac{h_m}{h_n + h_m} D_{i-1,j} \le D_{i,j}.$$

It turns out that (4.3.6) is true for every (i, j) with $\le i \le n$ and $0 \le j \le m$.

We are now ready for the proof of the existence of a limit solution of (4.3.1)

THEOREM 4.3.1 *Assume that* $(A1)_3$, $(A2)_3$, $(G1)_3$, $(G2)_3$, $(L1)_3$, *and* $(L2)_3$ *are satisfied. Assume that the condition* (4.3.2) *holds. Let* $\phi \in PC([-r, 0]; \overline{B_{\bar{r}}(0)})$ *be Lipschitz continuous with Lipschitz constant* $K = K_\phi > 0$ *and let* $\phi(0) \in \widehat{D}$. *Then there exists a limit solution* $x(t)$ *of* (4.3.1) *on* $[-r, T]$ *and* $x(t)$ *is Lipschitz continuous on* $[-r, T]$.

Proof. Let $\{t_i^n\}$, $\{t_j^m\}$ be two partitions of $[0, T]$, where $t_i^n = ih_n = i\frac{T}{n}$ for $i = 0, 1, 2, \cdots, n$ and $t_j^m = jh_m = j\frac{T}{m}$ for $j = 0, 1, 2, \cdots, m$.

Let $t \in (t_{i-1}^n, t_i^n] \cap (t_{j-1}^m, t_j^m]$. Then

$$|t_i^n - t_j^m| \le |t_i^n - t| + |t - t_j^m| \le h_n + h_m.$$

From Proposition 4.3.3

$$\begin{aligned} &\|x_n(t) - x_m(t)\| \\ &\quad = \|x_i^n - x_j^m\| \\ &\quad \le K_{16}((h_n + h_m)^2 + (h_n + h_m)T)^{\frac{1}{2}} \\ &\quad\quad + K_{13} \sum_{k=1}^{i} \sup_{t \in [-r, t_k^n]} \|x_n(t) - x_m(t)\| h_n \\ &\quad\quad + T(K_{16}((h_n + h_m)^2 + (h_n + h_m)T)^{\frac{1}{2}} + \epsilon_{n,m}). \end{aligned}$$

We define the function $F_{n,m}$ as follows

$$F_{n,m}(t) = \begin{cases} 0, \quad t = 0 \\ \sup_{s \in [-r, t_k^n]} \|x_n(s) - x_m(s)\|, \quad t \in (t_{k-1}^n, t_k^n], \\ \qquad\qquad\qquad\qquad\qquad\qquad k = 1, 2, \cdots, n. \end{cases}$$

Fix $t \in (0, T]$. Then $t \in (t_{k-1}^n, t_k^n]$ for some $k = 1, 2, \cdots, n$. Thus

$$F_{n,m}(t) = \max\Big\{ \sup_{s \in [-r, t]} \|x_n(s) - x_m(s)\|, \sup_{s \in [t, t_k^n]} \|x_n(s) - x_m(s)\| \Big\}.$$

If $s \in [t, t_k^n]$ then from Lemma 4.3.1 we have

$$\begin{aligned}
&\|x_n(s) - x_m(s)\| \\
&\quad \leq \|x_n(t) - x_m(t)\| + \|x_n(s) - x_n(t)\| + \|x_m(s) - x_m(t)\| \\
&\quad \leq \sup_{s \in [-r,t]} \|x_n(s) - x_m(s)\| + 2K_{11}|s - t| + K_{11}(h_n + h_m) \\
&\quad \leq \sup_{s \in [-r,t]} \|x_n(s) - x_m(s)\| + 2K_{11}h_n + K_{11}(h_n + h_m),
\end{aligned}$$

which yields

$$\begin{aligned}
&\sup_{s \in [t,t_k^n]} \|x_n(s) - x_m(s)\| \\
&\qquad \leq \sup_{s \in [-r,t]} \|x_n(s) - x_m(s)\| + 3K_{11}h_n + K_{11}h_m.
\end{aligned}$$

Thus we have

$$F_{n,m}(t) \leq \sup_{s \in [-r,t]} \|x_n(s) - x_m(s)\| + 3K_{11}h_n + K_{11}h_m$$

for every $t \in [0, T]$. Since

$$\begin{aligned}
\int_t^{t_i^n} F_{n,m}(\sigma)d\sigma &= \int_t^{t_i^n} \sup_{s \in [-r,t_i^n]} \|x_n(s) - x_m(s)\| d\sigma \\
&\leq \int_t^{t_i^n} 2\bar{r} d\sigma \leq 2\bar{r}h_n
\end{aligned}$$

for every $t \in (t_{i-1}^n, t_i^n]$ we obtain

$$\begin{aligned}
&\sum_{k=1}^{i} \sup_{t \in [-r,t_i^n]} \|x_n(t) - x_m(t)\| h_n \\
&\quad = \sum_{k=1}^{i} \sup_{t \in [-r,t_k^n]} \|x_n(t) - x_m(t)\|(t_k^n - t_{k-1}^n) \\
&\quad = \sum_{k=1}^{i} \int_{t_{k-1}^n}^{t_k^n} F_{n,m}(\sigma)d\sigma \\
&\quad = \int_0^t F_{n,m}(\sigma)d\sigma + \int_t^{t_i^n} F_{n,m}(\sigma)d\sigma \\
&\quad \leq \int_0^t \sup_{s \in [-r,\sigma]} \|x_n(s) - x_m(s)\| d\sigma + 3K_{11}Th_n + K_{11}Th_m \\
&\quad\quad + 2\bar{r}h_n.
\end{aligned}$$

It follows that

$$\|x_n(t) - x_m(t)\| \le \delta_{n,m} + K_{13} \int_0^t \sup_{s\in[-r,\sigma]} \|x_n(s) - x_m(s)\| d\sigma$$

for every $t \in (t_{i-1}^n, t_i^n] \cap (t_{j-1}^m, t_j^m]$, where

$$\begin{aligned}\delta_{n,m} = & K_{16}((h_n + h_m)^2 + (h_n + h_m)T)^{\frac{1}{2}} \\ & + T(K_{16}((h_n + h_m)^2 + (h_n + h_m)T)^{\frac{1}{2}} + \epsilon_{n,m}) \\ & + 3K_{11}K_{13}Th_n + K_{11}K_{13}Th_m + 2\bar{r}K_{13}h_n.\end{aligned}$$

Hence

$$\|x_n(t) - x_m(t)\| \le \delta_{n,m} + K_{13} \int_0^t \sup_{s\in[-r,\sigma]} \|x_n(s) - x_m(s)\| d\sigma$$

for every $t \in [0, T]$, which implies

$$\begin{aligned}& \sup_{s\in[-r,t]} \|x_n(s) - x_m(s)\| \\ & \qquad \le \delta_{n,m} + K_{13} \int_0^t \sup_{s\in[-r,\sigma]} \|x_n(s) - x_m(s)\| d\sigma\end{aligned}$$

for every $t \in [0, T]$. From Gronwall's inequality we have

$$\sup_{s\in[-r,t]} \|x_n(s) - x_m(s)\| \le \delta_{n,m} \exp\Big\{K_{13} \int_0^t d\sigma\Big\} \le \delta_{n,m} \exp\{K_{13}T\}$$

for every $t \in [0, T]$. This implies that $x_n(t)$ converges to a function $x(t)$ uniformly on $[-r, T]$. Also

$$\|x(t_1) - x(t_2)\| \le K_{11}|t_1 - t_2|$$

for every $t_i \in [-r, T]$, $i = 1, 2$, which proves the Lipschitz continuity of the function $x(t)$ on $[-r, T]$ with Lipschitz constant K_{11}.

In the following we establish a basic inequality involving limit solutions of problems of the type (4.3.1)

PROPOSITION 4.3.4 *Assume that the assumptions of Theorem 4.3.1 are satisfied. Let* ϕ, $\hat{\phi} \in PC([-r, 0]; \overline{B_{\bar{r}}(0)})$ *be Lipschitz continuous with Lipschitz constant* $K = K_\phi > 0, \widehat{K} = \widehat{K}_{\hat{\phi}} > 0$, *respectively. Let* $\phi(0)$, $\hat{\phi}(0) \in \widehat{D}$. *If* $x(t)$ *and* $y(t)$ *are limit solutions of* (4.3.1) *corresponding to*

ϕ and $\hat{\phi}$, respectively, then we obtain

$$\begin{aligned}
&\|x(t)-y(t)\| \\
&\quad \le \|x(s)-y(s)\| \\
&\quad\quad + L(\bar{r},\bar{r},\bar{r})\int_s^t \|x_\sigma - y_\sigma\|_{PC} d\sigma \\
&\quad\quad + \int_s^t [G(\sigma, x_\sigma, L_\sigma(x)) - G(\sigma, y_\sigma, L_\sigma(y)), x(\sigma)-y(\sigma)]_+ d\sigma
\end{aligned}$$

for $0 \le s \le t \le T$.

Proof. From the definition of the limit solution in Theorem 4.3.1 there exists a method of lines $\overline{x_i^n}(t)$ such that

$$\frac{x_i^n - x_{i-1}^n}{h_n} + A(t_i^n, (\overline{x_i^n})_{t_{i-1}^n})x_i^n \ni G(t_i^n, (\overline{x_i^n})_{t_i^n}, L_{t_i^n}(\overline{x_i^n}))$$

for $i = 1, 2, \cdots, n$, $x_0^n = \phi(0)$ and $\lim_{n\to\infty} x_n(t) = x(t)$, where $h_n = t_i^n - t_{i-1}^n$ and $x_n(t) = \overline{x_n^n}(t)$. Similarly there exists a method of lines $\overline{y_j^m}(t)$ such that

$$\frac{y_j^m - y_{j-1}^m}{k_m} + A(s_j^m, (\overline{y_j^m})_{s_{j-1}^m})y_j^m \ni G(s_j^m, (\overline{y_j^m})_{s_j^m}, L_{s_j^m}(\overline{y_j^m}))$$

for $j = 1, 2, \cdots, m$, $y_0^m = \hat{\phi}(0)$ and $\lim_{m\to\infty} y_m(t) = y(t)$, where $k_m = s_j^m - s_{j-1}^m$ and $y_m(t) = \overline{y_m^m}(t)$. From Proposition 4.3.2

$$\begin{aligned}
&\|x_i^n - y_j^m\| \\
&\quad \le \frac{h_n}{h_n+k_m}\|x_i^n - y_{j-1}^m\| + \frac{k_m}{h_n+k_m}\|x_{i-1}^n - y_j^m\| \\
&\quad\quad + \frac{h_n k_m}{h_n+k_m}(K_{12}|t_i^n - s_j^m| + L(\bar{r},\bar{r},\bar{r})\|(\overline{x_i^n})_{t_{i-1}^n} - (\overline{y_j^m})_{s_{j-1}^m}\|_{PC} \\
&\quad\quad\quad + [G(t_i^n, (\overline{x_i^n})_{t_i^n}, L_{t_i^n}(\overline{x_i^n})) \\
&\quad\quad\quad - G(s_j^m, (\overline{y_j^m})_{s_j^m}, L_{s_j^m}(\overline{y_j^m})), x_i^n - y_j^m]_+). \qquad (4.3.8)
\end{aligned}$$

Noting that

$$\begin{aligned}
[w,z]_+ &\le [w,z]_\lambda = \frac{\|w+\lambda z\| - \|w\|}{\lambda} \\
&\le [u,v]_\lambda + \|z-v\| + \frac{2\|w-u\|}{\lambda}
\end{aligned}$$

for every u, v, w, $z \in X$ we obtain

$$
\begin{aligned}
&[G(t_i^n,(\overline{x_i^n})_{t_i^n},L_{t_i^n}(\overline{x_i^n}))-G(s_j^m,(\overline{y_j^m})_{s_j^m},L_{s_j^m}(\overline{y_j^m})),x_i^n-y_j^m]_+\\
&\quad\le [G(t_i^n,(\overline{x_i^n})_{t_i^n},L_{t_i^n}(\overline{x_i^n}))-G(t_i^n,y_{t_i^n},L_{t_i^n}(y)),x_i^n-y(t_i^n)]_\lambda\\
&\quad\quad+\|G(s_j^m,(\overline{y_j^m})_{s_j^m},L_{s_j^m}(\overline{y_j^m}))-G(s_j^m,y_{s_j^m},L_{s_j^m}(y))\|\\
&\quad\quad+\|G(s_j^m,(\overline{y_j^m})_{s_j^m},L_{s_j^m}(\overline{y_j^m}))-G(t_i^n,y_{t_i^n},L_{t_i^n}(y))\|\\
&\quad\quad+\frac{2}{\lambda}\|y_j^m-y(s_j^m)\|+\frac{2}{\lambda}\|y(s_j^m)-y(t_i^n)\|
\end{aligned}
$$

and

$$
\begin{aligned}
&\|(\overline{x_i^n})_{t_{i-1}^n}-(\overline{y_j^m})_{s_{j-1}^m}\|_{PC}\\
&\quad=\|(x_n)_{t_{i-1}^n}-(y_m)_{s_{j-1}^m}\|_{PC}\\
&\quad\le\|(x_n)_{t_{i-1}^n}-(y_m)_{t_{i-1}^n}\|_{PC}+\|(y_m)_{t_{i-1}^n}-(y_m)_{s_{j-1}^m}\|_{PC}\\
&\quad=\|(x_n)_{t_{i-1}^n}-(y_m)_{t_{i-1}^n}\|_{PC}+\\
&\quad\quad+\sup_{\tau\in[-r,0]}\|y_m(t_{i-1}^n+\tau)-y_m(s_{j-1}^m+\tau)\|\\
&\quad\le\|(x_n)_{t_{i-1}^n}-(y_m)_{t_{i-1}^n}\|_{PC}+\sup_{\tau\in[-r,0]}K_{11}(|t_{i-1}^n-s_{j-1}^m|+k_m)\\
&\quad\le\|(x_n)_{t_{i-1}^n}-(y_m)_{t_{i-1}^n}\|_{PC}+K_{11}(|t_i^n-s_j^m|+h_n+2k_m).
\end{aligned}
$$

We now let $A_{i,j}=\|x_i^n-y_j^m\|$ to convert (4.3.8) to

$$
\begin{aligned}
A_{i,j}\le&\frac{h_n}{h_n+k_m}A_{i,j-1}+\frac{k_m}{h_n+k_m}A_{i-1,j}\\
&+\frac{h_nk_m}{h_n+k_m}(L(\bar r,\bar r,\bar r)\|(x_n)_{t_{i-1}^n}-(y_m)_{t_{i-1}^n}\|_{PC}\\
&+K_{17}|t_i^n-s_j^m|+\delta_i^n+\gamma_j^m+\rho(|t_i^n-s_j^m|)\\
&+L(\bar r,\bar r,\bar r)(h_n+2h_m)),
\end{aligned}
$$

where $K_{17}=K_{12}+K_{11}L(\bar r,\bar r,\bar r)$,

$$
\begin{aligned}
\delta_i^n &= [G(t_i^n,(\overline{x_i^n})_{t_i^n},L_{t_i^n}(\overline{x_i^n}))-G(t_i^n,y_{t_i^n},L_{t_i^n}(y)),x_i^n-y(t_i^n)]_\lambda=\\
&= [G(t_i^n,(x_n)_{t_i^n},L_{t_i^n}(x_n))-G(t_i^n,y_{t_i^n},L_{t_i^n}(y)),x_n(t_i^n)-y(t_i^n)]_\lambda,\\
\gamma_j^m &= G(s_j^m,(\overline{y_j^m})_{s_j^m},L_{s_j^m}(\overline{y_j^m}))-G(s_j^m,y_{s_j^m},L_{s_j^m}(y))\|\\
&\quad+\frac{2\|y_j^m-y(s_j^m)\|}{\lambda}\\
&= G(s_j^m,(y_m)_{s_j^m},L_{s_j^m}(y_m))-G(s_j^m,y_{s_j^m},L_{s_j^m}(y))\|\\
&\quad+\frac{2\|y_j^m-y(s_j^m)\|}{\lambda},
\end{aligned}
$$

and

$$\rho(t) = \sup_{|s-\sigma|\le t} \Big\{ \frac{2}{\lambda}\|y(s) - y(\sigma)\| + \|G(s, y_s, L_s(y) - G(\sigma, y_\sigma, L_\sigma(y))\| \Big\}$$

for every $t \in [0,T]$. Notice that ρ is a nondecreasing function on $[0,T]$. For $p \in \{0,1,2,\cdots,n\}$ and $q \in \{0,1,2,\cdots,m\}$ let $i = p, p+1, \cdots, n$ and $j = q, q+1, \cdots, m$. Then

$$\begin{aligned} |t_i^n - s_j^m| &\le |(t_i^n - s_j^m) - h_n| + h_n \\ &\le |(t_i^n - t_p^n) - (s_j^m - s_q^m) - h_n| + |t_p^n - s_q^m| + h_n \\ &\le D_{i-1,j} + |t_p^n - s_q^m| + h_n, \end{aligned}$$

where

$$D_{i-1,j} = \{((t_i^n - t_p^n) - (s_j^m - s_q^m))^2 + (t_i^n - t_p^n)h_n + (s_j^m - s_q^m)k_m\}^{\frac{1}{2}}.$$

Let $\delta \in (0, T/2)$ and assume that n and m are sufficiently large such that $\max\{h_n, k_m\} < \delta$. Then we obtain

$$\rho(|t_i^n - s_j^m|) \le \frac{\rho(T)}{\delta}||t_i^n - s_j^m| - h_n| + \rho(2\delta). \tag{4.3.9}$$

In fact, let $|t_i^n - s_j^m| \le 2\delta$. Since $\rho(t)$ is non-decreasing

$$\rho(|t_i^n - s_j^m|) \le \rho(2\delta).$$

Let $2\delta < |t_i^n - s_j^m| \le T$. Since $||t_i^n - s_j^m| - h_n| \ge \delta$,

$$\rho(|t_i^n - s_j^m|) \le \rho(T) \le \frac{\rho(T)}{\delta}||t_i^n - s_j^m| - h_n|.$$

Thus (4.3.9) holds. From (4.3.9) we have

$$\rho(|t_i^n - s_j^m|) \le \frac{\rho(T)}{\delta}(D_{i-1,j} + |t_p^n - s_q^m|) + \rho(2\delta).$$

Hence

$$\begin{aligned} A_{i,j} \le\ & \frac{h_n}{h_n + k_m}A_{i,j-1} + \frac{k_m}{h_n + k_m}A_{i-1,j} \\ &+ \frac{h_n k_m}{h_n + k_m}(L(\bar{r}, \bar{r}, \bar{r})\|(x_n)_{t_{i-1}^n} - (y_m)_{t_{i-1}^n}\|_{PC} \\ &\quad + \Big(\frac{\rho(T)}{\delta} + K_{17}\Big)(D_{i-1,j} + |t_p^n - s_q^m|) + \rho(2\delta) + \delta_i^n + \gamma_j^m \\ &\quad + (K_{17} + K_{11}L(\bar{r}, \bar{r}, \bar{r}))h_n + 2K_{11}L(\bar{r}, \bar{r}, \bar{r})h_m). \end{aligned} \tag{4.3.10}$$

Since

$$A(t_i^n, (\overline{x_i^n})_{t_{i-1}^n})x_i^n \ni G(t_i^n, (\overline{x_i^n})_{t_i^n}, L_{t_i^n}(\overline{x_i^n})) - \frac{x_i^n - x_{i-1}^n}{h_n}$$

we have

$$\begin{aligned}
|A(t_i^n, (\overline{x_i^n})_{t_{i-1}^n})x_i^n| &\le \|G(t_i^n, (\overline{x_i^n})_{t_i^n}, L_{t_i^n}(\overline{x_i^n}))\| + \frac{\|x_i^n - x_{i-1}^n\|}{h_n} \\
&\le \|G(t_i^n, (\overline{x_i^n})_{t_i^n}, L_{t_i^n}(\overline{x_i^n})) - G(t_i^n, \bar{0}, 0)\| \\
&\quad + \|G(t_i^n, \bar{0}, 0)\| + K_9 \\
&\le K_2(\|(\overline{x_i^n})_{t_i^n}\|_{PC} + \|L_{t_i^n}(\overline{x_i^n}))\|) + K_4 + K_9 \\
&\le K_2(\bar{r} + K_5) + K_4 + K_9 = K_{18}.
\end{aligned}$$

It follows from $(A2)_3$ that

$$\begin{aligned}
|A(t_l^n, (\overline{x_l^n})_{t_{l-1}^n})x_p^n| &\le |A(t_p^n, (\overline{x_p^n})_{t_{p-1}^n})x_p^n| + \\
&\quad + L(\|(\overline{x_l^n})_{t_{l-1}^n}\|_{PC}, \|(\overline{x_p^n})_{t_{p-1}^n}\|_{PC}, \|x_p^n\|) \\
&\qquad \times (|t_l^n - t_p^n|(1 + |A(t_p^n, (\overline{x_p^n})_{t_{p-1}^n})x_p^n|) \\
&\qquad + \|(\overline{x_l^n})_{t_{l-1}^n} - (\overline{x_p^n})_{t_{p-1}^n}\|_{PC}) \\
&\le K_{18} + L(\bar{r}, \bar{r}, \bar{r})(T(1 + K_{18}) + 2\bar{r}) = K_{19}.
\end{aligned}$$

Considering $l = p+1, p+2, \cdots, n$ we see that

$$\begin{aligned}
&\|x_l^n - x_p^n\| \\
&\quad = \|J_{h_n}(t_l^n, (\overline{x_l^n})_{t_{l-1}^n})(x_{l-1}^n + h_n G(t_l^n, (\overline{x_l^n})_{t_l^n}, L_{t_l^n}(\overline{x_l^n}))) - x_p^n\| \\
&\quad \le \|J_{h_n}(t_l^n, (\overline{x_l^n})_{t_{l-1}^n})(x_{l-1}^n + h_n G(t_l^n, (\overline{x_l^n})_{t_l^n}, L_{t_l^n}(\overline{x_l^n}))) \\
&\qquad - J_{h_n}(t_l^n, (\overline{x_l^n})_{t_{l-1}^n})x_p^n\| \\
&\qquad + \|J_{h_n}(t_l^n, (\overline{x_l^n})_{t_{l-1}^n})x_p^n - x_p^n\| \\
&\quad \le \|x_{l-1}^n - x_p^n\| + h_n\|G(t_l^n, (\overline{x_l^n})_{t_l^n}, L_{t_l^n}(\overline{x_l^n}))\| \\
&\qquad + h_n|A(t_l^n, (\overline{x_l^n})_{t_{l-1}^n})x_p^n| \\
&\quad \le \|x_{l-1}^n - x_p^n\| + K_{20}h_n,
\end{aligned}$$

where

$$K_{20} = \max\{K_{17} + K_{11}L(\bar{r}, \bar{r}, \bar{r}), K_{19} + K_[2(\bar{r} + K_5) + K_4\}.$$

Adding these inequalities for $l = p+1, p+2, \cdots, i$ we obtain

$$\|x_i^n - x_p^n\| \le (i-p)K_{20}h_n = K_{20}(t_i^n - t_p^n) \le K_{20}D_{i,q},$$

which yields

$$\begin{aligned}\|x_i^n - y_q^m\| &\leq \|x_p^n - y_q^m\| + \|x_i^n - x_p^n\| \\ &\leq \|x_p^n - y_q^m\| + K_{20}D_{i,q} \qquad (4.3.11)\end{aligned}$$

for $i = p, p+1, \cdots, n$.

Let $p \in \{0, 1, 2, \cdots, n\}$ and $q \in \{0, 1, 2, \cdots, m\}$. We claim that we obtain the estimate

$$\begin{aligned}\|x_i^n - y_j^m\| \leq\ & \|x_p^n - y_q^m\| + K_{20}D_{i,j} + \sum_{l=p}^{i} \delta_l^n h_n + \sum_{l=q}^{j} \gamma_j^m k_m \\ & + L(\bar{r}, \bar{r}, \bar{r}) \sum_{l=p}^{i} \|(x_n)_{t_{i-1}^n} - (y_m)_{t_{i-1}^n}\|_{PC} h_n \\ & + i h_n \Big(\Big(\frac{\rho(T)}{\delta} + K_{20} \Big) (D_{i,j} + |t_p^n - s_q^m|) + \rho(2\delta) \\ & \quad + K_{20}(h_n + k_m) \Big) \qquad (4.3.12)\end{aligned}$$

for $i = p, p+1, \cdots, n$ and $j = q, q+1, \cdots, m$. We know from (4.3.11) that (4.3.12) is true for $i = p, p+1, \cdots, n$ and $j = q$. Similarly we know that (4.3.12) holds also for $i = p$ and $j = q, q+1, \cdots, m$. To apply induction let assume that (4.3.12) holds for the pairs $(i-1, j)$ and $(i, j-1)$ for $i = p+1, p+2, \cdots, n$ and $j = q+1, q+2, \cdots, m$.

From (4.3.10) we have

$$\begin{aligned}A_{i,j} \leq\ & \frac{h_n}{h_n + k_m} \Big(\|x_p^n - y_q^m\| + K_{20}D_{i,j-1} + \sum_{l=p}^{i} \delta_l^n h_n + \sum_{l=q}^{j-1} \gamma_l^m k_m \\ & + L(\bar{r}, \bar{r}, \bar{r}) \sum_{l=p}^{i} \|(x_n)_{t_{l-1}^n} - (y_m)_{t_{l-1}^n}\|_{PC} h_n \\ & + i h_n \Big(\Big(\frac{\rho(T)}{\delta} + K_{20} \Big) (D_{i,j-1} + |t_p^n - s_q^m|) + \rho(2\delta) \\ & \quad + K_{20}(h_n + k_m) \Big) \Big) + \\ & + \frac{k_m}{h_n + k_m} \Big(\|x_p^n - y_q^m\| + K_{20}D_{i-1,j} + \sum_{l=p}^{i-1} \delta_l^n h_n + \sum_{l=q}^{j} \gamma_l^m k_m \\ & + L(\bar{r}, \bar{r}, \bar{r}) \sum_{l=p}^{i-1} \|(x_n)_{t_{l-1}^n} - (y_m)_{t_{l-1}^n}\|_{PC} h_n \\ & + (i-1) h_n \Big(\Big(\frac{\rho(T)}{\delta} + K_{20} \Big) (D_{i,j} + |t_p^n - s_q^m|) + \end{aligned}$$

$$
\begin{aligned}
&\qquad\qquad +\rho(2\delta)+K_{20}(h_n+k_m)\Big)\Big)\\
&+\frac{h_nk_m}{h_n+k_m}\Big(L(\bar r,\bar r,\bar r)\|(x_n)_{t^n_{i-1}}-(y_m)_{t^n_{i-1}}\|_{PC}\\
&\qquad +\Big(\frac{\rho(T)}{\delta}+K_{17}\Big)(D_{i-1,j}+|t^n_p-s^m_q|)+\rho(2\delta)+\delta^n_i\\
&\qquad +\gamma^m_j+K_{20}(h_n+k_m)\Big)\\
&\le \|x^n_p-y^m_q\|+K_{20}D_{i,j}+\sum_{l=p}^{i}\delta^n_l h_n+\sum_{l=q}^{j}\gamma^m_l k_m\\
&\quad +L(\bar r,\bar r,\bar r)\sum_{l=p}^{i}\|(x_n)_{t^n_{l-1}}-(y_m)_{t^n_{l-1}}\|_{PC}h_n\\
&\quad +ih_n\Big(\Big(\frac{\rho(T)}{\delta}+K_{20}\Big)(D_{i,j}+|t^n_p-s^m_q|)+\rho(2\delta)\\
&\qquad +K_{20}(h_n+k_m)\Big),
\end{aligned}
$$

where we have used the fact that

$$\frac{h_n}{h_n+k_m}D_{i,j-1}+\frac{k_m}{h_n+k_m}D_{i-1,j}\le D_{i,j}.$$

Hence we have shown that (4.3.12) holds for the pair (i,j). Therefore (4.3.12) holds for $i=p,p+1,\cdots,n$ and $j=q,q+1,\cdots,m$ with some $p\in\{0,1,2,\cdots,n\}$ and $q\in\{0,1,2,\cdots,m\}$.

Now we consider points $s\in(t^n_{p-1},t^n_p]\cap(s^m_{q-1},s^m_q]$ and $t\in(t^n_{i-1},t^n_i]\cap(s^m_{j-1},s^m_j]$. Letting $n\to\infty$ and $m\to\infty$ in (4.3.12) we obtain

$$
\begin{aligned}
&\|x(t)-y(t)\|\\
&\quad\le \|x(s)-y(s)\|+\limsup_{n\to\infty}\sum_{l=p}^{i}\delta^n_l h_n+\limsup_{m\to\infty}\sum_{l=q}^{j}\gamma^m_l k_m\\
&\qquad +L(\bar r,\bar r,\bar r)\limsup_{n,m\to\infty}\sum_{l=p}^{i}\|(x_n)_{t^n_{l-1}}-(y_m)_{t^n_{l-1}}\|_{PC}h_n+T\rho(2\delta)
\end{aligned}
$$

for every $\delta\in(0,\frac{T}{2})$. But since

$$
\begin{aligned}
\lim_{n\to\infty}\sum_{l=p}^{i}\delta^n_l h_n=\lim_{n\to\infty}\sum_{l=p}^{i}h_n[&G(t^n_l,(x_n)_{t^n_l},L_{t^n_l}(x_n))\\
&-G(t^n_l,y_{t^n_l},L_{t^n_l}(y)),x_n(t^n_l)-y(t^n_l)]_\lambda=
\end{aligned}
$$

$$= \int_s^t [G(\sigma, x_\sigma, L_\sigma(x)) \\ -G(\sigma, y_\sigma, L_\sigma(y)), x(\sigma) - y(\sigma)]_\lambda d\sigma$$

$$\lim_{m\to\infty} \sum_{l=q}^{j} \gamma_l^m k_m = \lim_{m\to\infty} \sum_{l=q}^{j} k_m \Big(\|G(s_l^m, (y_m)_{s_l^m}, L_{s_l^m}(y_m)) \\ -G(s_l^m, y_{s_l^m}, L_{s_l^m}(y))\| \\ + \frac{2\|y_m(s_l^m) - y(s_l^m)\|}{\lambda} \Big) = 0$$

and

$$\lim_{n,m\to\infty} \sum_{l=p}^{i} \|(x_n)_{t_{i-1}^n} - (y_m)_{t_{i-1}^n}\|_{PC} h_n = \int_s^t \|x_\sigma - y_\sigma\|_{PC} d\sigma$$

letting $\delta \to 0+$ in (4.3.12) we find

$$\begin{aligned} &\|x(t) - y(t)\| \\ &\quad \le \|x(s) - y(s)\| \\ &\quad\quad +L(\bar{r}, \bar{r}, \bar{r}) \int_s^t \|x_\sigma - y_\sigma\|_{PC} d\sigma \\ &\quad\quad + \int_s^t [G(\sigma, x_\sigma, L_\sigma(x)) - G(\sigma, y_\sigma, L_\sigma(y)), x(\sigma) - y(\sigma)]_\lambda d\sigma. \end{aligned}$$

Finally, letting $\lambda \to 0+$ above we obtain the desired inequality of this Proposition.

Now we consider the existence of an integral solution and a strong solution of (4.3.1).

THEOREM 4.3.2 *Assume that the assumptions of Theorem 4.3.1 are satisfied. Then there exists an integral solution $x(t)$ of (4.3.1) on $[-r, T]$.*

Proof. From Theorem 4.3.1 there exists a limit solution $x(t)$ of (4.3.1) on $[-r, T]$. We show that this $x(t)$ is an integral solution of (4.3.1) on $[-r, T]$. Let $x_n(t)$ be a method of lines with $\lim_{n\to\infty} x_n(t) = x(t)$ and $x_n(t) = \overline{x_n^n}(t)$, where

$$\overline{x_i^n}(t) = \begin{cases} \phi(t), & t \in [-r, 0] \\ x_j^n & t \in (t_{j-1}^n, t_j^n],\ j = 1, 2, \cdots, i-1 \\ x_i^n, & t \in (t_{i-1}^n, T] \end{cases}$$

and

$$\frac{x_i^n - x_{i-1}^n}{t_i^n - t_{i-1}^n} + A(t_i^n, (\overline{x_i^n})_{t_{i-1}^n})x_i^n \ni G(t_i^n, (\overline{x_i^n})_{t_i^n}, L_{t_i^n}(\overline{x_i^n}))$$

with $t_0^n = 0$, $t_n^n = T$ and $t_i^n = ih_n = i\dfrac{T}{n}$ for $i = 1, 2, \cdots, n$. Let $[u, v] \in A(\tau, x_\tau)$ for $\tau \in [0, T]$. Then there exists $z_i^n \in A(t_i^n, (\overline{x_{i-1}^n})_{t_{i-1}^n})x_i^n$ such that

$$(x_i^n - u) - (x_{i-1}^n - u) = h_n(G(t_i^n, (\overline{x_i^n})_{t_i^n}, L_{t_i^n}(\overline{x_i^n})) - z_i^n).$$

Then

$$\begin{aligned} h_n[v - z_i^n, x_i^n - u]_- &+ h_n[G(t_i^n, (\overline{x_i^n})_{t_i^n}, L_{t_i^n}(\overline{x_i^n})) - v, x_i^n - u]_+ \\ &\geq h_n[G(t_i^n, (\overline{x_i^n})_{t_i^n}, L_{t_i^n}(\overline{x_i^n})) - z_i^n, x_i^n - u]_- \\ &= [(x_i^n - u) - (x_{i-1}^n - u), x_i^n - u]_- \\ &\geq [x_i^n - u, x_i^n - u]_- - [x_{i-1}^n - u, x_i^n - u]_+ \\ &\geq \|x_i^n - u\| - \|x_{i-1}^n - u\|. \end{aligned} \tag{4.3.13}$$

Since

$$\begin{aligned} &\|x_i^n - u\| \\ &\quad = \|J_\lambda(t_i^n, (\overline{x_i^n})_{t_{i-1}^n})(x_i^n + \lambda z_i^n) - J_\lambda(\tau, x_\tau)(u + \lambda v)\| \\ &\quad \leq \|J_\lambda(t_i^n, (\overline{x_i^n})_{t_{i-1}^n})(x_i^n + \lambda z_i^n) - J_\lambda(t_i^n, (\overline{x_i^n})_{t_{i-1}^n})(u + \lambda v)\| \\ &\qquad + \|J_\lambda(t_i^n, (\overline{x_i^n})_{t_{i-1}^n})(u + \lambda v) - J_\lambda(\tau, x_\tau)(u + \lambda v)\| \\ &\quad \leq \|(x_i^n + \lambda z_i^n) - (u + \lambda v)\| \\ &\qquad + \|A_\lambda(t_i^n, (\overline{x_i^n})_{t_{i-1}^n})(x_i^n + \lambda z_i^n) - A_\lambda(\tau, x_\tau)(u + \lambda v)\| \end{aligned}$$

we obtain

$$\begin{aligned} &\frac{\|x_i^n - u\| - \|(x_i^n + \lambda z_i^n) - (u + \lambda v)\|}{\lambda} \\ &\quad \leq L(\|(\overline{x_i^n})_{t_{i-1}^n}\|_{PC}, \|x_\tau\|_{PC}, \|u + \lambda v\|) \\ &\qquad \times (|t_i^n - \tau|(1 + \|A_\lambda(\tau, x_\tau)(u + \lambda v)\|) \\ &\qquad + \|(x_n)_{t_{i-1}^n} - x_\tau\|_{PC}). \end{aligned}$$

Here we have used the fact that $(\overline{x_i^n})_{t_{i-1}^n} = (\overline{x_n^n})_{t_{i-1}^n} = (x_n)_{t_{i-1}^n}$. We now observe that

$$\|A_\lambda(\tau, x_\tau)(u + \lambda v)\| =$$

$$= \frac{1}{\lambda}\|(u+\lambda v) - J_\lambda(\tau, x_\tau)(u+\lambda v)\|$$
$$= \frac{1}{\lambda}\|\lambda v\| = \|v\|$$

and

$$\begin{aligned}
&\|(x_n)_{t_{i-1}^n} - x_\tau\| \\
&\quad = \sup_{t\in[-r,0]} \|x_n(t_{i-1}^n + t) - x(\tau + t)\| \\
&\quad \le \sup_{t\in[-r,0]} \|x_n(t_{i-1}^n + t) - x_n(\tau + t)\| \\
&\qquad + \sup_{t\in[-r,0]} \|x_n(\tau + t) - x(\tau + t)\| \\
&\quad \le \sup_{t\in[-r,0]} L_{11}(\mid t_{i-1}^n - \tau \mid + h_n) + \sup_{t\in[-r,T]} \|x_n(t) - x(t)\| \\
&\quad = K_{11}|t_{i-1}^n - \tau| + K_{11}h_n + \|x_n - x\|_T \\
&\quad = K_{11}|t_i^n - \tau| + 2K_{11}h_n + \|x_n - x\|_T.
\end{aligned}$$

Combining the above we arrive at

$$\begin{aligned}
&\frac{\|x_i^n - u\| - \|(x_i^n + \lambda z_i^n) - (u + \lambda v)\|}{\lambda} \\
&\quad \le L(\bar{r}, \bar{r}, \|u\| + \lambda\|v\|)(|t_i^n - \tau|(1 + \|y\| + K_{11}) \\
&\qquad + 2K_{11}h_n + \|x_n - x\|_T).
\end{aligned}$$

Letting $\lambda \to 0+$ we obtain

$$\begin{aligned}
&[z_i^n - v, x_i^n - u]_- \\
&\quad \le L(\bar{r}, \bar{r}, \|u\|)(|t_i^n - \tau|(1 + \|y\| + K_{11}) + 2K_{11}h_n + \|x_n - x\|_T).
\end{aligned}$$

Combining this with (4.3.14) we have

$$\begin{aligned}
&\|x_i^n - u\| - \|x_{i-1}^n - u\| \\
&\quad \le h_n L(\bar{r}, \bar{r}, \|u\|)(|t_i^n - \tau|(1 + \|y\| + K_{11}) + 2K_{11}h_n + \|x_n - x\|_T) \\
&\qquad + h_n[G(t_i^n, (\overline{x_i^n})_{t_i^n}, L_{t_i^n}(\overline{x_i^n})) - v, x_i^n - u]_+,
\end{aligned}$$

which yields

$$\begin{aligned}
&\|x_i^n - u\| - \|x_j^n - u\| \qquad\qquad (4.3.14) \\
&\quad \le \sum_{l=j+1}^{i} h_n([G(t_l^n, (\overline{x_l^n})_{t_l^n}, L_{t_l^n}(\overline{x_l^n})) - v, x_l^n - u]_+ +
\end{aligned}$$

$$+L(\bar{r},\bar{r},\|u\|)(|t_l^n-\tau|(1+\|y\|+K_{11})+2K_{11}h_n+\|x_n-x\|_T))$$

$$=\sum_{l=j+1}^{i} h_n([G(t_l^n,(\overline{x_l^n})_{t_l^n},L_{t_l^n}(\overline{x_l^n}))-v,x_l^n-u]_+ + C|t_i^n-\tau|$$

$$+2L(\bar{r},\bar{r},\|u\|)K_{11}h_n+L(\bar{r},\bar{r},\|u\|)\|x_n-x\|_T),$$

where $C=L(\bar{r},\bar{r},\|u\|)(1+\|y\|+K_{11})$.

Now let $t\in(t_{i-1}^n,t_i^n]$ and $s\in(t_{j-1}^n,t_j^n]$. Since

$$\lim_{n\to\infty}\sum_{l=j+1}^{i} 2L(\bar{r},\bar{r},\|u\|)K_{11}h_n^2$$
$$\leq 2L(\bar{r},\bar{r},\|u\|)K_{11}ih_n^2 \leq 2L(\bar{r},\bar{r},\|u\|)K_{11}Th_n=0$$

and

$$\lim_{n\to\infty}\sum_{l=j+1}^{i} L(\bar{r},\bar{r},\|u\|)\|x_n-x\|_T h_n$$
$$\leq L(\bar{r},\bar{r},\|u\|)\|x_n-x\|_T ih_n \leq L(\bar{r},\bar{r},\|u\|)\|x_n-x\|_T Th_n=0$$

letting $n\to\infty$ in (4.3.15) we have

$$\|x(t)-u\|-\|x(s)-u\|$$
$$\leq \int_s^t([G(\sigma,x_\sigma,L_\sigma(x))-v,x(\sigma)-u]_+ + C|\sigma-\tau|)d\sigma.$$

Since $x(t)=\phi(t)$ and $x(t)$ is continuous on $[0,T]$ $x(t)$ is an integral solution of (4.3.1) on $[-r,T]$.

THEOREM 4.3.3 *Assume that the assumptions of Theorem 4.3.1 are satisfied. Let X be reflexive. Then there exists a strong solution of* (4.3.1) *on* $[-r,T]$

Proof. From Theorem 4.3.1 there exists a limit solution $x(t)$ of (4.3.1) on $[-t,T]$ and $x(t)$ is Lipschitz continuous with Lipschitz constant $K_{11}>0$ on $[0,T]$. Thus $x(t)$ is absolutely continuous on $[0,T]$. Since X is reflexive $x(t)$ is differentiable a.e. on $(0,T)$.

Let $x(t)$ be differentiable at $t=t_0$ and $h>0$. Since from Theorem 4.3.2 $x(t)$ is an integral solution of (4.3.1) on $[-r,T]$ we obtain

$$\|x(t_0+h)-u\|-\|x(t_0)-u\|$$
$$\leq \int_{t_0}^{t_0+h}([G(\sigma,x_\sigma,L_\sigma(x))-v,x(\sigma)-u]_+ + C|\sigma-t_0|)d\sigma$$

for $[u, v] \in A(t_0, x_{t_0})$, where $C = L(\bar{r}, \bar{r}, \|u\|)(1 + \|y\| + K_{11})$. Dividing the above inequality by h and then letting $h \to 0+$ we see that

$$\left[\frac{dx}{dt}(t_0), x(t_0) - u\right]_+ \leq [G(t_0, x_{t_0}, L_{t_0}(x)) - v, x(t_0) - u]_+$$

for $[u, v] \in A(t_0, x_{t_0})$. Thus

$$\left[-\frac{dx}{dt}(t_0) + G(t_0, x_{t_0}, L_{t_0}(x)) - v, x(t_0) - u\right]_+ \leq 0$$

for $[u, v] \in A(t_0, x_{t_0})$. From the maximality of $A(t_0, x_{t_0})$ we have

$$\frac{dx}{dt}(t_0) + A(t_0, x_{t_0})x(t_0) \ni G(t_0, x_{t_0}, L_{t_0}(x)).$$

Hence $x(t)$ is a strong solution of (4.3.1) on $[-r, T]$.

4.4. Examples and Applications

We give an example to illustrate the application of one of the results in this section

Let Ω be a bounded open subset of $\mathbf{R}^n$ with sufficiently smooth boundary $\partial\Omega$, where $\mathbf{R} = (-\infty, \infty)$ and $n \geq 2$. We define

$$|\alpha| = \alpha_1 + \alpha_2 + \cdots + \alpha_n \text{ and } D^\alpha = D_1^{\alpha_1} D_2^{\alpha_2} \cdots D_n^{\alpha_n}$$

for a multi-index $\alpha = (\alpha_1, \alpha_2, \cdots, \alpha_n)$, where $D_i = \dfrac{\partial}{\partial x_i}$ for $i = 1, 2, \cdots, n$. Let $\mathbf{R}^{n_m}$ be the space of all real vectors of the form $\xi = \{\xi_\alpha \mid |\alpha| \leq m\}$. The Sobolev space $W^{m,2}(\Omega)$ is a separable Hilbert space with inner product $\langle \cdot, \cdot \rangle_{m,2} : W^{m,2}(\Omega) \times W^{m,2}(\Omega) \to \mathbf{R}$ defined by

$$\langle u, v \rangle_{m,2} = \sum_{|\alpha| \leq m} \langle D^\alpha u, D^\alpha v \rangle_{L^2(\Omega)}$$

for every $u, v \in W^{m,2}(\Omega)$, and thus, with norm $\| \cdot \|_{m,2}$ defined by

$$\|u\|_{m,2} = \langle u, u \rangle_{m,2}^{\frac{1}{2}} = \sum_{|\alpha| \leq m} \langle D^\alpha u, D^\alpha u \rangle_{L^2(\Omega)}^{\frac{1}{2}}$$

for every $u \in W^{m,2}(\Omega)$.

Also $W_0^{m,2}(\Omega) = \overline{C_0^\infty(\Omega)}$ is an another separable Hilbert space in $W^{m,2}(\Omega)$.

We consider the initial-boundary value problem

$$\begin{cases} \dfrac{\partial z(t,\omega)}{\partial t} + \mathbf{A}(t,\omega,z(t-r,\omega),z(t,\omega)) \\ \quad = g(t,z(t-r,\omega)), \displaystyle\int_0^t K(t,s,z(\sigma - r,\omega))ds, \quad a.e. t \in (0,T) \\ D^\alpha z(t,\omega) = 0, \quad a.e.\ t \in (0,T),\ \omega \in \partial\Omega,\ |\alpha| \le m, \\ z(t,\omega) = \phi(t,\omega), \quad t \in [-r,0],\ \omega \in \Omega, \end{cases} \tag{4.4.1}$$

where

$$\mathbf{A}(t,\omega,v,w) = \sum_{|\alpha|\le m} (-1)^{|\alpha|} D^\alpha b_\alpha(t,\omega,\xi(v)) A_\alpha(\omega,\xi(w))$$

is the elliptic differential operator with $\xi(w) = \{D^\alpha w \mid |\alpha| \le m\}$ and $\phi : [-r,0] \times \Omega \to \mathbf{R}$ for $r > 0$ is a given function.

Now we consider the following conditions.

$(B1)_4$ For every multi-index α, $A_\alpha : \Omega \times \mathbf{R}^{n_m} \to \mathbf{R}$ is bounded and Lipschitz continuous in the second variable. That is, there exist constants $M > 0$ and $M_0 > 0$ such that $|A_\alpha(\omega,\xi)| \le M$ and

$$|A_\alpha(\omega,\xi_1) - A_\alpha(\omega,\xi_2)| \le M_0|\xi_1 - \xi_2|$$

forevery ω, $\omega_i \in \Omega$ and ξ, $\xi_i \in \mathbf{R}^{n_m}$, $i = 1,2$; $(B2)_4$ For $\omega \in \Omega$ and $\xi_i \in \mathbf{R}^{n_m}$, $i = 1,2$

$$\sum_{|\alpha|\le m} (A_\alpha(\omega,\xi_1) - A_\alpha(\omega,\xi_2))((\xi_1)_\alpha - (\xi_2)_\alpha) \ge 0;$$

$(B3)_4$ For every multi-index α, the function $b_\alpha : [0,T] \times \Omega \times \mathbf{R}^{n_m}$ is defined, continuous and there exist constant $M_i > 0$ such that

$$|b_\alpha(t_1,\omega,\xi_1) - b_\alpha(t_2,\omega,\xi_2)| \le M_1|t_1 - t_2| + M_2|\xi_1 - \xi_2|$$

for every $t_i \in [0,T]$, $\omega \in \Omega$ and $\xi_i \in \mathbf{R}$, $i = 1,2$;

$(B4)_4$ The function $K : [0,T] \times [0,T] \times \mathbf{R} \to \mathbf{R}$ is defined and there exists a constant $M_3 > 0$ such that

$$|D^\alpha K(t,s,v_1) - D^\alpha K(t,s,v_2)| \le M_3(|D^\alpha v_1 - D^\alpha v_2|$$

for every $t, s \in [0,T]$ and $v_i \in \mathbf{R}$, $i = 1,2$, and

$$|D^\alpha K(t_1,s_1,v) - D^\alpha K(t_2,s_2,v)| \le |D^\alpha v|(|t_1 - t_2| + |s_1 - s_2|)$$

for every $t_i, s_i \in [0,T]$, $i = 1,2$, and $v \in \mathbf{R}$;

$(B5)_4$ The function $g : [0,T] \times \mathbf{R} \times \mathbf{R} \to \mathbf{R}$ is defined and there exists a constant $M_4 > 0$ such that

$$|D^\alpha g(t, v_1, v_2) - D^\alpha g(t, w_1, w_2)| \\ \leq M_4(|D^\alpha v_1 - D^\alpha w_1| + |D^\alpha v_2 - D^\alpha w_2|)$$

for every $t \in [0,T]$ and $v_i, w_i \in \mathbf{R}$, $i = 1, 2$, and

$$|D^\alpha g(t_1, v_1, v_2) - D^\alpha g(t_2, v_1, v_2)| \leq (|D^\alpha v_1| + |D^\alpha v_2|)|t_1 - t_2|$$

for every $t_i \in [0,T]$ and $v_i \in \mathbf{R}$, $i = 1, 2$.

THEOREM 4.4.1 *Assume that $(B1)_4$–$(B5)_4$ are satisfied. Let $\phi : [-r, 0] \times \Omega \to \boldsymbol{R}$ be Lipschitz continuous with Lipschitz constant $c > 0$. Then* (4.4.1) *has a unique solution on $[-r, T]$.*

Proof. Let $X = W_0^{m,2}(\Omega)$. We define an operator $\widehat{A}(t, v) : X \to X$ for every $t \in [0,T]$ and $v \in X$ by

$$\langle \widehat{A}(t, v)w, y\rangle_{m,2} \\ = \sum_{|\alpha| \leq m} \int_\Omega b_\alpha(t, \omega, \xi(v(\omega))) A_\alpha(\omega, \xi(w(\omega))) D^\alpha y(\omega) d\omega.$$

for every w, $y \in X$. Then $\widehat{A}(t, v)$ is continuous monotone and bounded on X, and satisfies the Lipschitz condition

$$\|\widehat{A}(t_1, v_1)w - \widehat{A}(t_2, v_2)w\|_{m,2} \leq M_5|t_1 - t_2| + M_6\|v_1 - v_2\|_{m,2} \quad (4.4.2)$$

for every $t_i \in [0,T]$, $v_i \in X$, $i = 1, 2$, $w \in X$ and for some positive constants $M_5 > 0$ and $M_6 > 0$.

In fact, from $(B1)_4$ for $t \in [0,T]$, v, $w_i \in X$, $i = 1, 2$, and $y \in X$

$$|\langle \widehat{A}(t, v)w_1 - \widehat{A}(t, v)w_2, y\rangle_{m,2}| \\ \leq \sum_{|\alpha| \leq m} \int_\Omega b_\alpha(t, \omega, \xi(v(\omega)))|A_\alpha(\omega, \xi(w_1(\omega))) - A_\alpha(\omega, \xi(w_2(\omega)))| \\ \times |D^\alpha y(\omega)| d\omega \\ \leq \sum_{|\alpha| \leq m} \int_\Omega b_\alpha(t, \omega, \xi(v(\omega))) M_0 |\xi(w_1(\omega)) - \xi(w_2(\omega))||D^\alpha y(\omega)| d\omega \\ \leq B_0\|w_1 - w_2\|_{m,2}\|y\|_{m,2}$$

for some $B_0 > 0$. Thus

$$\|\widehat{A}(t, v)w_1 - \widehat{A}(t, v)w_2\|_{m,2} \leq B_0\|w_1 - w_2\|_{m,2}.$$

[12] D. W. Brewer (1980). The asymptotic stability of a nonlinear functional differnetial equation of infinite delay, *Houston J. Math.* **6**, 321-330.

[13] D. W. Brewer (1982). Locally Lipschitz continuous functional differential equations and nonlinear semigroups, *Illinois J. Math.* **26**, 374-381.

[14] H. Brézis (1971). Monotonicity methods in Hilbert spaces and some applications to nonlinear partial differential equations, *Cont. Nonl. Funct. Anal.* edited by Zarantonello, (Academic Press: New York, USA), 101-156.

[15] H. Brézis (1973). *Opérateurs Maximaux Monotones et Semi-Groupes de Contractions dans les Espaces de Hilbert*, (North-Holland Publishing Company: Amsterdam).

[16] H. Brézis (1974). New results concerning monotone operators and nonlinear semigroups, *Anal. Nonlin. Probl., RIMS*, 2-27

[17] H. Brézis and A. Pazy (1970). Accretive sets and differential equations in Banach spaces, *Israel J. Math.* **8**, 367-383.

[18] B. C. Burch and J. A. Goldstein (1978). Nonlinear semigroups and problems in heat conduction, *Houston J. Math.* **4**, 311-328.

[19] B. Calvert and A. G. Kartsatos (1987). On the compactness of the nonlinear evolution operator in a Banach space, *Bull. London Math. Soc.* **19** (1987), 551-558.

[20] E. A. Coddington and N. Levinson (1955) *Theory of Ordinary Differential Equations*, (McGraw-Hill Book Company, Inc.: New York).

[21] C. Corduneanu and V. Lakshmikantham (1980). Equations with unbounded delay : a survey, *Nonlin. Anal. TMA* **4**, 831-877.

[22] M. G. Crandall (1971). Semigroups of nonlinear transformations in Banach spaces, *Controb. Nonl. Funct. Anal.*, (Academic Press: New York), 157-179.

[23] M. G. Crandall (1972). The semigroup approach to first order quaisilinear equations in several space variables, *Israel J. Math.* **12**, 108-132.

[24] M. G. Crandall (1973). A generalized domain for semigroup generators, *Proc. Amer. Math. Soc.* **37**, 434-440.

[25] M. G. Crandall and L. C. Evans (1975). On the relation of the operator $\frac{\partial}{\partial s}+\frac{\partial}{\partial \tau}$ to evolution governed by accretive operators, *Israel J. Math.* **21**, 261-278.

[26] M. G. Crandall and T. M. Liggett (1971). Generation of semi-groups of nonlinear transformations on general Banach spaces, *Amer. J. Math.* **93**, 265-298.

[27] M. G. Crandall, S. O. Londen and J. A. Nohel (1978). An abstract nonlinear Volterra integro-differential equation, *J. Math. Anal. Appl.* **64**, 701-735.

[28] M. G. Crandall and J. A. Nohel (1978). An abstract functional differential equation and a related nonlinear Volterra equation, *Israel J. Math.* **29**, 313-328.

[29] M. G. Crandall and A. Pazy (1972). Nonlinear evolution equations in Banach spaces, *Israel J. Math.* **11**, 57-94.

[30] M. G. Crandall and P. E. Souganidis (1989). On nonlinear equations of evolution, *Nonlin. Anal. TMA* **13**, 1375-1392.

[31] E. B. Davies (1980). *One-Parameter Semigroups*, (Academic Press: London).

[32] K. Deimling (1985). *Nonlinear Functional Analysis*, (Springer–Verlag: Berlin, Germany).

[33] M. B. Dhakne and B. G. Pachpatte (1988). On a general class of abstract functional integro-differential equations, *Indian J. Pure Appl. Math.* **19**, 728-746.

[34] M. B. Dhakne and B. G. Pachpatte (1988). On perturbed abstract functional integro-differential equations, *Acta. Math. Sci.* **8**, 263-282.

[35] J. Diestel (1977). Remarks on weak compactness in $L^1(\mu, X)$, *Glasgow Math. J.* **18**, 87-91.

[36] J. Diestel, W. M. Ruess and W. Schachermayer (1993). Weak compactness in $L^1(\mu, X)$, *Proc. Amer. Soc.* **118**, 447-453.

[37] N. Dunford and J. T. Schwartz (1958). *Linear Operators, Part I*, (Intersciences Publishing Inc.: New York).

[38] J. Dyson and R. Villella–Bressan (1975/76). Functional differential equations and nonlinear evolution operators, *Proc. Roy. Soc. Edinb.* **75A**, 223-234.

[39] J. Dyson and R. Villella–Bressan (1977). Nonlinear functional differential equations in L^1 spaces, *Nonlin. Anal. TMA* **1**, 383-395.

[40] J. Dyson and R. Villella–Bressan (1977). Some remarks on the asymptotic behavior of a non-autonomous, nonlinear functional differential equation, *J. Diff. Eqns.* **25**, 275-287.

[41] J. Dyson and R. Villella–Bressan (1979). A nonlinear boundary value problem for a functional differential equation, *J. Diff. Eqns.* **34**, 273-290.

[42] J. Dyson and R. Villella–Bressan (1979). Semigroups of translations associated with functional and functional differential equations, *Proc. Roy. Soc. Edinb.* **82A**, 171-188.

[43] J. Dyson and R. Villella–Bressan (1985). A semigroup approach to nonlinear non-autonomous neutral functional differential equations, *J. Diff. Eqns.* **59**, 206-228.

[44] J. Dyson and R. Villella–Bressan (1991). Propagation of solutions of a nonlinear functional differential equation, *Diff. Integ. Eqns.* **4**, 293-303.

[45] J. Dyson and R. Villella–Bressan (1995). Integral solutions of locally Lipschitz continuous functional differential equations, *Diff. Integ. Eqns.* **8**, 1157-1166.

[46] J. Dyson and R. Villella–Bressan (1995). On integral solutions of nonautonomous delay equations and their propagation, *Nonlin. Anal. TMA* **24**, 693-710.

[47] L. C. Evans (1977). Nonlinear evolution equations in an arbitrary Banach space, *Israel J. Math.* **26**, 1-42.

[48] W. E. Fitzgibbon (1977). Stability for abstract nonlinear Volterra equations involving finite delay, *J. Math. Anal. Appl.* **60**, 429-434.

[49] W. E. Fitzgibbon (1978). Semilinear functional differential equations in Banach spaces, *J. Diff. Eqns.* **29**, 1-14.

[50] W. E. Fitzgibbon (1985). Convergence theorems for semilinear Volterra equations with infinite delay, *J. Integ. Eqns.* **8**, 261-274.

[51] W. E. Fitzgibbon (1990). Asymptotic behavior of solutions to a class of Volterra integrodifferential equations, *J. Math. Anal. Appl.* **146**, 429-434.

[52] H. Flaschka and M. J. Leitman (1975). On semigroups of nonlinear operators and the solution of the functional differential equation $\dot{x}(t) = F(x_t)$, *J. Math. Anal. Appl.* **49**, 649-658.

[53] J. A. Goldstein (1985). *Semigroups of Linear Operators and Applications*, (Oxford University Press: New York).

[54] Z. Guan and A. G. Kartsatos (1995). Ranges of perturbed maximal monotone and m-accretive operators in Banach spaces, *Trans. Amer. Math. Soc.* **347**, 2403-2435.

[55] S. Gutman (1982). Compact perturbations of m-accretive operators in general Banach spaces, *SIAM J. Math. Anal.* **13**, 789-800.

[56] S. Gutman (1983). Evolutions governed by m-accretive plus compact operators, *Nonlin. Anal. TMA* **7**, 707-715.

[57] K. S. Ha (1979). Sur des semigroupes nonlinéaires dans les espaces $L^\infty(\Omega)$, *J. Math. Soc. Japan* **31**, 593-622.

[58] K. S. Ha (1987). *Theory of Nonlinear Partial Differential Equations with Nonlinear Semigroups* (in Korean), (Mineumsa: Seoul).

[59] K. S. Ha (1999). *Theory of Functional Differential Equations* (in Korean), (Arche: Seoul).

[60] K. S. Ha (2002). Existence of solutions for nonlinear functional differential equations in L^p spaces, *Differential Equations and Application, Proc. Sixth Int'l Conf. Nonl. Funct. Anal. Appl. 2000, in Masan and Jinju, Korea,* **2**, 43-54, (Nova Science Publshers, Inc.: New York).

[61] K. S. Ha and K. Shin (1996). Existence of solutions of nonlinear functional differential equations in L^p spaces, *Proc. World Congr. Nonl. Analysts '92* **3**, 2249-2259.

[62] K. S. Ha, K. Shin and B. J. Jin (1995). Existence of solutions of nonlinear functional integro-differential equations in Banach spaces, *Diff. Integ. Eqns.* **8**, 553-566.

[63] J. R. Haddock and W. E. Hornor (1988). Precompactness and convergence in norm of positive orbits in a certain fading memory space, *Funk. Ekva.* **31**, 349-361.

[64] J. K. Hale (1969). *Ordinary Differential Equations*, (Wiley-Interscience: New York).

[65] J. K. Hale (1971). *Functional Differential Equations*, (Springer–Verlag: Berlin).

[66] J. K. Hale (1974). Functional differential equations with infinite delays, *J. Math. Anal. Appl.* **48**, 276-283.

[67] J. K. Hale (1977). *Theory of Functional Differential Equations*, (Springer–Verlag: Berlin).

[68] E. Hille and R. S. Phillips (1957). *Functional Analysis and Semigroups*, (Amer. Math. Soc.: Rhode Island).

[69] J. G. Jeong and K. Shin (1996). Existence of solution of nonlinear functional differential equations in general Banach spaces, *Comm. Korean Math. Soc.* **11**, 1003-1013.

[70] B. J. Jin (1993). On the existence of solutions for nonlinear functional integro-differential equations in general Banach spaces, (Dissertation), Pusan National University.

[71] J. Kačur (1975). Application of Rothe's method to nonlinear evolution equations, *Mat. Čas* **25**, 63-81.

[72] J. Kačur (1978). Method of Rothe and nonlinear parabolic boundary value problems of arbitrary order, *Czech. Math. J.* **28**, 507-524.

[73] F. Kappel and W. Schappacher (1978). Autonomous nonlinear functional equations and averaging approximations, *Nonlin. Anal. TMA* **2**, 391-472.

[74] F. Kappel and W. Schappacher (1979). Non-linear functional differential equations and abstract integral equations, *Proc. Royal Soc. Edinb.* **81A**, 71-91.

[75] F. Kappel and W. Schappacher (1980). Some considerations to the fundamental theorey of infinite delay equations, *J. Diff. Eqns.* **37**, 141-183.

[76] A. G. Kartsatos (1978). Perturbations of m-accretive operators and quasi-linear evolution equations, *J. Math. Soc. Japan* **30**, 75-84.

[77] A. G. Kartsatos (1988). A direct method for existence of evolution operators associated with functional evolutions in general Banach spaces, *Funk. Ekva.* **31**, 89-102.

[78] A. G. Kartsatos (1988). Applications of nonlinear perturbation theorey to the existence of methods of lines for functional evolutions in reflexive Banach spaces, *Ann. Univ. Marie Curie Sklodowska* **42**, 41-51.

[79] A. G. Kartsatos (1991). The existence of bounded solutions on the real line of perturbed nonlinear evolution equations in general Banach spaces, *Nonlin. Anal. TMA* **17**, 1085-1092.

[80] A. G. Kartsatos (1993). On compact perturbations and compact resolvents of nonlinear m-accretive operators in Banach spaces, *Proc. Amer. Math. Soc.* **119**, 1189-1199.

[81] A. G. Kartsatos (1995). On the construction of the methods of lines for functional evolutions in general Banach spaces, *Nonlin. Anal. TMA* **25**, 1321-1331.

[82] A. G. Kartsatos (1996). New results in the perturbation theory of maximal monomone and m-accretive operators, *Trans. Amer. Math. Soc.* **248**, 1663-1707.

[83] A. G. Kartsatos (1996). Recent results invoving compact perturbations and compact resolvents of accretive operators in Banach spaces, *Proceedings of World Congress of Nonlinear Analysts '92* **3**, 2197-2222.

[84] A. G. Kartsatos and X. Liu (1997). On the construction and the convergence of the method of lines for quasi-nonlinear functional evolutions in general Banach spaces, *Nonlin. Anal. TMA* **29**, 385-414.

[85] A. G. Kartsatos and M. E. Parrott (1982). Existence of solutions and Galerkin approximations for nonlinear functional evolution equations, *Tôhoku Math. J.* **34**, 509-523.

[86] A. G. Kartsatos and M. E. Parrott (1983). Convergence of the Kato approximants for evolution equations involving functional perturbations, *J. Diff. Eqns.* **47**, 358-377.

[87] A. G. Kartsatos and M. E. Parrott (1983). Global solutions of functional evolution equations involving locally defined Lipschitzian perturbations, *J. London Math. Soc.* **27**, 306-316.

[88] A. G. Kartsatos and M. E. Parrott (1984). A simplified approach to the existence and stability problem of a functional evolution equation in a general Banach space, 'Infinite Dimensional Systems' edited by F. Kappel and W. Schappacher, *Lecture Notes in Math.* **1076**, 115-122, (Springer–Verlag: Berlin).

[89] A. G. Kartsatos and M. E. Parrott (1984). A method of lines for a nonlinear abstract functional evolution equation, *Trans. Amer. Math. Soc.* **286**, 73-89.

[90] A. G. Kartsatos and M. E. Parrott (1984). Functional evolution equations involving time dependent maximal monotone operators in Banach spaces, *Nonlin. Anal. TMA* **8**, 817-833.

[91] A. G. Kartsatos and M. E. Parrott (1988). The weak solution of a functional differential equation in a general Banach space, *J. Diff. Eqns.* **75**, 290-302.

[92] A. G. Kartsatos and K. Shin (1993). Solvability of functional evolutions via compactness methods in general Banach spaces, *Nonlin. Anal. TMA* **7**, 517-535.

[93] A. G. Kartsatos and K. Shin (1994). The method of lines and the approximation of zeros of m-accretive operators in general Banach spaces, *J. Diff. Eqns.* **113**, 128-149.

[94] A. G. Kartsatos and W. R. Zigler (1976). Rothe's methods and weak solutions of perturbed evolution equations in reflexive Banach spaces, *Math. Ann.* **219**, 159-166.

[95] T. Kato (1967). Nonlinear semigroups and evolution equations, *J. Math. Soc. Japan* **19**, 508-520.

[96] P. M. D. Kahn and B. B. Pachpatte (1987). On quasi-linear Volterra integro-differential equation in a Banach space, *Indian J. Pure Appl. Math.* **18**, 32-49.

[97] K. Kobayashi (1983), On difference approximation of time dependent nonlinear evolution equations in Banach spaces, *Mem. Sagami, Institute Tech.* **17**, 59-69.

[98] Y. Kobayashi (1975) Difference approximation of Cauchy problems for quasi-dissipative operators and generation of nonlinear semigroups, *J. Math. Soc. Japan* **27**, 640-665.

[99] Y. Komura (1967). Nonlinear semigroups in Hilbert spaces, *J. Math. Soc. Japan* **19**, 493-507.

[100] Y. Komura (1969). Differentiability of nonlinear semigroups, *J. Math. Soc. Japan* **21**, 375-402.

[101] V. Lakshmikantham and S. Leela (1981). *Nonlinear Differntial Equations in Abstract Spaces*, (Pergamon Press: New York).

[102] G. Lumer and R. S. Phillips (1961). Dissipative operators in a Banach space, *Pacific J. Math.* **11**, 679-698.

[103] G. Makay (1994). On the asymptotic stability of the solutions of functional differential equations with infinite delay, *J. Diff. Eqns.* **108**, 139-151.

[104] R. H. Martin Jr. (1976). *Nonlinear Operators and Differential Equations in Banach Spaces*, (John Wiley and Sons: New York).

[105] E. Mitidieri and I. I. Vrabie (1987) Existence for nonlinear functional differential equations, *Hiroshima Math. J.* **17**, 627-649.

[106] I. Miyadera (1977). *Nonlinear Semigroups* (in Japanese), (Kinokuniya: Tokyo).

[107] C. Morales (1986). Zero for strongly accretive set-valued mappings, *Comm. Math. Univ. Caro* **27**, 455-469.

[108] G. Morsanu (1988). *Nonlinear Evolution Equations and Applications*, (D. Reidel Publishing Company: Dordrecht).

[109] R. Nagel (1986). *One-Parameter Semigroups of Positive Operators*, Lecture Notes on Mathematics 1184, (Springer–Verlag: Berlin).

[110] J. Nečas (1974). Application of Rothe's method to abstract parabolic equations, *Czech. Math. J.* **24**, 496-500.

[111] S. K. Ntouyas, Y. G. Sficas and P. Ch. Tsamatos (1994). Existence results for initial value problems for neutral functional differential equations, *J. Diff. Eqns.* **114**, 527-537.

[112] B. G. Pachpatte (1973). A note on Gronwall–Bellman inequality, *J. Math. Anal. Appl.* **44**, 758-762.

[113] B. G. Pachpatte (1975). On some integro-differential equations in Banach spaces, *Bull. Australian Math. Soc.* **12**, 337-350.

[114] M. E. Parrott (1982). Representation and approximation of generalized solutions of a nonlinear functional differential equation, *Nonlin. Anal. TMA* **6**, 307-318.

[115] M. E. Parrott (1992). Linearized stability and irreducibility for a functional differential equation, *SIAM Math. Anal.* **23**, 649-661.

[116] D. Pascali and S. Sburlan (1978). *Nonlinear Mappings of Monotone Type*, (Sijthoff and Noordhoff International Publishers: Burcharest).

[117] N. H. Pavel (1987). *Nonlinear Evolution Operators and Semigroups*, Lecture Notes in Mathematics 1260, (Springer–Verlag: Berlin).

[118] A. Pazy (1983). *Semigroups of Linear Operators and Applications to Partial Differential Equations*, (Springer–Verlag: Berlin).

[119] A. T. Plant (1977). Nonlinear semigroups of translations in Banach space generated by functional differential equations, *J. Math. Anal. Appl.* **60**, 67-74.

[120] A. T. Plant (1977). Stability of nonlinear functional differential equations using weighted norms, *Houston J. Math.* **3**, 99-108.

[121] E. Rothe (1930). Zweidimensionale parabolische Randwert aufgaben als Grenzfall eindimensionaler Rabdwertaufgaben, *Math. Ann.* **102**, 650-670.

[122] W. M. Ruess (1993). The evolution operator approach to functional differential equations with delay, *Proc. Amer. Math. Soc.* **119**, 783-791.

[123] W. M. Ruess (1996). Existence of solutions to partial functional differential equations with delay, *Proceedings of Theory and Applications of Nonlinear Operators of Accretive and Monotone type* (ed. A. G. Kartsatos), (Marcel Dekker: New York), 259-288.

[124] W. M. Ruess (1999). Existence and stability of solutions to partial functional differential equations with delay, *Advances Diff. Eqns.* **4**, 843-876.

[125] W. M. Ruess and W. H. Summers (1994). Operator semigroups for functional differential equations with infinite delay, *Trans. Amer. Math. Soc.* **341**, 695-719.

[126] W. M. Ruess and W. H. Summers (1996). Almost periodicity and stability for solutions to functional differential equations with infinte delay, *Diff. Integ. Eqns.* **9**, 1225-1252.

[127] K. Sawano (1982). Some considerations on the fundamental theorems for functional differential equations with infinite delay, *Funk. Ekva.* **25**, 97-104.

[128] N. Shioji (1996). Local existence theorems for nonlinear differential equations and compactness of integral solutions in $L^p(0,T;X)$, *Nonlin. Anal. TMA* **26**, 799-811.

[129] R. Schöneberg (1978). Zero of nonlinear monotone operators in Hilbert space, *Canadian Math. Bull.* **21**, 213-219.

[130] N. Tanaka (1988). Nonlinear nonautonomous differential equations, *RIMS, Kokyuroku* **647**, 36-56.

[131] N. Tanaka (1988). On the existence of solutions for functional evolution equations, *Nonlin. Anal. TMA* **12**, 1087-1104.

[132] C. C. Travis and G. F. Webb (1974). Existence and stability for partial functional differential equations, *Trans. Amer. Math. Soc.* **200**, 395-418.

[133] C. C. Travis and G. F. Webb (1976). Partial differential equations with deviating arguments in the time variable, *J. Math. Anal. Appl.* **56**, 397-409.

[134] I. I. Vrabie (1982). An existence result for a class of nonlinear evolution equations in Banach spaces, *Nonlin. Anal. TMA* **6**, 711-722.

[135] I. I. Vrabie (1987). *Compactness Methods for Nonlinear Evolutions*, Pitman Monographs and Surveys in Pure and Applied Mathematics **32**, (Longman Scientific and Technical: London).

[136] G. F. Webb (1974). Autonomous nonlinear functional differential equations and nonlinear semigroups, *J. Math. Anal. Appl.* **46**, 1-12.

[137] G. F. Webb (1976). Asymptotic stability for abstract nonlinear functional differential equations, *Proc. Amer. Math. Soc.* **54**, 225-230.

[138] G. F. Webb (1976). Functional differential equations and nonlinear semigroups in L^p-spaces, *J. Diff. Eqns.* **20**, 71-89.

[139] G. F. Webb and M. Badii (1981). Nonlinear non-autonomous functional differential equations in L^p spaces, *Nonlin. Anal. TMA* **5**, 203-223.

[140] K. Yosida (1948). On the differentiability and the representation of one-parameter semi-groups of linear operators, *J. Math. Soc. Japan* **1**, 15-21.

[141] K. Yosida (1974). *Functional Analysis*, (Springer–Verlag: Berlin).

Index